# STUDENT'S SOLUTIONS MANUAL

### MILTON LOYER
*Penn State University*

# ELEMENTARY STATISTICS
## USING THE
## TI-83/84 PLUS CALCULATOR
### SECOND EDITION

## Mario F. Triola
*Dutchess Community College*

Boston San Francisco New York
London Toronto Sydney Tokyo Singapore Madrid
Mexico City Munich Paris Cape Town Hong Kong Montreal

Reproduced by Pearson Addison-Wesley from electronic files supplied by the author.

Copyright © 2008 Pearson Education, Inc.
Publishing as Pearson Addison-Wesley, 75 Arlington Street, Boston, MA 02116.

All rights reserved. No part of this publication may be reproduced, stored in a retrieval system, or transmitted, in any form or by any means, electronic, mechanical, photocopying, recording, or otherwise, without the prior written permission of the publisher. Printed in the United States of America.

ISBN-13: 978-0-321-46258-9
ISBN-10: 0-321-46258-0

3 4 5 6 BB 10 09 08

# TABLE OF CONTENTS

Chapter 1: Introduction to Statistics.................................................................1

Chapter 2: Summarizing and Graphing Data....................................................12

Chapter 3: Statistics for Describing, Exploring, and Comparing Data.................27

Chapter 4: Probability....................................................................................42

Chapter 5: Discrete Probability Distributions....................................................60

Chapter 6: Normal Probability Distributions.....................................................73

Chapter 7: Estimates and Sample Sizes.........................................................108

Chapter 8: Hypothesis Testing......................................................................128

Chapter 9: Inferences from Two Samples......................................................158

Chapter 10: Correlation and Regression........................................................186

Chapter 11: Multinomial Experiments and Contingency Tables........................213

Chapter 12: Analysis of Variance..................................................................231

Chapter 13: Nonparametric Statistics............................................................246

Chapter 14: Statistical Process Control.........................................................276

# INTRODUCTION

## by Milton Loyer

This *Students' Solutions Manual* contains detailed solutions to selected exercises in the text *Elementary Statistics Using the TI-83/84 Plus Calculator*, Second Edition, by Mario Triola. Specifically, the manual contains the solutions to the odd-numbered exercises for each section, and the solutions to all of the end-of-chapter exercises: the Statistical Literacy and Critical Thinking, the Review Exercises, and the Cumulative Review Exercises.

To aid in the comprehension of the calculations, worked problems typically include intermediate steps of algebraic or computer/calculator notation. In addition, the manual makes extensive use of screen captures from the actual set-up and solution display screens of the TI-83/84 Plus calculator. When appropriate, additional hints and comments are included and prefaced by NOTE.

Many statistical problems are best solved using particular formats. Because recognizing and following these patterns promote understanding, this manual identifies and employs such formats whenever practicable.

For best results, read the text carefully <u>before</u> attempting the exercises, and attempt the exercises <u>before</u> consulting the solutions. This manual has been prepared to provide a check and extra insights for exercises that have already been completed, and to provide guidance for solving exercises that have already been attempted but have not been successfully completed.

I would like to thank Mario Triola for writing an excellent elementary statistics textbook and for inviting me to prepare this solutions manual.

# Chapter 1

# Introduction to Statistics

## 1-2 Types of Data

1. A parameter is a numerical value describing a population, while a statistic is a numerical value describing a sample.

3. Discrete data can take on only a countable number of specified values, while continuous data can take on an infinite number of values over some range.

5. Statistic, since the value 2.58 was obtained from a sample of households.

7. Parameter, since the value 706 was obtained from the population of all passengers.

9. Discrete, since the number of letters arriving at the target must be an integer.

11. Continuous, since weight can be any value on a continuum.

13. Nominal, since the numbers are for identification only. Numbers on jerseys are merely numerical names for the players.

15. Nominal, since the numbers are for identification only. Even though SS numbers are assigned chronologically within regions and can be placed in numerical order, there is no meaningful way to compare 208-34-3338 and 517-94-1438. If all the numbers had been assigned chronologically beginning with 000-00-0001, like the order of finishers in a race, then SS numbers would illustrate the ordinal level of measurement..

17. Interval, since differences are meaningful but ratios are not.

19. Ordinal, since the ratings are relative positions in a hierarchy.

IMPORTANT NOTE for exercises 21-24: The population and sample are determined by the intent of the researcher, which must be clearly defined at the outset of any project. Unfortunately these exercises state only what the researcher did, and do not specifically identify the intent of the researcher. Consequently there may be differences of interpretation in some of the exercises, but some general principles apply. (1) The sample is a subset of the population of interest and must have the same units as the population. If the population of interest is all households, for example, then the sample must be a selection of households and not a selection of adults – as households with more adults would have a higher chance of being included in the study, thus creating a bias. (2) The problem of nonresponse must be addressed. If 500 persons are randomly selected and asked a personal question, for example, but only 400 choose to answer the question, what is the sample? Depending on the situation, the sample could be either the 500 people randomly selected (and the 100 "no response" answers be reported as part of the sample data) or the just 400 people who actually gave data in the form of a specific answer.

21. a. The sample is the 25 Senators selected at random.
    b. The population is all 100 US Senators.
    Yes. If the selections are made at random there is no bias in the design of the experiment that would cause the sample not to be representative.

23. a. The sample is the 1059 randomly selected adults.
    b. The population is all US adults.
    Yes. If the adults were selected at random there is no reason why they would not be representative of the population.
    NOTE: This is a sample of adults, not of homes. The appropriate inference is that about 39% of the adults live in a home with a gun, not that there is a gun in about 39% of the homes. Since adults and not homes were selected at random, homes with more adults are more likely to be represented in the survey – and that bias would prevent the survey from being representative of all homes.

25. Temperature ratios are not meaningful because a temperature of 0° does not represent the absence of temperature in the same sense that $0 represents the absence of money. The zero temperature in the exercise (whether Fahrenheit or Centigrade) was determined by a criterion other than "the absence of temperature."

27. This is an example of ordinal data. It is not interval data because differences are not meaningful – i.e., the difference between the ratings +4 and +5 does not necessarily represent the same differential in the quality of food as the difference between 0 and +1.

## 1-3 Critical Thinking

1. A voluntary response sample is one in which the participants choose whether or not to participate. This is generally unsuitable for statistical purposes because it is persons with strong feelings and/or a personal interest that tend to respond – and the opinions of such people are not necessarily representative of the entire population.

3. No. This is a voluntary response sample. Most of the data in a voluntary response sample comes from people with strong feelings or a personal interest in the topic, and there is usually a lack of data from "typical" people.

5. People who choose to play basketball are already taller than the general population when they choose to participate in the sport. It is not the participation that makes them taller.

7. If the population of Orange County includes significantly more minority drivers than white drivers, one would expect more speeding tickets to be issued to minorities than to whites – even if the percentage of white drivers who violated the speed limit was greater than the percentage of minority drivers who did so. It is also possible that police tend to target minority drivers – so that the numbers of tickets issued to the various racial/ethnic groups does not correspond to the actual amount of speed limit violations occurring. The fact that more speeding tickets are given to minorities does not warrant the conclusion that minority persons are more likely to speed.

9. The fact that the study was financed by a company with vested interests in the outcome might have influenced the conducting and/or reporting to concentrate on aspects of the study favorable to the company. The fact that a product contains one ingredient associated with positive health benefits may not outweigh other negative aspects of the product that were not reported.

11. There are at least two reasons why her conclusions should not be applied to the general population of all women. First, the target sample was biased because she sent the 100,000 questionnaires to members of various women's groups – and women who join such groups are not necessarily representative of the entire population, and certainly do not include the "non-

joiner" types. Secondly, the 4500 replies she received are a voluntary response sample (and a very small 4.5% response rate at that) – which means that the responses will include an over representation of those with strong feelings and/or a personal interest in the survey.

13. Using the telephone directory as a source of survey subjects eliminates persons without telephones and persons with unlisted numbers, and it also over-includes persons who live in a house with more than one telephone listing (as some households have different numbers and/or listings for each spouse as well as the children). A more subtle problem is that telephone listings are usually by household, not by individual – which means that a person in a one-adult household has a greater chance of being included in the survey than a person in a multi-adult household.

15. People who were killed in motorcycle crashes, when a helmet may have saved their lives, were not present to testify.

17. No, this is not likely to be a good estimate of the average of all wages earned in the United States. States with small populations but high wages (due to the cost of living) like Alaska and Hawaii, for example, contribute 2/50 = 4% of the average for the 50 states – but since they do not contain 4% of the US wage earners, they inflate the estimate to make it higher than the true average of all US wage earners. In general, the average of averages does not give the overall average. Imagine, for example, two states: one where all 4 wage earners earned $100,000 each, and one where all 96 wage earners earned $50,000 each. The true average wage for the country would be [(4)(100,000) + (96)(50,000)]/100 = 5,200,000/100 = $52,000, but the average of the state averages would be [100,000 + 50,000]/2 = 150,000/2 = $75,000.

19. No, the results will not be representative of all households in the state. This approach violates the basic principle that the population and the sample have the same units – if the population of interest is all *households*, then you cannot survey *children*. This creates a bias because households with fewer children have a smaller chance of being included in the data, and households with no children will be missed entirely.

21. a. 3/20 = (3/20)(100%) = 15%
    b. 56.7% = 56.7/100 = 0.567
    c. (34%)(500) = (34/100)(500) = 170
    d. 0.789 = (0.789)(100%) = 78.9%

23. a. (52%)(1038) = (52/100)(1038) = 540
    b. (52/1038)(100%) = 5%

25. NOTE: The number who <u>actually committed</u> a campus crime while under the influence of alcohol or drugs cannot be determined from the survey, but the number who <u>say</u> that have committed such crimes can be determined as shown below. This distinction is an important one.
    (8%)(1875) = (8/100)(1875) = 150 say they've committed a campus crime
    (62%)(150) = (62/100)(150) = 93 say they've committed a campus crime while under the influence of alcohol or drugs

27. If the researcher started with 20 mice in each group, and none of the mice dropped out of the experiment for any reason, then the proportions of success could only be fractions like 0/20, 1/20, 2/20, etc. – and the success rates in percents could only be multiples of 5 like 0%, 5%, 10%, etc. A success rate of 53%, for example, would not be possible.

4   CHAPTER 1   Introduction to Statistics

**1-4   Design of Experiments**

1. In a random sample, every individual member has an equal chance of being selected; while in a simple random sample of size n, every possible sample of size n has an equal chance of being selected.

3. Blinding occurs when the subject (or evaluator) does not know whether the real vaccine or a placebo has been given. Double-blinding occurs when both the subject and the evaluator don't know whether the real vaccine or a placebo has been given. Blinding is important in situations when the data involve subjective responses (e.g., whether there is less pain, more mobility, etc.) that may be influenced (deliberately or subconsciously) by knowing whether the real vaccine or the placebo has been given.

5. Observational study, since the researcher (Emily) merely measured whether the therapist could identify the chosen hand and did nothing to modify the therapist.

7. Observational study, since specific characteristics are measured on unmodified tablets.

9. Prospective, since the data are to be collected in the future from a group sharing a common factor.

11. Cross-sectional, since the data are collected for a specific point in time.

13. Systematic, since every $5^{th}$ driver was stopped.

15. Stratified, since the population of interest was divided into activity subgroups from which the actual sampling was done.

17. Cluster, since all the wait staff at selected restaurants were surveyed.

19. Convenience, since the sample is those who happen to be in the author's classes.

21. Random, since each adult has an equal chance of being selected.
    NOTE: This is a complex situation. The above answer ignores the fact that there is not a 1 to 1 correspondence between adults and phone numbers. An adult with more than one number will have a higher chance of being selected. An adult who shares a phone number with other adults will have a lower chance of being selected. And, of course, adults with no phone numbers have no chance of being selected at all.

23. Cluster, since all the students in selected classes were interviewed.
    NOTE: Ideally the division of the population should place each subject in one and only one cluster, and every subject should have an equal chance to be included in the sample. In this scenario there are at least two problems: (1) students carrying more classes will have a greater chance of being included in the survey and (2) a student may be included in more than one cluster. While some techniques may eliminate those two problems, they could introduce others – for example, selecting the 10 classes from among all 1 pm MWF classes, means that a student not taking a class then has zero chance of being included in the survey.

25. Yes, this results in a random sample because each return has an equal chance of being included in the sample.
    No, this does not result in a simple random sample of size 48 because some groupings of 48 returns are not possible – for example, a grouping with 24 Monday returns and 24 Tuesday returns would not be possible.

27. Whether the sample is a random sample depends on how the first selection is made. If the engineer chooses the first one at random from 1 to 10,000 and every 10,000$^{th}$ one thereafter, then every M&M has an equal chance of being selected (namely 1 in 10,000) and the sample is a random sample. If the engineer determines to start with #1 and choose every 10,000$^{th}$ one thereafter, then some M&M's have no chance of being selected (e.g., #2) and the sample is not a random sample.
No, no matter how the first selection is made the sample will not be a simple random sample of size n. All possible groupings of size n are not possible – any grouping containing #1 and #2, for example, could not occur.

29. Yes, this results in a random sample because every student has an equal chance of being selected.
Yes, this results in a simple random sample of size 6 because all possible groupings of size 6 have an equal chance to occur.

31. Assume that "students who use this book" refers to a particular semester, or other point in time. Before attempting any sampling procedure below, two preliminary steps are necessary: (1) obtain from the publisher a list of each school that uses the book and (2) obtain from each school the number of students enrolled in courses requiring the text. Note that using only the number of books sold to the school by the publisher for the semester of will miss those students purchasing used copies turned in from previous semesters, buying textbooks on the Internet, etc. "Students" in parts (a)-(e) below is understood to mean students using this book.
   a. Random: Conceptually (i.e., without actually doing so) make a list of all N students by assigning to each school one place on the list for each of its students. Pick 100 random numbers from 1 to N. Each number identifies a particular school and a conceptual student. For each particular school identified select a student at random (or more than one student, if that school was identified more than once). In this manner every student has the same chance of being selected. In fact every possible grouping of 100 students has the same chance of being selected, and the sample is also a simple random sample.
   b. Systematic. Place the n schools in a list. Pick a random number between 1 and (n/10) to determine a starting selection. Select that school and every (n/10)$^{th}$ school thereafter. At each of the 10 schools thus systematically selected, randomly (or systematically) select and survey 10 students.
   c. Convenience. Randomly (or conveniently) select and survey 100 students from your school.
   d. Stratified. Divide the schools into four categories: 2 year, 4 year public, 4 year private, proprietary. Randomly select one school from each category. For each school randomly (or in some stratified manner) select and survey the number of students equal to the percent of the N students that are in that category.
   e. Cluster. Select a school at random and survey all the students, or cluster sample again by selecting a class within that school and surveying all the students in that class. Repeat as necessary until the desired sample of 100 is reached.

33. The study may be described as
   a. prospective because the data are to be collected in the future, after the patients are assigned to different treatment groups.
   b. randomized because the patients were assigned to one of the three treatment groups at random.
   c. double-blind because neither the patient nor the evaluating physician knew which treatment had been administered.

d. placebo-controlled because there was a treatment with no drug that allowed the researchers to distinguish between real effects of the drugs in question and (1) the psychological effects of being "treated" and/or (2) the effects (e.g., natural improvement over time) of not receiving any drug.

## 1-5 Introduction to the TI-83/84 Plus Calculator

1. An APP is a software application that can be downloaded (from a computer or another calculator) to a TI-83/84 Plus calculator using the cable that comes with the calculator.

3. The "–" key indicates subtraction, and it is an operation that must be preceded by and followed by a value. The "(-)" indicates negation, and it is used before a value to indicate that indicate the negative of that value.

NOTE: The characters and numbers in the following problem appear on the TI-83/84 Plus keyboard, and the answers were obtained by pressing the keys in the order in which they appear in the statement of the problem (and then pressing the ENTER key). The answers below give the keys used and not what actually appears on the display – e.g., "3 × 4" on the keyboard produces "4∗3" on the display, and "2nd x2" on the keyboard produces "$\sqrt{\ }$(" on the display.

5. "4 × (15 – 20) ENTER" gives the answer -20

7. "5 – (-)2 ENTER" gives the answer 7

9. "9 ÷ (-)2 ENTER" gives the answer -4.5

11. "2nd $x^2$ 234 ENTER" gives the answer 15.29705854

13. 3; 2nd $x^2$ is the function $\sqrt{\ }$ and the answer is $\sqrt{9}$

15. 25; the answer is $5^2$

17. {0 223 0 176 0...; the result is a display of the values in L1.

19. {0 446 0 352 0...; the result is a display of twice each value in L1

21. "2 ^ 45 ENTER" gives the answer 3.518437209E13 = 35,184,372,090,000

23. "0.5 ^ 10 ENTER" gives the answer 9.765625E-4 = .0009765625

25. NOTE: Using the TI Connectivity cable and software to accomplish a screen capture, the resulting four screen displays for parts b-e are as follows.

a. A list of the applications available, including TRIOLAXE.
b. A list of file names: MHEALTH, FHEALTH, BTEMP,…
c. Below are the names: TAR, NICOT, CO.
d. The top line reads NAMES  OPS  MATH.
   Below are the names of available lists: L1, L2, L1, L4, L5, L6, CO,…
e. {16  16  16  9  1  8...; which is the beginning of the list of TAR values in Data Set 3.

## Statistical Literacy and Critical Thinking

1. A large sample is not necessarily a good sample. As in many other situations, quality is more important than quantity. More important than the size of the sample is whether it was selected in such as way as to be representative of the population of interest.

2. a. Height data are quantitative because height is an actual numerical measurement and not merely a categorization.
   b. Height data are continuous because they can take on any value over a specified range.
   c. The population is all runners who finish the New York Marathon.

3. This sample would not necessarily be representative of the general population. It is a voluntary response sample, which typically includes only those with certain motivation or characteristics – in this case that would probably be people who either (1) think they have a high IQ and /or enjoy such tests or (2) need the $50.

4. This would not be a good estimate for at least two reasons. First, she appears to be depending on values provided by the owners. An owner might not know a car's true value or might tend to exaggerate its value. Secondly, she is counting each state equally. Alaska has very few cars, but they probably include many relatively expensive (even considering that state's overall high cost of items) four-wheel drive utility vehicles. California has thousands of times as many cars as Alaska, and they would probably tend to be less expensive smaller vehicles. The average for 10 cars from each state would likely not be a good estimate for the average of all the cars in the nation.

## Review Exercises

1. The results cannot be considered representative of the population of the United States for at least two reasons. First, only AOL Internet subscribers were polled – and they are not necessarily representative racially, socio-economically, geographically, etc. Secondly, the responders constitute a voluntary response sample – and such responders tend to be those with strong opinions and/or a personal stake in the issue.

2. Let N be the total number of full-time students and n be the desired sample size.
   a. Random. Obtain a list of all N full-time students, and number the students from 1 to N. Select n random numbers from 1 to N, and survey each student whose number on the list is one of the random numbers selected.
   b. Systematic. Obtain a list of all N full-time students, and number the students from 1 to N. Let m be the largest integer less than the fraction N/n, select a random number between 1 and m, and survey that student and every $m^{th}$ student thereafter.
   c. Convenience. Select a location by which most of the students usually pass and survey the first n full-time students that pass by.

8   CHAPTER 1   Introduction to Statistics

    d. Stratified. Obtain a list of all N full-time students and the gender of each. Divide the list by gender and randomly survey n/2 students of each gender.

    e. Cluster. Obtain a list of all the classes meeting at a popular time, select at random one of the classes (or as many of the classes as needed to obtain a sample of n students), and survey all the students in that class.

3. a. Ratio, since differences are valid and there is a meaningful 0.
   b. Ordinal, since there is a hierarchy but differences are not valid.
   c. Nominal, since the categories have no meaningful inherent order.
   d. Interval, since differences are valid but the 0 is arbitrary.

4. a. Discrete, since the number of shares held must be an integer.
      NOTE: Even if partial shares are allowed (e.g., 5½ shares), the number of shares must be some fractional value and cannot be any value on a continuum – e.g., a person cannot own π shares.
   b. Ratio, since differences between values are valid and there is a meaningful 0.
   c. Stratified, since the population of interest (all stockholders) was divided into subpopulations (by states) from which the actual sampling was done.
   d. Statistic, since it is calculated from a sample and not from the entire population.
   e. There is no unique correct answer, but the following are reasonable possibilities. (1) The proportion of stockholders holding above that certain number of shares that would make them "influential." (2) The proportion of stockholders holding below that certain number of shares that would make them "insignificant." (3) The numbers of shares held by the largest stockholders.
   f. There is no unique correct answer, but the following are reasonable possibilities. (1) The results would be from a voluntary response sample, and hence not necessarily be representative of all stockholders. (2) If the questionnaire did not include information on the number of shares owned, the views of small stockholders could not be distinguished from the views of large stockholders (whose views should carry more weight).

5. NOTE: The second part of the problem involves deciding "whether the sampling scheme is likely to result in a sample that is representative of the population." This is subjective, since the term "representative" has not been well-defined. Mathematically, any random sample is representative in the sense that there is no bias and a series of such samples can be expected to average out to the true population values – even though any one particular sample may not be representative. And so we expect a random sample of size n=50 to be representative. But how about a random sample of size n=2? It has all the mathematical properties of a random sample of size n=50, but common sense suggests that any one sample of size n=2 is not necessarily likely to be representative of the population.
   a. Systematic, since the selections were made at regular intervals. Yes, there is no reason why every 500th stockholder should have some characteristic that would introduce a bias.
   b. Convenience, since those selected were the ones who happened to attend. No, those who attend would tend to be the more interested and/or more well-off stockholders.
   c. Cluster, since the stockholders were organized into groups (by stockbroker) and all the stockholders in the selected groups were surveyed. Yes, since every stockholder has the same chance of being included in the sample (assuming each stockholder works through a single stockbroker), the sample will be a random sample – and considering the large number of stockbrokers involved (see the NOTE at the beginning of the problem), it is reasonable to expect the sample to be representative of the population.

d. Random, since each stockholder has the same chance of being selected. Yes, since every stockholder has the same chance of being included in the sample, the scheme is likely to result in a representative sample.

e. Stratified, since the stockholders were divided into subpopulations (zip codes) from which the actual sampling was done. No, since all the zip codes were given equal weights (5 stockholders from each) – because "significant" zip codes (with large numbers of large stockholders) are counted equal with "insignificant" zip codes (with few numbers of small stockholders), the sample will probably be biased in favor of small stockholders.

6. a. Blinding occurs when those involved in an experiment (either as subjects or evaluators) do not know whether they are dealing with a treatment or a placebo. It might be used in this experiment by (1) not telling the subjects whether they are receiving Sleepeze or the placebo and/or (2) not telling any post-experiment interviewers or evaluators which subjects received Sleepeze and which ones received the placebo. Double-blinding occurs when neither the subjects nor the evaluators know whether they are dealing with a treatment or a placebo.

b. The data reported will probably involve subjective assessments (e.g., "on a scale of 1 to 10, how well did it work?") that may be influenced by whether the subject was known to have received Sleepeze or the placebo.

c. In a completely randomized design, subjects are assigned to the treatments (in this case Sleepeze or the placebo) at random.

d. In a rigorously controlled design, the subjects are carefully placed into treatment groups (in this case Sleepeze or the placebo) according to characteristics that might affect the outcome of the experiment. For example: if the Sleepeze group contains an athletic male under 20, then the placebo group should contain an athletic male under 20 – since gender, physical condition, and age are all characteristics that might reasonably affect the outcome of the experiment. This makes the treatment groups as similar as possible, so that any observed differences are likely to be due to the treatments and not to extraneous factors that might affect the experiment.

e. Replication involves performing the experiment on a sample of subjects large enough to ensure (1) that atypical responses of a few subjects will not give a distorted view of the true situation and (2) that repeating the experiment would likely produce similar results.

7. a. The figure 39.82% is a parameter, since it refers to the entire population.
  b. Discrete, since the numbers of votes cast must be integers.
  c. $(39.82\%)(1{,}865{,}908) = (39.82/100)(1{,}865{,}908) = 743{,}005$

8. a. To contain 100% less fat would be to contain no fat at all – 100% of the fat has been removed. It is not physically possible to contain 125% less fat.
  b. $(19\%)(237) = (19/100)(237) = 45$
  c. $(30/186)(100\%) = 16.1\%$

# 10  CHAPTER 1  Introduction to Statistics

**Cumulative Review Exercises**

NOTE: Throughout the text intermediate mathematical steps will be shown to aid those who may be having difficulty with the calculations. In practice, most of the work can be done continuously on the TI-83/84 Plus calculator. Even when the calculations cannot be done continuously, DO NOT WRITE AN INTERMEDIATE VALUE ON YOUR PAPER AND THEN RE-ENTER IT IN THE CALCULATOR. That practice can introduce round-off and copying errors. Store any intermediate values in the calculator so that you can recall them with infinite accuracy and without copying errors. In general, the degree of accuracy appropriate depends upon the particular problem – and guidelines for this will be given as needed in subsequent chapters. Unless there is reason to otherwise, answers in this section are given with 3 decimal accuracy. For this section only, in the spirit of a cumulative review, the solutions are given both in regular algebraic format and as they would be entered in the TI-83/84 Plus calculator – from this point on, however, we employ the equal sign (=) to represent use of the ENTER key.

1. $\dfrac{3.0630+3.0487+2.9149+3.1358+2.9753}{5} = \dfrac{15.1377}{5} = 3.02754$ grams

   $(3.0630 + 3.0487 + 2.9149 + 3.1358 + 2.9753) \div 5 = 3.02754$ grams

2. $\dfrac{98.20-98.60}{0.62} = \dfrac{-0.40}{0.62} = -0.645$

   $(98.20 - 98.60) \div 0.62 = 0.6451612903$

3. $\dfrac{98.20-98.60}{\dfrac{0.62}{\sqrt{106}}} = \dfrac{-0.40}{0.0602} = -6.642$

   $(98.20 - 98.60) \div (0.62 \div 2\text{nd } x^2\ 106) = -6.642342026$

4. $\left[\dfrac{1.96 \cdot 0.25}{0.03}\right]^2 = [16.333]^2 = 266.778$

   $(1.96 \times 0.25 \div 0.03)\ \wedge\ 2 = 266.7777778$

5. $\dfrac{(50-45)^2}{45} = \dfrac{(5)^2}{45} = \dfrac{25}{45} = 0.556$

   $(50 - 45)\ \wedge\ 2 \div 45 = .5555555556$

   NOTE: The hierarchy of operations makes it unnecessary to place the (50-45)^2 in parentheses.

6. $\dfrac{(2-4)^2+(3-4)^2+(7-4)^2}{3-1} = \dfrac{(-2)^2+(-1)^2+(3)^2}{2} = \dfrac{4+1+9}{2} = \dfrac{14}{2} = 7$

   $((2 - 4)\wedge 2 + (3 - 4)\wedge 2 + (7 - 4)\wedge 2\ ) \div (3 - 1) = 7$

7. $\sqrt{\dfrac{(2-4)^2+(3-4)^2+(7-4)^2}{3-1}} = \sqrt{\dfrac{(-2)^2+(-1)^2+(3)^2}{2}} = \sqrt{\dfrac{4+1+9}{2}} = \sqrt{\dfrac{14}{2}} = \sqrt{7} = 2.646$

2nd $x^2$ ((2 − 4)^2 + (3 − 4)^2 + (7 − 4)^2 ) ÷ (3-1) = 2.645751311
NOTE: Three "(" symbols will appear on the display after "2nd $x^2$". Whenever "2nd $x^2$" is entered on the TI-83/84 Plus, a "(" symbol appears automatically. Using the ENTER key automatically applies the indicated operation to what the user has entered.

8. $\dfrac{8(151{,}879)-(516.5)(2176)}{\sqrt{8(34{,}525.75)-516.5^2}\sqrt{8(728{,}520)-2176^2}} = \dfrac{91128}{\sqrt{9433.75}\sqrt{1093184}} = 0.89735$

(8 × 151879 − 516.5 × 2176) ÷ (2nd $x^2$ 8 × 34525.75 − 516.5^2) 2nd $x^2$ 8 × 728520 − 2176^2)) = .8973523904
NOTE: An additional "(" symbol will appear on the display after each "2nd $x^2$". Whenever "2nd $x^2$" is entered on the TI-83/84 Plus, a "(" symbol appears automatically.

9. 0.5^10 = 9.765625E-4 = .0009765625

10. 2^40 = 1.099511628E12 = 1,099,511,628,000
    NOTE: The final 3 digits are not really zero, but most calculators will not be able to give all 13 digits of the exact answer.

11. 7^12 = 1.38412872E10 = 13,841,287,200
    NOTE: The final 2 digits are not really zero, but most calculators will not be able to give all 11 digits of the exact answer.

12. 0.8^50 = 1.427247693E-3 = 0.00001427247693

# Chapter 2

# Summarizing and Graphing Data

## 2-2 Frequency Distributions

1. A frequency distribution is a table with two columns – one column that identifies data values either individually or in groups, and one column that tells how many data values meet that criterion. It is useful for summarizing data and giving information about the distribution of the data – the minimum, maximum, shape, and concentration of the data.

3. The given class intervals overlap. The classes of a frequency distribution should be defined so that each data value falls into one and only class. Because the value 10, for example, is in two different classes, the frequency distribution constructed using the given class intervals would not be well-defined.

5. Subtracting the first two lower class limits indicates that the class width is 40 – 35 = 5. Since there is a gap of 1.0 between the upper class limit of one class and the lower class limit of the next, class boundaries are determined by increasing or decreasing the appropriate class limits by (1.0)/2 = 0.5  The class boundaries and class midpoints are given in the table below.

| temperature (°F) | class boundaries | class midpoint | frequency |
|---|---|---|---|
| 35 – 39 | 34.5 – 39.5 | 37 | 1 |
| 40 – 44 | 39.5 – 44.5 | 42 | 3 |
| 45 – 49 | 44.5 – 49.5 | 47 | 5 |
| 50 – 54 | 49.5 – 54.5 | 52 | 11 |
| 55 – 59 | 54.5 – 59.5 | 57 | 7 |
| 60 – 64 | 59.5 – 64.5 | 62 | 7 |
| 65 – 69 | 64.5 – 69.5 | 67 | 1 |
| | | | 35 |

NOTE: Although they often contain extra decimal places and may involve consideration of how the data were obtained, class boundaries are the key to tabular and pictorial data summaries. Once the class boundaries are obtained, everything else falls into place. In exercise #5 it is readily seen that the width of the first class is 39.5 – 34.5 = 5.0 and the midpoint of the first class is (34.5+39.5)/2 = 37.0. In this manual class boundaries will typically be calculated first and that information used to determine other values. Because the sum of the frequencies is an informative number and is used in many subsequent calculations, it will be shown as an integral part of each frequency distribution.

7. Subtracting the first two lower class limits indicates that the class width is 65.0-60.0 = 5.0 Since there is a gap of 0.1 between the upper class limit of one class and the lower class limit of the next, class boundaries are determined by increasing or decreasing the appropriate class limits by (0.1)/2 = 0.05  The class boundaries and class midpoints are given in the table on the following page.

Frequency Distributions  SECTION 2-2   13

| height (in) | class boundaries | class midpoint | frequency |
|---|---|---|---|
| 60.0 – 64.9 | 59.95 – 64.95 | 62.45 | 4 |
| 65.0 – 69.9 | 64.95 – 69.95 | 67.45 | 25 |
| 70.0 – 74.9 | 69.95 – 74.95 | 72.45 | 9 |
| 75.0 – 79.9 | 74.95 – 79.95 | 77.45 | 1 |
| 80.0 – 84.9 | 79.95 – 84.95 | 82.45 | 0 |
| 85.0 – 89.9 | 84.95 – 89.95 | 87.45 | 0 |
| 90.0 – 94.9 | 89.95 – 94.95 | 92.45 | 0 |
| 95.0 – 99.9 | 94.95 – 99.95 | 97.45 | 0 |
| 100.0 –104.9 | 99.95 –104.95 | 102.45 | 0 |
| 105.0 –109.9 | 104.95 –109.95 | 107.45 | 1 |
|  |  |  | 40 |

9. Yes; in a relatively symmetric manner the frequencies start out low, increase to a maximum, and then decrease.

11. The height of the tallest man is stated to be between 105 and 110 inches – i.e., between 8'9" and 9'2". This is almost certainly an error. When this value is deleted the distribution of the heights is approximately normal.

NOTE: For relative frequency distributions, as with frequency distributions, the sum is included as an integral part of the table. For relative frequency distributions this should always be 1.000 (or 100.0%) and serves as a check for the calculations. [Because of rounding in the individual classes, the listed relative frequencies might actually sum to 0.999 or 1.001 – but the true sum with infinite accuracy is always 1.000, and deviations from that indicate that an error has been made.] Relative frequencies may be expressed as decimals or percents – while the choice is arbitrary, individual instructors may specify the use of one or the other.

13. The relative frequency for each class is found by dividing its frequency by 35, the sum of the frequencies.

| temperature (°F) | relative frequency |
|---|---|
| 35 – 39 | 2.9% |
| 40 – 44 | 8.6% |
| 45 – 49 | 14.3% |
| 50 – 54 | 31.4% |
| 55 – 59 | 20.0% |
| 60 – 64 | 20.0% |
| 65 – 69 | 2.9% |
|  | 100.0% |

NOTE: For cumulative tables, this manual uses upper class boundaries in the "less than" column. Consider exercise #15, for example, to understand why this is done. Conceptually, temperatures occur on a continuum and the integer values reported are assumed to be the nearest whole number representation of the precise measure. An exact temperature of 39.7, for example, would be reported as 40 and fall in the second class. The values in the first class, therefore, are better described as "less than 39.5" (using the upper class boundary) than as "less than 40" (using the lower class limit of the next class). This distinction is crucial in the construction of pictorial representations in the next section.

In addition, that the final cumulative frequency must equal the total number (i.e., the sum of the frequency column in the frequency distribution) serves as a check for the calculations. The sum of the cumulative frequencies, however, has absolutely no meaning and is not included in the table.

15. The cumulative frequencies are the combined number in each class and all previous classes.

| temperature (°F) | cumulative frequency |
|---|---|
| less than 39.5 | 1 |
| less than 44.5 | 4 |
| less than 49.5 | 9 |
| less than 54.5 | 20 |
| less than 59.5 | 27 |
| less than 64.5 | 34 |
| less than 69.5 | 35 |

17. The frequency distribution of the last digits shows unusually high numbers of 0's and 5's. This is typical for data that have been rounded off to "convenient" values. It appears that the heights were reported and not actually measured.

| digit | frequency |
|---|---|
| 0 | 9 |
| 1 | 2 |
| 2 | 1 |
| 3 | 3 |
| 4 | 1 |
| 5 | 15 |
| 6 | 2 |
| 7 | 0 |
| 8 | 3 |
| 9 | 1 |
|   | 37 |

19. Since most of the days have no rainfall the data values accumulate in the lower end of the table. The frequency distribution does not appear to approximate a normal distribution.

| rainfall (in) | frequency |
|---|---|
| 0.00 – 0.19 | 44 |
| 0.20 – 0.39 | 6 |
| 0.40 – 0.59 | 1 |
| 0.60 – 0.79 | 0 |
| 0.80 – 0.99 | 0 |
| 1.00 – 1.19 | 0 |
| 1.20 – 1.39 | 1 |
|   | 52 |

21. Most of the data values occur at the lower end of the table, and there are very high values that are not balanced by very low ones. The frequency distribution does not appear to approximate a normal distribution.

| BMI | frequency |
|---|---|
| 15.0 – 20.9 | 10 |
| 21.0 – 26.9 | 15 |
| 27.0 – 32.9 | 11 |
| 33.0 – 38.9 | 2 |
| 39.0 – 44.9 | 2 |
|   | 40 |

23. The weights of the pennies appear to be normally distributed, but with an unusually heavy concentration of data values at the center and very few values at the extremes.

| weight (lbs) | frequency |
|---|---|
| 2.9500 – 2.9999 | 2 |
| 3.0000 – 3.0499 | 3 |
| 3.0500 – 3.0999 | 22 |
| 3.1000 – 3.1499 | 7 |
| 3.1500 – 3.1999 | 1 |
|  | 35 |

25. The two relative frequency distributions, each constructed with 200 as the lower class limit of the first class and a class width of 20, are given below.

0.0109 CANS

| weight (lbs) | relative frequency |
|---|---|
| 200 – 219 | 6.9% |
| 220 – 239 | 5.1% |
| 240 – 259 | 10.3% |
| 260 – 279 | 48.0% |
| 280 – 299 | 22.7% |
|  | 100.0% |

0.0111 CANS

| weight (lbs) | relative frequency |
|---|---|
| 200 – 219 | 3.4% |
| 220 – 239 | 2.9% |
| 240 – 259 | 6.9% |
| 260 – 279 | 20.6% |
| 280 – 299 | 49.7% |
| 300 – 319 | 16.0% |
| 320 – 339 | 0.0% |
| 340 – 359 | 0.0% |
| 360 – 379 | 0.0% |
| 380 – 399 | 0.0% |
| 400 – 419 | 0.0% |
| 420 – 439 | 0.0% |
| 440 – 459 | 0.0% |
| 460 – 479 | 0.0% |
| 480 – 499 | 0.0% |
| 500 – 519 | 0.6% |
|  | 100.0% |

For both distributions most of the data values occur near the center of the table, and the relative frequencies taper off toward each end.

Both distributions have approximately the same "off-balance normal" shape – and as might be expected, the data values are slightly higher for the stronger 0.0111 cans.

27. Let n = the number of data values and let x = the number of classes.
Solve the given formula x = 1 + (log n)/(log 2) for n to get $n = 2^{x-1}$.

Use the values x = 5.5, 6.5, 7.5,…
to get cut-off values for n shown below.

| x | $n = 2^{x-1}$ |
|---|---|
| 5.5 | 22.63 |
| 6.5 | 45.25 |
| 7.5 | 90.51 |
| 8.5 | 181.02 |
| 9.5 | 362.04 |
| 10.5 | 724.04 |
| 11.5 | 1448.15 |
| 12.5 | 2896.31 |

Assuming n is at least 16, use the cut-off values to complete the table as follows.

| n | ideal # of classes |
|---|---|
| 16 – 22 | 5 |
| 23 – 45 | 6 |
| 46 – 90 | 7 |
| 91 – 181 | 8 |
| 182 – 362 | 9 |
| 363 – 724 | 10 |
| 725 – 1448 | 11 |
| 1449 – 2896 | 12 |

NOTE: The cut-off values indicate that
For n < 22.63, x rounds to 5; for n > 22.63, x rounds to 6.
For n < 45.25, x rounds to 6; for n > 45.25, x rounds to 7.
etc.

16    CHAPTER 2    Summarizing and Graphing Data

**2-3 Histograms**

1. A histogram reveals the basic shape of the distribution. It also gives a preliminary indication of the typical values and the amount of spread of the values.

3. Histograms and other summaries are appropriate to organize the data when they represent too much information for the mind to assimilate as individual values – and data sets with only five values, for example, typically do not need to be summarized to get a feel for the data. In addition, five values would generally not be enough to provide an adequate visual or tabular representation of the underlying distribution.

IMPORTANT NOTE FOR THE REMAINDER OF THE MANUAL: The next several exercises illustrate the practical implications of using the TI-83/84 Plus, or any other graphing calculator. Very briefly, the necessary compact format and lack of labeling on the graphics displayed by the calculator are not ideal for the formal presentations often desired in practice.

The histogram presented for exercises 5-8 cannot be interpreted without reference to the settings in the window that was used to create it. Even then, careful counting of the markings on the x and y axes must be employed to answer the questions asked in the exercises. Ideally, statistical graphics summarize the data, give the reader a feel for the data, and allow the reader to answer relatively technical questions about the data – all from a single figure. Figures 2-2 and 2-3 in the text, for example, are such figures.

Furthermore, images on the screen of a graphing calculator can not be preserved unless the calculator has built-in printing capabilities or can be connected to a printer. For the TI-83/84 Plus calculator, this can be accomplished with a TI Connectivity cable and software. Undoubtedly, some instructors may require such equipment so that assignments can be turned in for grading, etc.

In general, this manual uses the TI-83/84 Plus to generate the values necessary to produce the required graphics, but then follows the techniques given in the text to produce the type of well-labeled and informative presentation typically demanded by employers (and instructors). When appropriate, the manual will indicate the TI-83/84 Plus commands used to generate the values.

Exercise 9 indicates the dilemma one faces when preparing a solutions manual. The histogram presented is what would be expected from an employer and is typical of the type of graphic presented in the remainder of the manual. Leading the reading through the steps necessary to produce to compact TI-83/84 Plus histogram would necessitate the following comments and screen captures.

#9.  Place the 37 data values in list L1
   Use STAT PLOT on list L1 as shown in the left screen below.
   Use WINDOW and enter the values show in the middle screen below.
   Use GRAPH to generate the right screen below.

   Press TRACE and use the arrows to see the information associated with each of the individual bars.

Rather than giving the compact and unlabeled right screen above, this manual presents the type of figure given in the text. As suggested by the author in the graph-oriented Chapter 14, the manual uses Minitab software to generate figures that are intended to be the final product of an exercise and the TI-83/84 Plus calculator to generate figures used within an exercise for diagnostic purposes only.

In addition, the TI-83/84 Plus formulas from now on will generally be given in the manual as they appear on the display instead of using the notation on the keys – e.g., "$\sqrt{\phantom{x}}$" instead of "2nd $x^2$", "$\rightarrow$" instead of "STO$\rightarrow$", "L1" instead of "2nd STAT L1 ENTER", etc.

5. The Yscl=1 setting indicates the dots on the left border are one unit apart. The number of crew members is $2 + 10 + 5 + 1 = 18$. This represents 9 persons (1 coxswain and 8 rowers) from each school.

7. Rowers need to be muscular in order to power the boat. Since the coxswain directs the rowers and represents "unproductive" weight, it would be advantageous for a crew of rowers to have a light coxswain. The two lighter weights likely belong to the two coxswains.

9. The histogram is given below. See the extended NOTE on the previous page.

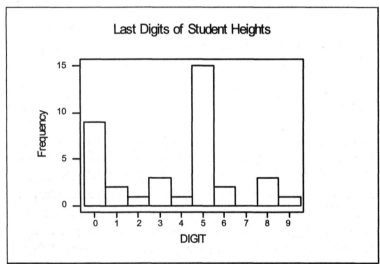

The digits 0 and 5 occur disproportionately more than the others. This is typical for data that have been rounded off to "convenient" values. It appears that the heights were reported and not actually measured.

11. The histogram is given below.

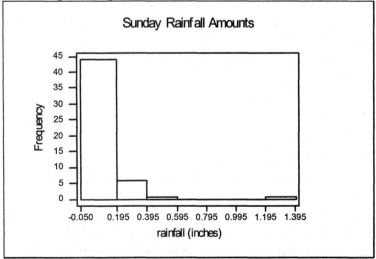

No, the data do not appear to have a distribution that is approximately normal.

18   CHAPTER 2   Summarizing and Graphing Data

13. The histogram is given below.

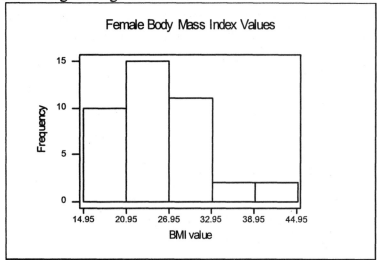

Yes, the data appear to have a distribution that is approximately normal.

15. The histogram is given below.

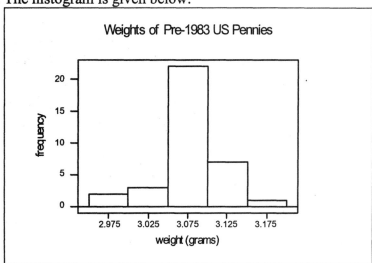

Yes, the data appear to have a distribution that is approximately normal.

17. The two relative frequency histograms are given below. For the sake of comparison, the same horizontal and vertical axes have been used for both histograms.

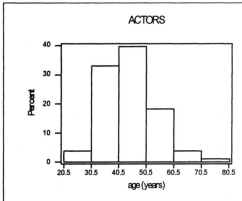
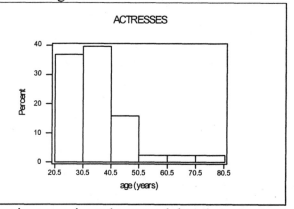

Each set of ages appears to have a distribution that is approximately normal, but there are two significant differences between the two sets: the ages for the actors are larger and more spread out than those for the actresses. Yes, the two genders seem to win Oscars at different ages –

with males winning most often in their 40's, and females winning most often in their 30's.

19. The back-to-back relative frequency histogram is given on the following page. Because it is an outlier and not helpful to the intended comparison, the 504 value for the 0.0111 thickness cans is not included.

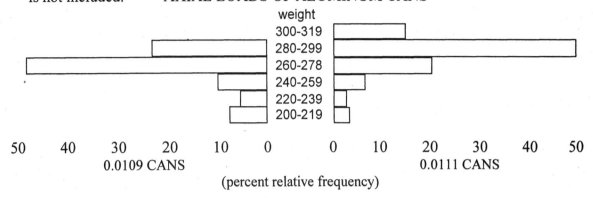

## 2-4 Statistical Graphics

1. The main objective in graphing data is to better understand the information contained in the data. An appropriate graph can present the information efficiently and reveal nuances that otherwise might go unnoticed.

3. A time series graph requires data from a single variable that has been collected for different points in time. It can reveal trends or patterns that occur over time.

5. The dotplot is given below.

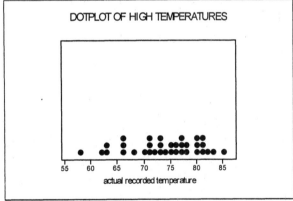

The dotplot suggests that the high temperatures during the period are fairly evenly distributed, with slightly fewer occurrences at the extremes.

7. The frequency polygon is given below.

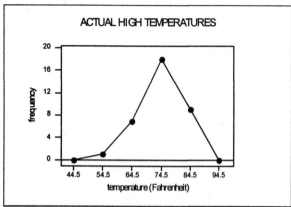

NOTE: The midpoints are not integer values, and the polygon must begin and end at zero at the midpoints of the adjoining classes that contain no data values.

9. The stemplot is given below.
```
 9 | 55
10 |
11 | 00055
12 | 000000005555
13 | 0000000066668
14 | 000088
15 | 00
```
The stemplot suggests that the distribution of the heights is approximately normal.
NOTE: The values suggest that the heights are being estimated and not measured precisely – hence the preponderance of heights ending in 0 and 5. This is to be expected, for how would it be possible to get an exact measurement of the height of a geyser? In the 130's and 140's, the only non-zero final digits are 6 and 8 – which may be a function of the algorithm used to estimate the height.

11. The ogive is given below.

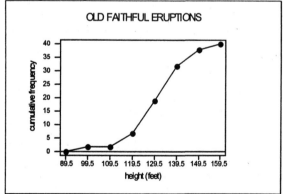

Using the figure: move up from 119.5 on the horizontal scale to intersect the graph, then move left to intersect the vertical scale at 7. This indicates there were approximately 7 data values which would have been recorded as being below 120. This agrees with the actual data values.
NOTE: Ogives always begin on the vertical axis at zero and end at n, the total number of data values. All cumulative values are plotted at the upper class boundaries.

13. See the figure below, with the bars arranged in order of magnitude.

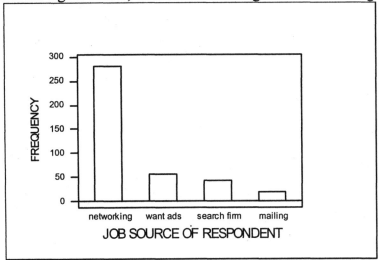

Networking appears to be the most effective job-seeking approach.

15. See the figure below. The corresponding central angles were determined as follows.
    vehicles: 2375/5524 = 43%, and 43% of 360° is 155°
    objects: 884/5524 = 16%, and 16% of 360° is 58°
    violence: 829/5524 = 15%, and 15% of 360° is 54°
    falls: 718/5524 = 13%, and 13% of 360° is 47°
    exposure: 522/5524 = 10%, and 10% of 360° is 36°
    fire: 166/5524 = 3%, and 3% of 360° is 11°

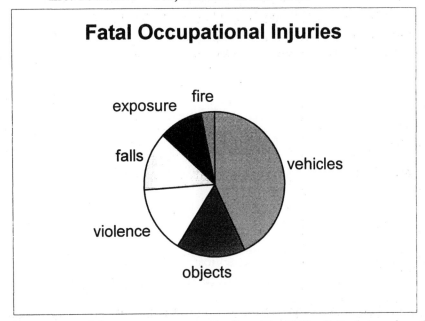

Since some of the areas are very similar, it's difficult to determine the exact order of importance from the pie chart.

17. See the figure below. To be complete the figure should have a title, and each axis should be labeled both numerically and with the name of the variable.
Note: When there are multiple occurrences of the same (or very similar) data pairs, some computer generated plots show only a single symbol. This does not give a true picture of the data or the relationship between the variables. In the figure below, the numbers 2 or 3 are superimposed on the computer generated circles to indicate multiple occurrences.

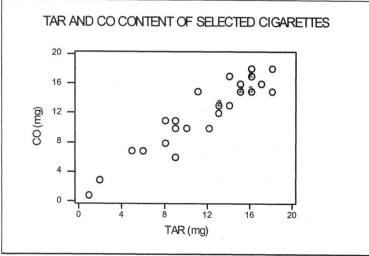

Cigarettes high in tar tend to be high in CO. The points cluster around a straight line from (0,0) to (18,18), indicating that the mg of CO tends to be about equal to the mg of tar.

19. See the figure below.

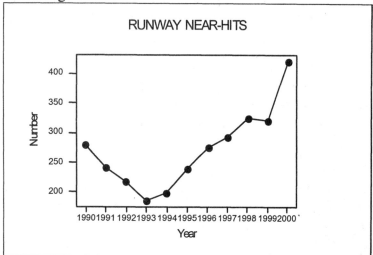

Yes, there appears to be a trend. The number of near-hits appears to have been decreasing prior to 1993 and increasing since 1993. Perhaps some change occurred about that time that may have contributed to the reversal in the trend.

IMPORTANT NOTE: Chapter 1 contained the warning that not starting the horizontal axis at zero "tends to produce a misleading subjective impression, causing readers to incorrectly believe that the difference is much greater than it really is" and that consequently "to correctly interpret a graph we should analyze the numerical information given in the graph." The visual impression above is that the occurrence of near-hits was almost eliminated in 1993 and that the numbers have risen dramatically since then. The time-series graph on the next page, which starts the horizontal axis at zero, gives a very different impression – that except for a spike in 2000, the numbers seem relatively constant.

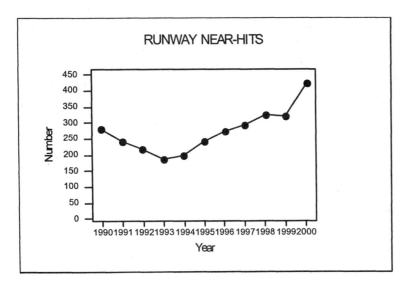

21. According to the figure, 422,000 started and 10,000 returned.
    10,000/422,000 = 2.37%

23. The figure indicates that of the 400,000 − 80,000 = 320,000 men who started from Vilna toward Moscow, only 100,000 arrive in Moscow. The scale gives the length d on the chart that corresponds to 100 miles. The straight-line distance from Vilna to Moscow is about 4.9d = 490 miles, while the distance following the route is about 5.3d = 530 miles.

25. The final form of the back-to-back stemplot is given below, an example of adapting a standard visual form in order to better communicate the data. While such decisions are arbitrary, we chose to display "outward" from the central stem but to keep both ages in increasing order from left to right.

```
                       actresses' ages |   | actors' ages
       1244555566667777788889999999 | 2 | 9
        001112233333444555555677888899 | 3 | 001222444556667778888999
                       011111223569 | 4 | 0001111112222223333444555567778999
                                 04 | 5 | 112222346677
                                013 | 6 | 00222
                                  4 | 7 | 6
                                  0 | 8 |
```

Female Oscar winners tend to be younger than male Oscar winners. If one assumes that acting ability does not peak differently according to gender, the data may reveal a difference in the standard by which males and females are judged.

**Statistical Literacy and Critical Thinking**

1. Since it provides a visual image, a histogram is generally a more effective method for investigating the distribution of a data set.

2. If the data sets contain the same number of observations, the choice is arbitrary – but frequency distributions may be preferred since they give the added information of the actual counts. Since the actual number of observations per category is a function of the total number of observations as well as the nature of the distribution, relative frequency distributions should be used when comparing data sets with different total numbers of observations.

# 24 CHAPTER 2 Summarizing and Graphing Data

3. A time-series graph is designed to identify trends and patterns in the data over time. A histogram is designed to compare different categories. For the scenario suggested, the time-series graph would be more appropriate. In addition, if there are data from each of the 50 years, a line graph connecting 50 points would be much more readable than a series of 50 bars in a histogram.

4. A histogram constructed for data that are approximately normally distributed should have the following three features.
    (1) The tallest bar in the histogram should be near the center.
    (2) The heights of the other bars should decrease as their distance from the center increases.
    (3) The figure should be approximately symmetric.

## Review Exercises

1. The requested frequency distribution is given below.

    | ages of actors | frequency |
    |---|---|
    | 21-30 | 3 |
    | 31-40 | 25 |
    | 41-50 | 30 |
    | 51-60 | 14 |
    | 61-70 | 3 |
    | 71-80 | 1 |
    | | 76 |

    The ages of the actors are greater than those of the actresses.

2. The requested histogram is given below, with the same class boundaries used in Figure 2-2.

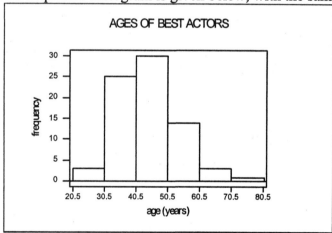

The ages of the actors are greater than those of the actresses.

3. The requested dotplot is given below.

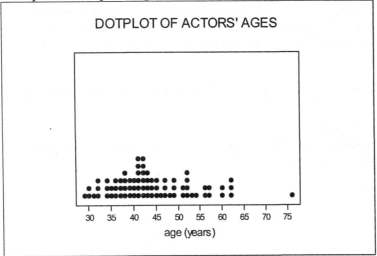

The ages of the actors are greater than those of the actresses.

4. The requested stemplot is given below.

Actors' Ages
2 | 9
3 | 00122244455666777888899
4 | 000111111222222333344455567778999
5 | 112222346677
6 | 00222
7 | 6

The ages of the actors are greater than those of the actresses.

5. The requested scatterplot is given below, using the first 10 entries by row for each gender and placing the ages of the actresses on the x axis.

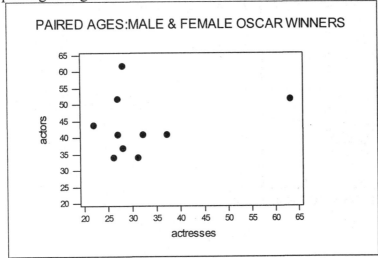

There does not appear to be an association in any given year between the ages of the best actress and the best actor.

26  CHAPTER 2  Summarizing and Graphing Data

6. The requested time-series graph is given below, using the entries in order row by row.

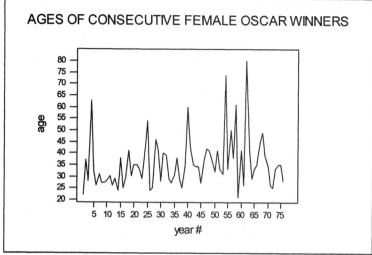

There is no apparent trend or pattern over time.

**Cumulative Review Exercises**

1. No, these numbers do not measure or count anything.

2. The outcome results are at the nominal level of measurement – the numbers that result from the spins are merely names for the slots on the roulette wheel. The frequency results are at the ratio level of measurement – the numbers of occurrences are actual counts, and differences in counts are meaningful and there is a natural zero.

3. Since there were 380 spins and 38 slots on the roulette wheel, one would expect each slot to occur about 10 times. Each of the first 7 categories represents 5 slots and should have a frequency of about 50, while the last category represents 3 slots and should have a frequency of about 30. The frequencies reported agree with these expectations. Based on these results, there is no reason to suspect that the wheel is not fair.

4. Yes. On a fair wheel, each of the 38 slots is equally likely and the average of the resulting numbers should be close to $(1+2+3+\ldots+35+36+0+00)/38 = 666/38 \approx 17.5$. If the average of 500 spins is close to 5.0, then the wheel is biased in favor of the lower numbers and betting on lower numbers would be a wise strategy.

5. No. These results cannot be used to make valid conclusions about the population of all car owners for at least two reasons. (1) The sample is a voluntary response sample, which is not necessarily representative of the original population. (2) The owners stated the amount they spent for their cars – which assumes that they correctly remember the prices and honestly report them, and neither of those assumptions can be counted on.

# Chapter 3

# Statistics for Describing, Exploring, and Comparing Data

## 3-2 Measures of Center

1. The mean, median, mode, and midrange are measures of "center" in the sense that they each attempt to determine (by various criteria – i.e., by using different approaches) what might be designated as a typical or representative value.

3. No; it makes no sense to calculate the mean of those numbers. The numbers are merely numerical names for the players and do not represent the measurement of any quantity or characteristic. If the individual numbers have no quantitative meaning, then neither would any value calculated from them.

NOTE: As it is common in mathematics and statistics to use symbols instead of words to represent quantities that are used often and/or that may appear in equations, this manual employs the following symbols for the various measures of center.

mean = $\bar{x}$         mode = M
median = $\tilde{x}$         midrange = m.r.

This manual will generally report means, medians and midranges accurate to one more decimal place than found in the original data. In addition, these two conventions will be employed.

(1) When there are an odd number of data, the median will be one of the original values. The manual follows the example given in this section and reports the median as given in the original data. And so the median of 1,2,3,4,5 is reported as $\tilde{x} = 3$.

(2) When the mean falls <u>exactly</u> between two values accurate to one more decimal place than the original data, the round-off rule in this section gives no specific direction. This manual follows the commonly accepted convention of always rounding up. And so the mean of 1,2,3,3 is reported as $\bar{x} = 9/4 = 2.3$ [i.e., $9/4 = 2.25$, rounded up to 2.3].

In addition, the median is the middle score <u>when the scores are arranged in order</u>, and the midrange is halfway between the first and last score <u>when the scores are arranged in order</u>. It is usually helpful to arrange the scores in order. This will not affect the mean, and it may also help in identifying the mode. Finally, no measure of center can have a value lower than the smallest score or higher than the largest score. Remembering this helps to protect against gross errors, which most commonly occur when calculating the mean.

5. Arranged in order, the n=8 scores are: 49 49 52 53 58 62 68 75
   a. $\bar{x} = (\Sigma x)/n = (466)/8 = 58.3$ sec        c. M = 49 sec
   b. $\tilde{x} = (53 + 58)/2 = 55.5$ sec                 d. m.r. = (49 + 75)/2 = 62.0 sec
   A sample of statistics students would not be representative of all adults in terms of age, scientific background, etc.

NOTE: This manual generally recommends doing smaller data sets by hand in order to gain an understanding of the material, and using the TI-83/84 Plus on the larger data sets. Exercise #5 may be done using the commands on the left to obtain the screen displays on the right.

STAT EDIT  4 [ClrList] L1
STAT EDIT  1 [Edit] (enter values in L1)
STAT EDIT  2 [SortA] L1
STAT EDIT  1 [Edit] (view sorted values)
STAT CALC  1 [1-Var Stats] L1

```
1-Var Stats
x̄=58.25
Σx=466
Σx²=27772
Sx=9.467991188
σx=8.856494792
↓n=8
```

```
1-Var Stats
↑n=8
minX=49
Q₁=50.5
Med=55.5
Q₃=65
maxX=75
```

7. Arranged in order, the n=25 scores are: 1 1 1 1 1 1 1 1 1 1 1 2 2 2 2 2 2 2 2 3 3 3 3 3 4
   a. $\bar{x} = (\Sigma x)/n = (47)/25 = 1.9$          c. M = 1
   b. $\tilde{x} = 2$          d. m.r. = (1 + 4)/2 = 2.5
   Yes, the measures of center can be calculated. No, the measures of center do not make sense – except for the mode of 1, which indicates that the smooth-yellow phenotype occurs more than any of the others.

9. Arranged in order, the n=14 scores are:
   120 120 125 130 130 130 130 135 138 140 140 143 144 150
   a. $\bar{x} = (\Sigma x)/n = (1875)/14 = 133.9$ mm      c. M = 130 mm
   b. $\tilde{x} = (130 + 135)/2 = 132.5$ mm      d. m.r. = (120 + 150)/2 = 135 mm
   Given that the same person was measured, the data vary considerably. If these results are typical, it appears that blood pressure readings either (1) are dependent on the person taking the reading or (2) fluctuate widely from moment to moment or (3) are influenced by a lack of precision in the present equipment and technology.

11. Arranged in order, the n=11 scores are: 0.64 0.68 0.72 0.76 0.84 0.84 0.84 0.84 0.90 0.90 0.92
    a. $\bar{x} = (\Sigma x)/n = (8.88)/11 = 0.807$ mm     c. M = 0.84 mm
    b. $\tilde{x} = 0.84$ mm     d. m.r. = (0.64 + 0.92)/2 = 0.780 mm
    No, a sample from one location would not be representative of the genetic variety possible.

13. Arranged in order, the scores are as follows.
    Roof: 1 2 2 2 2 2 2 2 3 3 4 4 4 4 6 7 7 8 11
    Hat: 2 2 2 2 3 3 3 3 3 3 3 3 3 4 4 4 5 5

    Roof
    n = 20
    $\bar{x} = (\Sigma x)/n = 78/20 = 3.9$ letters
    $\tilde{x} = (3 + 3)/2 = 3.0$ letters

    Hat
    n = 20
    $\bar{x} = (\Sigma x)/n = 62/20 = 3.1$ letters
    $\tilde{x} = (3 + 3)/2 = 3.0$ letters

    Based on the means, the words appear to be longer in *Cat on a Hot Tin Roof*.

15. Arranged in order, the scores are as follows.
    One-day: -3 -2 -2 -1 0 0 0 0 1 1 1 2 2 8
    Five-day: -9 -6 -4 -3 -2 -2 -1 -1 0 2 4 5 6 6

    One-day
    n = 14
    $\bar{x} = (\Sigma x)/n = 7/14 = 0.5$ °F
    $\tilde{x} = (0 + 0)/2 = 0.0$ °F

    Five-day
    n = 14
    $\bar{x} = (\Sigma x)/n = -5/14 = -0.4$ °F
    $\tilde{x} = (-1 + -1)/2 = -1.0$ °F

    On the average both forecasts seem to have about the same degree of accuracy, but the errors in the five-day forecasts are understandably more variable.

17. Arranged in order, the scores are as follows.
    JVB: 6.5 6.6 6.7 6.8 7.1 7.3 7.4 7.7 7.7 7.7
    Prov: 4.2 5.4 5.8 6.2 6.7 7.7 7.7 8.5 9.3 10.0

    JVB
    n = 10
    $\bar{x} = (\Sigma x)/n = 71.5/10 = 7.15$ min
    $\tilde{x} = (7.1 + 7.3)/2 = 7.20$ min
    m.r. = (6.5 + 7.7)/2 = 7.10 min
    M = 7.7 min

    Prov
    n = 10
    $\bar{x} = (\Sigma x)/n = 71.5/10 = 7.15$ min
    $\tilde{x} = (6.7 + 7.7)/2 = 7.20$ min
    m.r. = (4.2 + 10.0)/2 = 7.10 min
    M = 7.7 min

    The two data sets have identical measures of center, but the Providence (individual lines) waiting times seem to be more variable.

Measures of Center   SECTION 3-2   29

19. Arranged in order, the scores are as follows.
   Before: 3.0398  3.0406  3.0586  3.0762  3.0775  3.1038  3.1043  3.1086  3.1274  3.1582
   After:  2.4823  2.4848  2.4907  2.4950  2.4998  2.5024  2.5113  2.5127  2.5163  2.5298

   Before (n=10)
   $\bar{x} = (\Sigma x)/n = 30.8950/10 = 3.08950$ g
   $\tilde{x} = (3.0775 + 3.1038)/2 = 3.09065$ g

   After (n=10)
   $\bar{x} = (\Sigma x)/n = 25.0251/10 = 2.50251$ g
   $\tilde{x} = (2.4998 + 2.5024)/2 = 2.50110$ g

   Yes, the weights after 1983 are considerably lower. Apparently zinc is lighter than copper.

21. Use the LIST [2$^{nd}$ STAT] menu to access existing lists and create lists of the one-day and five-day errors as follows.
    LACTHI – LPH1 STO→ ERR1             LACTHI – LPHI5 STO→ ERR2
    Use 1-VAR STATS with ERR1 and ERR2 to generate the following screens.

    The desired information may be displayed as follows.

    One-day
    n = 35
    $\bar{x} = (\Sigma x)/n = -21/35 = -0.6$ °F
    $\tilde{x} = 0$ °F

    Five-day
    n = 35
    $\bar{x} = (\Sigma x)/n = -25/35 = -0.7$ °F
    $\tilde{x} = -1$ °F

    The results are very close to those obtained using only the first 14 days. No, the conclusions do not change when the larger data sets are used. Yes, we have more confidence in the results from the larger data set.

23. Use 1-VAR STATS with the lists CPPRE and CPPST to generate the following screens.

    The desired information may be displayed as follows.

    Before
    n = 35
    $\bar{x} = (\Sigma x)/n = 107.6174/35 = 3.07478$ g
    $\tilde{x} = 3.0763$ g

    After
    n = 37
    $\bar{x} = (\Sigma x)/n = 92.4668/37 = 2.49910$ g
    $\tilde{x} = 2.5004$ g

    The results are very close to those obtained using only the first 10 coins. No, the conclusions do not change when the larger data sets are used.

25. The x values below are the class midpoints from the given frequency table.

    | x  | f  | f·x  |
    |----|----|------|
    | 37 | 1  | 37   |
    | 42 | 3  | 126  |
    | 47 | 5  | 235  |
    | 52 | 11 | 572  |
    | 57 | 7  | 399  |
    | 62 | 7  | 434  |
    | 67 | 1  | 67   |
    |    | 35 | 1870 |

    $\bar{x} = \Sigma(f \cdot x)/\Sigma f = 1870/35 = 53.4$ °F

    The answer may also be found using 1-VAR STATS L1,L2 as described in the text. [see exercise #25 in section 3-3]

    This is close to the 53.8 found using the original data.

30    CHAPTER 3   Statistics for Describing, Exploring, and Comparing Data

27. The x values below are the class midpoints from the given frequency table, the class limits of which indicate the original data were integers.

| x | f | f·x |
|---|---|---|
| 43.5 | 25 | 1087.5 |
| 47.5 | 14 | 665.0 |
| 51.5 | 7 | 360.5 |
| 55.5 | 3 | 166.5 |
| 59.5 | 1 | 59.5 |
|  | 50 | 2339.0 |

$\bar{x} = \Sigma(f \cdot x)/\Sigma f = 2339.0/50 = 46.8$ mph

The answer may also be found using 1-VAR STATS L1,L2 as described in the text. [see exercise #27 in section 3-3]

This is close to the 46.7 mph found using the original data. The mean speed of 46.8 mph of those ticketed by the police is more than 1.5 times the posted speed limit of 30 mi/hr. But this indicates nothing about the mean speed of all drivers, a figure which may or may not be higher than the posted limit.

29. a. Arranged in order, the original 54 scores are:
26  29  34  40  46  48  60  62  64  65  76  79  80  86  90  94  105  114  116  120  125  132
140  140  144  148  150  150  154  166  166  180  182  202  202  204  204  212  220  220  236  262  270  316
332  344  348  356  360  365  416  436  446  514
$\bar{x} = \Sigma x/n = 9876/54 = 182.9$ lbs

b. Trimming the highest and lowest 10% (or 5.4 = 5 scores), the remaining 44 scores are:
48  60  62  64  65  76  79  80  86  90  94  105  114  116  120  125  132  140  140  144  148  150
150  154  166  166  180  182  202  202  204  204  212  220  220  236  262  270  316  332  344  348  356  360
$\bar{x} = \Sigma x/n = 7524/44 = 171.0$ lbs

c. Trimming the highest and lowest 20% (or 10.8 = 11 scores), the remaining 32 scores are:
79  80  86  90  94  105  114  116  120  125  132  140  140  144  148  150  150  154  166  166  180  182
202  202  204  204  212  220  220  236  262  270
$\bar{x} = \Sigma x/n = 5093/32 = 159.2$ lbs

Since physical considerations limit the lower values more than the upper values, increased trimming reduces the mean.

31. Treating the 5+ values as 5.0, $\bar{x} = \Sigma x/n = 17.1/5 = 3.42$. Since continuing the experiment would only increase $\Sigma x$, one can conclude that the actual mean battery life in the sample is greater than 3.42 years.

33. Let $\bar{x}_h$ stand for the harmonic mean.
$\bar{x}_h = n/[\Sigma(1/x)] = 2/[1/40 + 1/60] = 2/[0.0417] = 48.0$ mph

35. R.M.S. $= \sqrt{\Sigma x^2 / n} = \sqrt{[(111.2)^2 + (108.7)^2 + (109.3)^2 + (104.8)^2 + (112.6)^2]/5} = 109.35$ volts


## 3-3 Measures of Variation

1. The standard deviation is considered a measure of variation because it gives an indication of how much spread there is in the data. Very loosely, the standard deviation measures the distance a typical score is from the mean

3. Very loosely, the standard deviation measures the distance a typical score is from the mean. A student with a score of 85 would be 3.5 times farther away from the mean than is the typical student. Yes, in this context a score of 85 could be considered "unusual."

NOTE: Although not given in the text, the symbol R will be used for the range throughout this manual. Remember that the range is the difference between the highest and the lowest scores, and not necessarily the difference between the last and first scores as they are listed. Since calculating the range involves only the subtraction of 2 original pieces of data, the manual follows the example in this section of reporting that measure of variation with the same accuracy as the original data. In general, the standard deviation and the variance will be reported with one more decimal place than the original data.

When finding the square root of the variance to obtain the standard deviation, use <u>all</u> the decimal places of the variance – and not the rounded value reported as the answer. To do this, keep the value on the calculator display or place it in the memory. Do not copy down all the decimal places and then re-enter them to find the square root – as that not only is a waste of time, but also could introduce copying and/or round-off errors.

5. preliminary values: $n = 8$   $\Sigma x = 466$   $\Sigma x^2 = 27772$
   $R = 75 - 49 = 26$ sec
   $s^2 = [n(\Sigma x^2) - (\Sigma x)^2]/[n(n-1)] = [8(27772) - (466)^2]/[8(7)] = 5020/56 = 89.6$ sec$^2$
   $s = 9.5$ sec
   A sample of statistics students would not be representative of all adults in terms of age, scientific background, etc. Being a rather homogeneous group, they would likely exhibit less variability than is present in the general population of adults.

NOTE: The above format will be used in this manual for calculation of the variance and standard deviation for small data sets. To reinforce some of the key symbols and concepts, the following detailed mathematical analysis is given for the data of the above problem. In addition, the screen display from the STAT CALC 1-Var Stats program is given in the box. Even when working a problem by hand, it is recommended that you use this program to verify the values for $\Sigma x$ and $\Sigma x^2$.

| x | x-$\bar{x}$ | (x-$\bar{x}$)$^2$ | x$^2$ |
|---|---|---|---|
| 49 | -9.25 | 85.5625 | 2401 |
| 49 | -9.25 | 85.5625 | 2401 |
| 52 | -6.25 | 39.0625 | 2704 |
| 53 | -5.25 | 27.5625 | 2809 |
| 58 | -0.25 | 0.0625 | 3364 |
| 62 | 3.75 | 14.0625 | 3844 |
| 68 | 9.75 | 95.0625 | 4624 |
| 75 | 16.75 | 280.5625 | 5625 |
| 466 | 0 | 627.5000 | 27772 |

$\bar{x} = (\Sigma x)/n = 466/8 = 58.25$ sec

by formula 3-4, $s^2 = \Sigma(x-\bar{x})^2/(n-1)$
   $= 627.5000/7$
   $= 89.6429$ sec$^2$

by formula 3-5, $s^2 = [n(\Sigma x^2) - (\Sigma x)^2]/[n(n-1)]$
   $= [8(27772) - (466)^2]/[8(7)]$
   $= 5020/56$
   $= 89.6429$ sec$^2$

```
1-Var Stats
x̄=58.25
Σx=466
Σx²=27772
Sx=9.467991188
σx=8.856494792
↓n=8
```

NOTE: As shown in the preceding detailed mathematical analysis, formulas 3-4 and 3-5 give the same answer. The following three observations give hints for using these two formulas.
• When using formula 3-4, constructing a table having the first 3 columns shown above helps to organize the calculations and makes errors less likely. In addition, verify that $\Sigma(x-\bar{x}) = 0$ [except for a possible discrepancy at the last decimal due to rounding] before proceeding – if such is not the case, an error has been made and further calculation is fruitless.
• When using formula 3-5, the quantity $[n(\Sigma x^2) - (\Sigma x)^2]$ must be non-negative (i.e., greater than or equal to zero) – if such is not the case, an error has been made and further calculation is fruitless.
• In general, formula 3-5 is to be preferred over formula 3-4 because (1) it does not involve round-off error or messy calculations when the mean does not "come out even" and (2) the quantities $\Sigma x$ and $\Sigma x^2$ can be found directly (on most calculators, with one pass through the data) without having to construct a table or make a second pass through the data after finding the mean.
• The 1-VAR STATS function on the TI-83/84 Plus may be used with any data set. This manual uses the format outlined above for small data sets, but uses only the 1-VAR STATS function for large data sets and the data sets given in Appendix B of the text.

7. preliminary values: n = 25    Σx = 47    Σx² = 109
   R = 4 – 1 = 3
   $s^2 = [n(\Sigma x^2) - (\Sigma x)^2]/[n(n-1)] = [25(109) - (47)^2]/[25(24)] = 516/600 = 0.9$
   s = 0.9
   Yes, the formulas can be applied to these values to obtain measures of variation – but no, the resulting values will have no meaning because the data are at the nominal level.

9. preliminary values: n = 14    Σx = 1875    Σx² = 252179
   R = 150 – 120 = 30 mm
   $s^2 = [n(\Sigma x^2) - (\Sigma x)^2]/[n(n-1)] = [14(252179) - (1875)^2]/[14(13)] = 14881/182 = 81.8$ mm²
   s = 9.0 mm
   If the subject's blood pressure remains constant and is accurately measured, all the values would be the same and there would be no variability. Each value would equal the mean, and the standard deviation would be zero.

11. preliminary values: n = 11    Σx = 8.88    Σx² = 7.2568
    R = 0.92 – 0.64 = 0.28 mm
    $s^2 = [n(\Sigma x^2) - (\Sigma x)^2]/[n(n-1)] = [11(7.2568) - (8.88)^2]/[11(10)] = 0.9704/110 = 0.009$ mm²
    s = 0.094 mm
    No, a sample from one location would not be representative of the genetic variety possible. Being genetically homogeneous, the fruit flies in the sample would likely yield a smaller standard deviation than would result from a more representative group.

13. <u>Cat on a Hot Tin Roof</u>                              <u>Cat in the Hat</u>
    n = 20, Σx = 78, Σx² = 434                           n = 20, Σx = 62, Σx² = 208
    R = 11 – 1 = 10 letters                              R = 5 – 2 = 3 letters
    $s^2 = [n(\Sigma x^2) - (\Sigma x)^2]/[n(n-1)]$      $s^2 = [n(\Sigma x^2) - (\Sigma x)^2]/[n(n-1)]$
       $= [20(434) - (78)^2]/[20(19)]$                     $= [20(208) - (62)^2]/[20(19)]$
       = 2596/380 = 6.8 letters²                           = 316/380 = 0.8 letters²
    s = 2.6 letters                                      s = 0.9 letters
    Yes, there is significant less variation in word length in the Dr. Seuss book.

15. <u>One-day</u>                                           <u>Five-day</u>
    n = 14, Σx = 7, Σx² = 93                             n = 14, Σx = -5, Σx² = 269
    R = 8 – (-3) = 11 °F                                 R = 6 – (-9) = 15 °F
    $s^2 = [n(\Sigma x^2) - (\Sigma x)^2]/[n(n-1)]$      $s^2 = [n(\Sigma x^2) - (\Sigma x)^2]/[n(n-1)]$
       $= [14(93) - (7)^2]/[14(13)]$                       $= [14(269) - (-5)^2]/[14(13)]$
       = 1253/182 = 6.9 degrees²                           = 3741/182 = 20.6 degrees²
    s = 2.6 °F                                           s = 4.5 °F
    There appears to be a greater variation in the errors for the five-day forecasts. Yes, together with the fact that the forecasts do not differ significantly in mean error, this does suggest that the one-day forecasts are more accurate.

17. <u>Jefferson Valley</u>                                  <u>Providence</u>
    n = 10, Σx = 71.5, Σx² = 513.27                      n = 10, Σx = 71.5, Σx² = 541.09
    R = 7.7 – 6.5 = 1.2 min                              R = 10.0 – 4.2 = 5.8 min
    $s^2 = [n(\Sigma x^2) - (\Sigma x)^2]/[n(n-1)]$      $s^2 = [n(\Sigma x^2) - (\Sigma x)^2]/[n(n-1)]$
       $= [10(513.27) - (71.5)^2]/[10(9)]$                  $= [10(541.09) - (71.5)^2]/[10(9)]$
       = 20.45/90 = 0.23 min²                              = 298.65/90 = 3.32 min²
    s = 0.48 min                                         s = 1.82 min
    Exercise #17 of section 3-2 indicated that the mean waiting time was 7.15 minutes at each

bank. The Jefferson Valley (single line) waiting times, however, are considerably less variable. The range measures the differences between the extremes. The longest and shortest waits at Jefferson Valley differ by a little over a minute (R=1.2), while the longest and shortest waits at Providence differ by almost 6 minutes (R=5.8). The standard deviation measures the typical deviation from the mean. A Jefferson Valley customer usually receives service within about ½ minute (s=0.48) of 7.15 minutes, while a Providence customer usually receives service within about 2 minutes (s=1.82) of the mean.

19. <u>Before 1983</u>
    n = 10, $\Sigma x$ = 30.8950, $\Sigma x^2$ = 95.46318394
    R = 3.1582 – 3.0398 = 0.1184 g
    $s^2 = [n(\Sigma x^2) - (\Sigma x)^2]/[n(n-1)]$
      $= [10(95.46318394) - (30.8950)^2]/[10(9)]$
      $= 0.13081440/90 = 0.00145$ $g^2$
    s = 0.03812 g

    <u>After 1983</u>
    n = 10, $\Sigma x$ = 25.0251, $\Sigma x^2$ = 62.62760433
    R = 2.5298 – 2.4823 = 0.0475 g
    $s^2 = [n(\Sigma x^2) - (\Sigma x)^2]/[n(n-1)]$
      $= [10(62.62760433) - (25.0251)^2]/[10(9)]$
      $= 0.02041329/90 = 0.00023$ $g^2$
    s = 0.01506 g

    The weights of the pre-1983 pennies appear to vary about twice as much as the weights of the post-1983 pennies. The differences in variability could be caused by production factors – either zinc can be worked with more precisely than copper, or more precise penny-making machines have been developed since 1983. But it seems more likely that the differences in variability would be caused by wear – the older pennies have been through a wider variety of use, some being very worn from over 25 years of constant use and some being almost new because they have been out of heavy circulation for one reason or another.

NOTE: Exercises #21-#25 repeat earlier exercises, but use the full data sets instead of the first several values. In general, the variance and standard deviation of any sample size are good estimates of the population variance and standard deviation – although the larger the sample, the more confidence we have in the estimate. Increasing the sample size should not necessarily increase or decrease the variance or standard deviation. The range, however, is dependent upon the sample size because it is based only on the extremes and does not attempt to determine a "typical" spread based on all the data. In general, we expect larger samples to produce a larger range – in fact, it is not possible for the range to decrease as more data values are added to the sample.

21. Create lists of the one-day and five-day errors as follows.
        LACTHI – LPH1 → ERR1        LACTHI – LPHI5 → ERR2
    Use 1-VAR STATS with ERR1 and ERR2 to generate the following screens.

    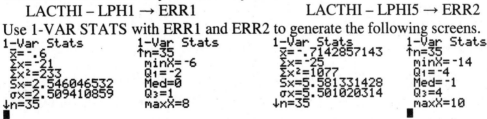

    The desired information may be displayed as follows.
    <u>One-day</u>
    n = 35
    R = 8 – (-6) = 14 °F
    s = 2.5 °F

    <u>Five-day</u>
    n = 35
    R = 10 – (-14) = 24 °F
    s = 5.6 °F

    There appears to be a grater variation in the errors for the five-day forecasts. Yes, together with the fact that the forecasts do not differ significantly in mean error, this does suggest that the one-day forecasts are more accurate. The conclusions are the same as in exercise #15.

34   CHAPTER 3   Statistics for Describing, Exploring, and Comparing Data

23. Use 1-VAR STATS with the lists CPPRE and CPPST to generate the following screens.

The desired information may be displayed as follows.

Before 1983
n = 35
R = 3.1582 – 2.9593 = 0.1989 g
s = 0.03910 g

After 1983
n = 37
R = 2.5298 – 2.4529 = 0.0769 g
s = 0.01648 g

The weights of the pre-1983 pennies appear to vary about twice as much as the weights of the post-1983 pennies. This is the same conclusion that was reached in exercise #19, and the individual results are almost identical.

NOTE: The following TI-83/84 Plus screens for exercises #25 and #27 were generated using 1-VAR STATS L1,L2 with the class midpoints in L1 and the class frequencies in L2.

#25
L1: 37,42,...
L2: 1,3,...

#27
L1: 43.5,47.5,...
L2: 24,14,...

25. The screen indicates s = 7.0°F. This is close to the 6.9 °F found using the original data.
    NOTE: The value for s may be found by hand as follows.

    | x | f | f·x | f·x² |
    |---|---|-----|------|
    | 37 | 1 | 37 | 1369 |
    | 42 | 3 | 126 | 5292 |
    | 47 | 5 | 235 | 11045 |
    | 52 | 11 | 572 | 29744 |
    | 57 | 7 | 399 | 22743 |
    | 62 | 7 | 434 | 26908 |
    | 67 | 1 | 67 | 4489 |
    |   | 35 | 1870 | 101590 |

    $s^2 = [n\Sigma(f \cdot x^2) - (\Sigma f \cdot x)^2]/[n(n-1)]$
    $= [35(101590) - (1870)^2]/[35(34)]$
    $= 58750/1190$
    $= 49.36$
    $s = 7.0\,°F$

27. The screen indicates s = 4.1 mph. This is close to the 4.0 mph found using the original data.

29. The Range Rule of Thumb suggests s ≈ range/4. Answers will vary, but one reasonable estimate for the range would be R = 70 – 26 = 44 years. This suggests s ≈ 44/4 = 11 years.

31. The Range Rule of Thumb suggests s ≈ range/4. Solving for the range, R ≈ 4s. If the distribution is approximately symmetric about the mean, then reasonable values for the upper and lower usual limits would be $\bar{x} \pm 2s$. In this instance,
    minimum usual value = $\bar{x} - 2s$ = 2838 – 2(504) = 1830 kwh
    maximum usual value = $\bar{x} + 2s$ = 2838 + 2(504) = 3846 kwh
    Yes, in this context a consumption of 578 kwh would be considered unusual.

33. a. The range from 169 to 183 is $\bar{x} \pm 1s$. The empirical rule suggests that about 68% of the data values should fall within those limits.

b. The range from 155 to 197 is $\bar{x} \pm 3s$. The empirical rule suggests that about 99.7% of the data values should fall within those limits.

35. The coefficient of variation for each sample is as follows.
    men: n=6   $\bar{x} = 69.167$   s = 2.137   CV = s/$\bar{x}$ = 2.137/69.167 = 0.031 = 3.1%
    eggs: n=9   $\bar{x} = 22.144$   s = 1.130   CV = s/$\bar{x}$ = 1.130/22.144 = 0.051 = 5.1%
    The values are similar, but the lengths of the eggs are slightly more variable for their context.
    NOTE: The coefficient of variation provides a direct comparison of the variability of two data sets even for objects that differ greatly in size and are measured in different units.

37. Even though the standard deviation measures the "typical" deviation and the deviation of the outlier will be "averaged" with the other deviations to get a "representative" value, the outlier will have a large effect and will increase the standard deviation.

## 3-4 Measures of Relative Standing

1. Since the sign of a z score is determined by subtracting $\bar{x}$ from the x value, a negative z score indicates that the x value is less than (i.e., below) the mean. Since the z score tells how many standard deviations a value is from the mean, an x value that has a z score of -2 is 2 standard deviations below the mean.

3. If 15 is the first quartile, then about 25% of the values fall below 15 – and about 75% of the values are above 15.

5. a. $|x - \bar{x}| = |182 - 176| = |6| = 6$ cm
   b. 6/7 = 0.86 standard deviations
   c. z = (x – $\bar{x}$)/s = (182 – 176)/7 = 6/7 = 0.86
   d. Since –2 < 0.86 < 2, Darwin's height is not considered unusual.

7. a. $|x - \bar{x}| = |67 - 71.5| = |-4.5| = 4.5$ in
   b. 4.5/2.1 = 2.14 standard deviations
   c. z = (x – $\bar{x}$)/s = (67 – 71.5)/2.1 = -4.5/2.1 = -2.14
   d. Since -2.14 < -2, McKinley's height is considered unusual.

9. In general z = (x – $\bar{x}$)/s
   a. $z_{97.5}$ = (97.5 – 98.20)/0.62 = -0.7/0.62 = -1.13
   b. $z_{98.60}$ = (98.60 – 98.20)/0.62 = 0.40/0.62 = 0.65
   c. $z_{98.20}$ = (98.20 – 98.20)/0.62 = 0/0.62 = 0

11. z = (x – $\bar{x}$)/s = (308 – 268)/15 = 40/15 = 2.67
    Yes, since 2.67 > 2, a pregnancy of 308 days is unusual. "Unusual" is just what the word implies – out of the ordinary. Other exercises in this section noted that there are unusual heights and IQ's, and so there can be unusual pregnancies.

13. Convert the two test scores to z scores.
    psychology: z = (x – $\bar{x}$)/s = (85 – 90)/10 = -5/10 = -0.50
    economics: z = (x – $\bar{x}$)/s = (45 – 55)/5 = -10/5 = -2.00
    Since -0.50 > -2.00, the psychology test score was relatively better.

For exercises #15-#29, refer to the list of ordered scores at the right.

15. Let b = # of scores below x; n = total number of scores
    In general, the percentile of score x is (b/n)·100.
    The percentile score of 25 is (4/76)·100 = 5.

17. Let b = # of scores below x; n = total number of scores
    In general, the percentile of score x is (b/n)·100.
    The percentile score of 40 is (57/76)·100 = 75.

19. To find $P_{10}$, L = (10/100)·76 = 7.6, rounded up to 8.
    Since the 8$^{th}$ score is 25, $P_{10}$ = 25.

21. To find $P_{25}$, L = (25/100)·76 = 19, a whole number.
    The mean of the 19$^{th}$ and 20$^{th}$ scores, $P_{25}$ = (28+28)/2 = 28.0.

23. To find $P_{33}$, L = (33/100)·76 = 25.08, rounded up to 26.
    Since the 26$^{th}$ score is 29, $P_{33}$ = 29.

25. To find $P_{01}$, L = (1/100)·76 = 0.76, rounded up to 1.
    Since the 1$^{st}$ score is 21, $P_{01}$ = 21.

27. a. The interquartile range is $Q_3 - Q_1$.
    For $Q_3 = P_{75}$, L = (75/100)·76 = 57 – a whole number.
    The mean of the 57$^{th}$ and 58$^{th}$ scores, $Q_3$ = (39+40)/2 = 39.5.
    For $Q_1 = P_{25}$, L = (25/100)·76 = 19 – a whole number.
    The mean of the 19$^{th}$ and 20$^{th}$ scores, $Q_1$ = (28+28)/2 = 28.0.
    The interquartile range is 39.5 – 28.0 = 11.5
    b. The midquartile is $(Q_1+Q_3)/2$ = (28.8 + 39.5)/2 = 33.75.
    c. The 10-90 percentile range is $P_{90} - P_{10}$.
    For $P_{90}$, L = (90/100)·76 = 68.4, rounded up to 69.
    The 69$^{th}$ score, $P_{90}$ = 49.
    For $P_{10}$, L = (10/100)·76 = 7.6, rounded up to 8.
    The 8$^{th}$ score, $P_{10}$ = 25.
    The 10-90 percentile range is 49 – 25 = 24.
    d. Yes, $Q_2 = P_{50}$ by definition. Yes, they are always equal.
    e. For $Q_2 = P_{50}$, L = (50/100)·76 = 38 – a whole number.
    The mean of the 38$^{th}$ and 39$^{th}$ scores, $Q_2$ = (33 + 34)/2 = 33.5.
    No; in this case $Q_2 = 33.5 \neq 33.75 = (Q_1 + Q_3)/2$, which demonstrates that the median does not necessarily equal the midquartile.

29. a. $D_1 = P_{10}$, $D_5 = P_{50}$, $D_8 = P_{80}$
    b. For d = 1,2,3,4,5,6,7,8,9, $L_d$ = (10d/100)·76 = 7.6d.
    For $D_1$, L = 7.6 and so $D_1 = x_8 = 25$
    For $D_2$, L = 15.2 and so $D_2 = x_{15} = 27$
    For $D_3$, L = 22.8 and so $D_3 = x_{23} = 29$
    For $D_4$, L = 30.4 and so $D_4 = x_{31} = 31$
    For $D_5$, L = 38 and so $D_5 = (x_{38} + x_{39})/2 = (33 + 34)/2 = 33.5$
    For $D_6$, L = 45.6 and so $D_6 = x_{46} = 35$
    For $D_7$, L = 53.2 and so $D_7 = x_{54} = 38$
    For $D_8$, L = 60.8 and so $D_8 = x_{61} = 41$
    For $D_9$, L = 68.4 and so $D_9 = x_{69} = 49$

| # | age | # | age |
|---|---|---|---|
| 1 | 21 | 39 | 34 |
| 2 | 22 | 40 | 34 |
| 3 | 24 | 41 | 34 |
| 4 | 24 | 42 | 35 |
| 5 | 25 | 43 | 35 |
| 6 | 25 | 44 | 35 |
| 7 | 25 | 45 | 35 |
| 8 | 25 | 46 | 35 |
| 9 | 26 | 47 | 35 |
| 10 | 26 | 48 | 35 |
| 11 | 26 | 49 | 36 |
| 12 | 26 | 50 | 37 |
| 13 | 27 | 51 | 37 |
| 14 | 27 | 52 | 38 |
| 15 | 27 | 53 | 38 |
| 16 | 27 | 54 | 38 |
| 17 | 28 | 55 | 38 |
| 18 | 28 | 56 | 39 |
| 19 | 28 | 57 | 39 |
| 20 | 28 | 58 | 40 |
| 21 | 29 | 59 | 41 |
| 22 | 29 | 60 | 41 |
| 23 | 29 | 61 | 41 |
| 24 | 29 | 62 | 41 |
| 25 | 29 | 63 | 41 |
| 26 | 29 | 64 | 42 |
| 27 | 30 | 65 | 42 |
| 28 | 30 | 66 | 43 |
| 29 | 31 | 67 | 45 |
| 30 | 31 | 68 | 46 |
| 31 | 31 | 69 | 49 |
| 32 | 32 | 70 | 50 |
| 33 | 32 | 71 | 54 |
| 34 | 33 | 72 | 60 |
| 35 | 33 | 73 | 61 |
| 36 | 33 | 74 | 63 |
| 37 | 33 | 75 | 74 |
| 38 | 33 | 76 | 80 |

c. For q = 1,2,3,4, $L_q = (20q/100) \cdot 76 = 15.2q$. These correspond to the even-numbered values computed in part (b) for the deciles.
  Quintile$_1$ = $P_{20}$ = $D_2$ = 27
  Quintile$_2$ = $P_{40}$ = $D_4$ = 31
  Quintile$_3$ = $P_{60}$ = $D_6$ = 35
  Quintile$_4$ = $P_{80}$ = $D_8$ = 41

## 3-5 Exploratory Data Analysis (EDA)

1. The boxplot gives the following information about the data set.
   minimum = 2     $Q_1$ = 5     median = 10
   maximum = 20    $Q_3$ = 12
   In addition, the data do not have a symmetric distribution about the mean – there is a concentration of scores from 10 to 12.

3. Sigma has less variation in repair times – and this is not merely a function of a few extreme values for Newport, as Newport also has a much larger the interquartile range $Q_3$-$Q_1$. Because the Sigma repair times have less variability, and costs are largely a function of the hourly rate, Sigma should have more predictable costs than Newport.

NOTE: Following the pattern established is previous section, the manual finds the 5-number by hand for small data sets and uses the TI-83/84 Plus for larger data sets and/or data sets from **Appendix B** in the text. In addition, the boxplots presented are the fully-labeled type that convey the relevant information to the reader in a single figure.

5. Order the n=11 scores to find the 5-number summary as follows.
      minimum = $x_1$ = 1316
      $Q_1$ = $x_3$ = 1511
      median = $x_6$ = 1910
      $Q_3$ = $x_9$ = 2060
      maximum = $x_{11}$ = 2496
   The boxpolot is given at the right.

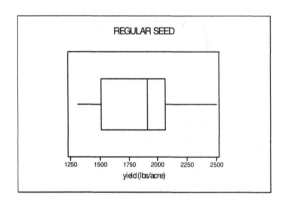

7. Using 1-VAR STATS with the list CQPRE gives the 5-number summary shown below. The boxplot is given at the right.
```
1-Var Stats
↑n=40
 minX=6.0002
 Q1=6.13305
 Med=6.19435
 Q3=6.26605
 maxX=6.3669
```

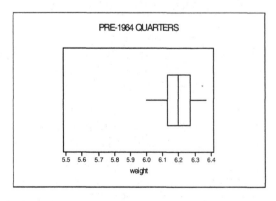

9. Using 1-VAR STATS with the list BLEN gives
the 5-number summary shown below.
The boxplot is given at the right.
```
1-Var Stats
↑n=54
 minX=36
 Q₁=50
 Med=60.75
 Q₃=66.5
 maxX=76.5
```

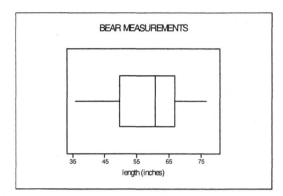

While the median is toward the higher end of the main box (indicating a slight negative skew), the distribution is approximately symmetric.

11. Using 1-VAR STATS with the list MBMI gives
the 5-number summary shown below.
The boxplot is given at the right.
```
1-Var Stats
↑n=40
 minX=19.6
 Q₁=23.65
 Med=26.2
 Q₃=27.6
 maxX=33.2
```

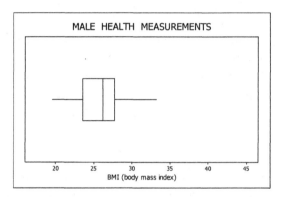

13. Choose Brand 1. Brand 1 and Brand 3 batteries both have a longer median life than Brand 2, but the Brand 1 values have less variability than those for Brand 3. In addition to making it easier to anticipate and schedule battery replacements, the small variability in values for Brand 1 may also indicate better quality control and hence an overall better product.

15. Using 1-VAR STATS with the list RNMON gives the 5 number summary below.
```
1-Var Stats
↑n=52
 minX=0
 Q₁=0
 Med=0
 Q₃=.01
 maxX=1.41
```
  IQR = $Q_3 - Q_1$ = 0.01 − 0.00 = 0.01
  1.5(IQR) = 1.5(0.01) = 0.015
  3(IQR) = 3(0.01) = 0.03
There are no outliers at the lower end, since there are no values below $Q_1$ = 0.00.
The outliers at the upper end are as follows.
  mild: $Q_3 + 1.5(IQR) < x \le Q_3 + 3(IQR)$     extreme: $x > Q_3 + 3(IQR)$
      0.01 + 0.015 < x ≤ 0.01 + 0.03           x > 0.01 + 0.03
          0.025 < x ≤ 0.04                     x > 0.04
The only mild outlier is 0.03.
The extreme outliers are 0.05, 0.11, 0.12, 0.12, 0.41, 0.43, 0.47, 0.49, 0.59, 0.92, 1.41.

## Statistical Literacy and Critical Thinking

1. No, reducing the standard deviation of the repair times means that there is less variability from time to time – not that the times are smaller. Suppose, for example that the repair times were previously about 2 hours each. The redesigned procedure could call for a minimum time of 2 hours spent on each repair, no matter how simple the problem. This would eliminate all the very short times and the remaining times would be more tightly clustered at 2 hours and above – a reduction in the standard deviation of the repair times, but a very inefficient increase in the typical repair time.

2. No; even though the mean life is greater than 10 years as desired, the standard deviation cannot be ignored. If the mean life is just barely greater than 10, about half the batteries would have a life less than 10. And if the standard deviation is large, many of those lower values could be considerably less than 10.

3. The additional value will have a considerable effect on the mean and the standard deviation. The mean can be expected to increase by about ($1,000,000)/51 = $19,600. Because it measures "typical deviation" from the mean, the standard deviation can also be expected to increase – since scores less than the old mean will now be about $19,600 farther from the new mean, and the individual deviations for scores above the old mean will also likely increase. The median will change very little – moving from the average of the $25^{th}$ and $26^{th}$ ordered scores to the $26^{th}$ ordered score.

4. No, the results are not likely to yield a mean that is fairly close to the mean value of all cars owned by Americans. Some possible reasons for this are as follows.
   #1. The subscribers of any one particular ISP are not necessarily representative of the general population. Persons contracting with an ISP, for example, may tend to be more affluent than the general population.
   #2. The sample is a voluntary response sample, and those who choose to respond voluntarily are not necessarily representative of the group that was asked to respond.
   #3. The values were self-reported. People may not have an accurate or objective idea of the value of their cars – and some people may deliberately exaggerate the value of their cars to make themselves look good.

## Review Exercises

1. The n=20 ordered circumferences are as follows.
   1.8  1.8  1.9  2.4  3.1  3.4  3.7  3.7  3.8  3.9  4.0  4.1  4.9  5.1  5.1  5.2  5.3  5.5  8.3  13.7

   a. $\bar{x} = (\Sigma x)/n = 90.7/20 = 4.54$ ft

   b. $\tilde{x} = (3.9 + 4.0)/2 = 3.95$ ft

   c. M = 1.8 ft, 3.7 ft, 5.1ft (tri-modal)

   d. m.r. = (1.8 + 13.7)/2 = 7.75 ft

   e. R = 13.7 – 1.8 = 11.9 ft

   f. $s^2 = [n(\Sigma x^2) - (\Sigma x)^2]/[n(n-1)]$
   $= [20(544.85) - (90.7)^2]/[20(19)] = 2670.51/380 = 7.03$
   s = 2.65 ft

g. $s^2 = 7.03$ ft$^2$ [from part (f)]

h. $Q_1 = (x_5+x_6)/2 = (3.1 + 3.4)/2 = 3.25$ ft

i. $Q_3 = (x_{15}+x_{16})/2 = (5.1 + 5.2)/2 = 5.15$ ft

j. $P_{10} = (x_2 + x_3)/2 = (1.8 + 1.9)/2 = 1.85$ ft

2. a. $z = (x - \bar{x})/s$
   $z_{13.7} = (13.7 - 4.535)/2.65 = 9.165/2.65 = 3.46$

   b. Yes; since 13.7 is more than 2 standard deviations above the mean it is considered an unusual circumference.

   c. The Range Rule of Thumb suggests $s \approx$ range/4. Solving for the range, $R \approx 4s$. If the distribution is approximately symmetric about the mean, then reasonable values for the upper and lower usual limits would be $\bar{x} \pm 2s$. In this instance,
   minimum usual value = $\bar{x} - 2s = 4.535 - 2(2.65) = -0.765$ ft
   maximum usual value = $\bar{x} + 2s = 4.535 + 2(2.65) = 9.835$ ft
   Other than the 13.7 already identified, no other listed circumferences are considered unusual.   3. The requested frequency distribution is given below.

   | circumference | frequency |
   |---|---|
   | 1.0 – 2.9 | 4 |
   | 3.0 – 4.9 | 9 |
   | 5.0 – 6.9 | 5 |
   | 7.0 – 8.9 | 1 |
   | 9.0 – 10.9 | 0 |
   | 11.0 – 12.9 | 0 |
   | 13.0 – 14.9 | 1 |
   |  | 20 |

4. The requested histogram is given below.

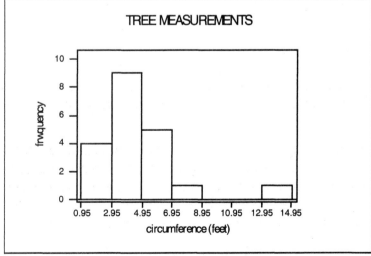

The distribution is approximately bell-shaped, with a definite positive skew.

5. Order the n=20 scores to find the 5-number
   summary as follows.
   minimum = $x_1$ = 1.8 ft
   $Q_1 = (x_5+x_6)/2$
      = (3.1 + 3.4)/2 = 3.25 ft
   median = $(x_{10}+x_{11})/2$
      = (3.9 + 4.0)/2 = 3.95 ft
   $Q_3 = (x_{15}+x_{16})/2$
      = (5.1 + 5.2)/2 = 5.15 ft
   maximum = $x_{20}$ = 13.7 ft
   The boxpolot is given at the right.

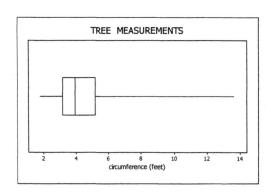

6. To make a comparison, convert each test score to a standard score (i.e., to a z score).
   management: $z = (x - \bar{x})/s = (72 - 80)/12 = -8/12 = -0.67$
   production: $z = (x - \bar{x})/s = (19 - 20)/5 = -1/5 = -0.20$
   Since -0.20 > -0.67, the score of 19 on the production test is the better score.

7. Answers will vary, but the following are reasonable possibilities.
   a. The estimated mean age is 4 years.
   b. The estimated maximum and minimum ages are 20 years and 0 years, giving an estimated range of R = 20 – 0 = 20. The Range Rule of Thumb suggests s ≈ R/4. And so the estimated standard deviation is 20/4 = 5 years.

8. The Range Rule of Thumb suggests s ≈ range/4. Solving for the range, R ≈ 4s. If the distribution is approximately symmetric about the mean, then reasonable values for the upper and lower usual limits would be $\bar{x} \pm 2s$. In this instance,
   minimum usual value = $\bar{x} - 2s$ = 69.0 – 2(2.5) = 64 in
   maximum usual value = $\bar{x} + 2s$ = 69.0 + 2(2.5) = 74 in
   Since 64 ≤ 72 ≤ 74, a height of 72 inches is within limits for usual heights and would not be considered unusual in this context.

## Cumulative Review Exercises

1. a. The given values are continuous, since circumference can assume any value on a continuum.
   b. The given values are at the ratio level of measurement – since differences are meaningful and there is a natural zero.

2. a. The mode is the appropriate measure of "typicalness" for data at the nominal level. If the data is at the nominal level, numbers would be valid for identification only and could not be meaningfully ordered or otherwise mathematically manipulated. The mean, median and midrange all require ordering and/or other mathematical manipulations.
   b. This is convenience sampling, since he is contacting people who happen to be handy according to some criterion.
   c. This is cluster sampling, since all the subjects at randomly selected locations are sampled.
   d. Consistency is the lack of variability. Since she is dealing with variability, the standard deviation is the most relevant of the given statistics. It should be lowered. While the interquartile range (IQR = Q3 – Q1) also measures variability, it essentially ignores the extremely high and the extremely low scores – which are the very ones causing most of the problem. The standard deviation measures variability and considers all the scores.

# Chapter 4

# Probability

## 4-2 Fundamentals

1. "The probability of winning the grand prize in the Illinois lottery is 1/20,358,520" means that there are 20,358,520 equally likely possibilities (which may be the number of people who entered the lottery, the number of possible numeric outcomes, etc.) and that only one of them corresponds to winning. Yes, under these conditions such a win would be unusual – it has only 1 chance in 20,358,520 of occurring.

3. Yes, a probability of 0.001 = 1/1000 indicates that the event in question has only 1 chance in 1,000 of occurring.

5. a. "You will surely pass" indicates a probability of 1.00 – a certainty.
   b. "A 10% chance of rain" indicates a probability of 0.10.
   c. "A snowball's chance in hell" indicates a probability of 0.00 – an impossibility.

7. The probability of any event must be a value from 0 to 1 inclusive. The following listed values cannot be probabilities.
   -1, since it is less than 0
   2, since it is greater than 1
   5/3 = 1.667, since it is greater than 1
   $\sqrt{2}$ = 1.414, since it is greater than 1

9. a. Exactly one girl is the set with 3 simple events A = {BBG,BGB,GBB}.
      P(A) = 3/8 = 0.375
   b. Exactly two girls is the set with 3 simple events B = {BGG,GBG,GGB}.
      P(B) = 3/8 = 0.375
   c. All three girls is the set with 1 simple event C = {GGG}.
      P(C) = 1/8 = 0.125

11. There were 428 + 152 = 580 plants in the experiment.
    Let G = getting an offspring pea that is green.
    The estimate is P(G) = 428/580 = 0.738, which is very close to the expected value of 0.750.

13. There were 295 + 30 = 325 births in the experiment.
    Let G = getting a baby girl.
    The estimate is P(G) = 295/325 = 0.908.
    Yes, it appears that the technique is effective in increasing the likelihood of a girl.

15. a. There are 13R + 25O + 8Y + 8Br + 27Bl + 19G = 100 total M&M's, 27 of which are blue.
       The estimate is P(Blue) = 27/100 = 0.27.
    b. Yes, the estimate from part (a) agrees with the claim that 24% of all M&M's are blue. Even if precisely 24% of all M&M's are blue, we do not expect to get exactly 24 blue M&M's in every group of 100.

17. a. Let B = the birthdate is correctly identified. P(B) = 1/365 = 0.003.
    b. Yes. Since 0.003 ≤ 0.05, it would be unusual to guess correctly on the first try.
    c. Most people would probably believe that Mike's correct answer was the result of having inside information – and not making a lucky guess.
    d. Fifteen years is a big error. If the guess was made in all seriousness, Kelly would likely think that Mike was not a knowledgeable person and/or be personally insulted – and a second date would be unlikely. If the guess is perceived as being made in jest, Kelly might well appreciate his sense of humor and/or handling of a potentially awkward situation – and a second date would be likely.

19. Let A = a 20-24 year old has an accident this year. The estimate is P(A) = 136/400 = 0.340. If a driver who will be in that age bracket next year is selected, and the probabilities do not change from year to, the approximate probability that he or she will be in an accident next year is 0.340. Since 0.340 > 0.05, it would not be unusual for a driver that age to be involved in a crash next year. Yes, the result is high enough to be of concern.

21. There were 117 + 617 = 734 total patients in the experiment.
    Let H = a patient experiences a headache. The estimate is P(H) = 117/734 = 0.159.
    No. Since 0.159 > 0.05, experiencing a headache from taking the drug is not an unusual occurrence. Yes, the probability of experiencing a headache is high enough to be of concern – especially since a headache is one symptom that can interfere with the intended purpose for taking the drug in the first place.

23. a. The estimated probability of a wrong result for a person who does not use marijuana is 24/178 = 0.135.
    b. No. Since 0.135 > 0.05, it is not unusual to obtain a wrong test result for a person who does not use marijuana.

25. a. Listed by birth order, the sample space contains 4 simple events: BB  BG  GB  GG.
    b. P(GG) = 1/4 = 0.25
    c. P(BG or GB) = 2/4 = 0.50

27. a. Listed with the brown/brown parent's contribution first, the sample space has 4 simple events: brown/brown  brown/blue  brown/brown  brown/blue
    b. P(blue/blue) = 0/4 = 0, an impossibility
    c. P(brown eyes) = P(blue/brown or brown/blue or brown/brown) = 4/4 = 1, a certainty

29. The odd against winning are P(not winning)/P(winning) = (423/500)/(77/500) = 423/77, or 423:77 – which is approximately 5.5:1, or 11:2.

31. Of the 38 slots, 18 are odd. Let W = winning because an odd number occurs
    a. P(W) = 18/38 = 9/19 = 0.474
    b. odds against W = P(not W)/P(W) = (20/38)/(18/38) = 20/18 = 10/9, or 10:9
    c. If the odds against winning are 1:1, a win gets back your bet plus $1 for every $1 bet. If you bet $18 and win, you get back $18 + $18 and your profit is $18.
    d. If the odds against winning are 10:9, a win gets get back your bet plus $10 for every $9 bet. If you bet $18 and win, you get back $18 + $20 and your profit is $20.

44   CHAPTER 4   Probability

33. If the odds against A are a:b, the definition of odds indicates that P(not A)/P(A) = a/b.
Since it is also true that P(not A) + P(A) = 1, algebra indicates that
P(not A) = a/(a+b)
P(A) = b/(a+b).
If a=9 and b=1 as given in the problem, P(A) = b/(a+b) = 1/(9+1) = 1/10 = 0.100

35. No matter where the two flies land, it is possible to cut the orange in half to include flies on the same half. Since this is a certainty, the probability is 1. [Compare the orange to a globe. Suppose fly A lands on New York City, and consider all the circles of longitude. Wherever fly B lands, it is possible to slice the globe along some circle of longitude that places fly A and fly B on the same half.]
NOTE: If the orange is marked into two predesignated halves before the flies land, the probability is different – once fly A lands, fly B has ½ a chance of landing on the same half. If both flies are to land on a specific one of the two predesignated halves, the probability is different still – fly A has ½ a chance of landing on the specified half, and only ½ the time will fly B pick the same half: the final answer would be ½ of ½, or ¼.

## 4-3 Addition Rule

1. Two events are disjoint if they cannot occur at the same time. The occurrence of one means that the other cannot occur – in other words, the occurrence of one prevents the occurrence of the other.

3. Close friends and relatives are a convenience sample and may not be representative of the general population.

5. a. No, it is possible to elect a female president.
   b. No, it is possible to select a male who smokes cigars.
   c. Yes, receiving Lipitor means that a person is not in the group that received no medication.

7. In general $P(\bar{A}) = 1 - P(A)$.
   a. $P(\bar{A}) = 1 - P(A)$
      $= 1 - 0.05$
      $= 0.95$
   b. Let R = a woman has red/green color blindness.
      $P(\bar{R}) = 1 - P(R)$
      $= 1 - 0.0025$
      $= 0.9975$

NOTE: Throughout the manual we use the first letter (or other natural designation) of each category to represent that category. And so in exercises #9– #12,
P(P) = P(selecting an accident in which the pedestrian was intoxicated)
P(D) = P(selecting an accident in which the driver was intoxicated)
P(PD) = P(selecting an accident in which pedestrian and driver were intoxicated)
$P(P\bar{D})$ = P(selecting an accident in which pedestrian was intoxicated and driver was not)
etc.
If there is ever cause for ambiguity, the notation will be clearly defined. Since mathematics and statistics use considerable notation and formulas, it is important to clearly define what various letters stand for.

9. Make a table like the one at the right.
   Let P = the pedestrian was intoxicated.
   Let D = the driver was intoxicated.
   There are two approaches.

|  |  | PEDESTRIAN | | |
|---|---|---|---|---|
|  |  | yes | no | |
| DRIVER | yes | 59 | 79 | 138 |
|  | no | 266 | 581 | 847 |
|  |  | 325 | 660 | 985 |

- Use broad categories and correct for double-counting – the "formal addition rule".
  
  P(P or D) = P(P) + P(D) − P(P and D)
  
  = 325/985 + 138/985 − 59/985
  
  = 404/985
  
  = 0.410

- Use individual mutually exclusive categories – the "intuitive addition rule".
  
  P(P or D) = P(PD or P$\bar{D}$ or $\bar{P}$D)
  
  = P(PD) + P(P$\bar{D}$) + P($\bar{P}$D)
  
  = 59/985 + 266/985 + 79/985
  
  = 404/985
  
  = 0.410

NOTE: In general, using broad categories and correcting for double-counting is a more powerful technique that "lets the formula do the work" and requires less analysis by the solver. Except when more detailed analysis is instructive, this manual uses the first approach.

11. Refer to exercise #9.
    
    P(P or $\bar{D}$) = P(P) + P($\bar{D}$) − P(P and $\bar{D}$)
    
    = 325/985 + 847/985 − 266/985
    
    = 906/985 = 0.920

13. Make a table like the one at the right.
    
    P($\bar{A}$) = 1 − P(A)
    
    = 1 − 40/100
    
    = 60/100 = 0.60

|  |  | GROUP | | | | |
|---|---|---|---|---|---|---|
|  |  | O | A | B | AB | |
| TYPE | Rh+ | 39 | 35 | 8 | 4 | 86 |
|  | Rh− | 6 | 5 | 2 | 1 | 14 |
|  |  | 45 | 40 | 10 | 5 | 100 |

15. Refer to exercise #13.
    
    P(A or Rh−) = P(A) + P(Rh−) − P(A and Rh−)
    
    = 40/100 + 14/100 − 5/100
    
    = 49/100 = 0.49

17. Refer to exercise #13.
    
    P(not Rh+) = 1 − P(Rh+)
    
    = 1 − 86/100
    
    = 14/100 = 0.14

19. Refer to exercise #13.
    
    P(AB or Rh+) = P(AB) + P(Rh+) − P(AB and Rh+)
    
    = 5/100 + 86/100 − 4/100
    
    = 87/100 = 0.87

21. The table is given below.

|  |  | FLOWER | | |
|---|---|---|---|---|
|  |  | purple | white | |
| POD | green | 5 | 3 | 8 |
|  | yellow | 4 | 2 | 6 |
|  |  | 9 | 5 | 14 |

46   CHAPTER 4   Probability

23. Make a table like the one at the right.
    Let A = a person is age 18-21
        N = a person does not respond
    P(A or N) = P(A) + P(N) – P(A and N)
              = 84/359 + 31/359 – 11/359
              = 104/359 = 0.290

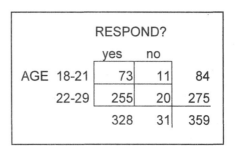

25. No, as illustrated by the following example.
    Let A = has prostate cancer.
    Let B = is a female.
    Let C = has testicular cancer.

27. The expression can be derived using algebra as follows.
    P(A or B or C)
        = P[(A or B) or C]
        = P(A or B)                  + P(C) – P[(A or B) and C]
        = [P(A) + P(B) – P(A and B)] + P(C) – P[(A and C) or (B and C)]
        = P(A) + P(B) – P(A and B)   + P(C) – [P(A and C) + P(B and C) – P(A and B and C)]
        = P(A) + P(B) – P(A and B)   + P(C) – P(A and C) – P(B and C) + P(A and B and C)
        = P(A) + P(B) + P(C) – P(A and B) – P(A and C) – P(B and C) + P(A and B and C)

**4-4 Multiplication Rule: Basics**

1. Two events are independent if the occurrence of one does not affect the occurrence of the other. Mathematically, independence means that the probability that one event occurs is not affected by whether or not the other event has already occurred.

3. If only the 24 unselected students are available for the second selection, he is sampling without replacement. When sampling without replacement, the second outcome is not independent of the first – the first outcome affects the first in that whoever was chosen in the first selection cannot be chosen in the second selection.

5. a. Technically, dependent. The first outcome affects the second outcome in that whichever quarter was chosen in the first selection cannot be chosen in the second selection.
    NOTE: If selecting from a small set of pre-2001 quarters (e.g., the 8 that you have your possession), this dependence is significant. If selecting from "all" pre-2001 quarters, this dependence is negligible and for all practical purposes the events are independent.
   b. Technically, dependent. The first outcome affects the second outcome in that whichever viewer was chosen in the first selection cannot be chosen in the second selection.
    NOTE: If selecting from a small set of such viewers (e.g., the 8 that are in a given room), this dependence is significant. If selecting from "all" such viewers, this dependence is negligible and for all practical purposes the events are independent.
   c. In general these events are dependent, but the answer depends on the situation.
    • Dependent, if the invitation is made with visual contact – since physical appearance is one factor in deciding whether or not to accept a date with a person when there is visual contact.
    • Independent, if the invitation is made without visual contact (e.g., over the telephone) – in which case physical appearance has no opportunity to influence the decision on whether or not to accept the date.

7. Let T = getting the 2-choice true/false question correct by random guessing.
   Let M = getting the 4-choice multiple choice question correct by random guessing.
   P(T and M) = P(T)·P(M|T)
   $\qquad\qquad$ = (1/2)(1/4) = 1/8 = 0.125
   Random guessing does not appear to be a very good strategy.

9. Let O = selecting a hunter who was wearing orange.
   P(O) = 6/123 = 0.048, for the first selection from the sample
   a. $P(O_1$ and $O_2) = P(O_1) \cdot P(O_2|O_1)$
      $\qquad\qquad$ = (6/123)(6/123) = 0.00238
   b. $P(O_1$ and $O_2) = P(O_1) \cdot P(O_2|O_1)$
      $\qquad\qquad$ = (6/123)(5/122) = 0.00200
   c. Selecting without replacement makes more sense in this context. Selecting with replacement could lead to re-interviewing the same hunter, which would lead to no new information.

11. Let O = selecting a CD that is OK.
    P(O) = 4900/5000, for the first selection
    P(accepting the entire batch) = $P(O_1$ and $O_2$ and $O_3$ and $O_4)$
    $\qquad\qquad$ = $P(O_1) \cdot P(O_2|O_1) \cdot P(O_3|O_1$ and $O_2) \cdot P(O_4|O_1$ and $O_2$ and $O_3)$
    $\qquad\qquad$ = (4900/5000)(4899/4999)(4898/4998)(4897/4997) = 0.922

13. Let G = a girl is born.
    P(G) = 0.50, for each couple
    P(all 12 girls) = $P(G_1$ and $G_2$ and...and $G_{12})$
    $\qquad\qquad$ = $P(G_1) \cdot P(G_2) \cdot ... \cdot P(G_{12})$
    $\qquad\qquad$ = $(0.50)(0.50)...(0.50) = (0.50)^{12} = 0.000244$
    If there are actually 12 girls born to the 12 couples in the experiment, then the gender-selection method appears to be effective. Either the method is effective or a very rare event (that has a 1 in 4096 chance of occurrence) has taken place.

15. Let A = the alarm clock of that type functions properly.
    P(A) = 0.975
    a. $P(\overline{A}) = 1 - P(A)$
       $\qquad$ = 1 − 0.975 = 0.025

    b. $P(\overline{A}_1$ and $\overline{A}_2) = P(\overline{A}_1) \cdot P(\overline{A}_2)$
       $\qquad\qquad$ = (0.025)(0.025) = 0.000625
    c. P("awakened with two clocks") = 1 − 0.000625 = 0.999375
       NOTE: This part could also be worked as follows.
       $P(A_1$ or $A_2) = P(A_1) + P(A_2) - P(A_1$ and $A_2)$
       $\qquad\qquad$ = 0.975 + 0.975 − (0.975)(0.975)
       $\qquad\qquad$ = 0.999375
    d. Yes, the second alarm clock results in greatly improved reliability.
       NOTE: Parts (c) and (d) actually give the probability that "at least one alarm clock functions properly" and not the probability that someone is "awakened with two clocks." If a person fails to arise because he sleeps through an alarm, then an additional clock may not increase the likelihood of awakening.

17. Make a table like the one at the right.
Let D = the driver was intoxicated.
Let P = the pedestrian was intoxicated.
$P(D_1 \text{ and } D_2) = P(D_1) \cdot P(D_2|D_1)$
$= (138/985)(137/984) = 0.0195$

|  |  | PEDESTRIAN | | |
|---|---|---|---|---|
|  |  | yes | no |  |
| DRIVER | yes | 59 | 79 | 138 |
|  | no | 266 | 581 | 847 |
|  |  | 325 | 660 | 985 |

19. Refer to exercise #17.
Let N = an accident in which neither the pedestrian nor the driver was intoxicated.
a. $P(N) = P(\overline{P} \text{ and } \overline{D}) = 581/985 = 0.590$
b. $P(N_1 \text{ and } N_2) = P(N_1) \cdot P(N_2|N_1)$
$= (581/985)(580/984) = 0.34767$
c. $P(N_1 \text{ and } N_2) = P(N_1) \cdot P(N_2|N_1)$
$= (581/985)(581/985) = 0.34792$
d. The answers in parts (c) and (d) are both 0.348 when rounded to 3 significant digits. The answer in part (d) is slightly larger because replacing the type-N accident after the first selection makes it more likely that another type-N accident will be chosen on the second selection.

21. Let D = a birthday is different from any yet selected.
$P(D_1) = 366/366 = 1$ NOTE: With nothing to match, it <u>must</u> be different.
$P(D_2) = P(D_2|D_1) = 365/366$
$P(D_3) = P(D_3|D_1 \text{ and } D_2) = 364/366$
...
$P(D_5) = P(D_5|D_1 \text{ and } D_2 \text{ and } D_3 \text{ and } D_4) = 362/366$
...
$P(D_{25}) = P(D_{25}|D_1 \text{ and } D_2 \text{ and } D_3 \text{ and}\ldots\text{and } D_{24}) = 342/366$

a. P(all different) = $P(D_1 \text{ and } D_2 \text{ and } D_3)$
$= P(D_1) \cdot P(D_2) \cdot P(D_3)$
$= (366/366)(365/366)(364/366) = 0.992$
b. P(all different) = $P(D_1 \text{ and } D_2 \text{ and } D_3 \text{ and } D_4 \text{ and } D_5)$
$= P(D_1) \cdot P(D_2) \cdot P(D_3) \cdot P(D_4) \cdot P(D_5)$
$= (366/366)(365/366)(364/366)(363/366)(362/366) = 0.973$
c. P(all different) = $P(D_1 \text{ and } D_2 \text{ and}\ldots\text{and } D_{25})$
$= P(D_1) \cdot P(D_2) \cdot \ldots \cdot P(D_{25})$
$= (366/366)(365/366)\ldots(342/366) = 0.432$

NOTE: The above calculations assume that all birthdays (even February 29) are equally likely. If one omits February 29 and considers there to be only 365 assumed equally likely birthdays, the answers to (a) and (b) do not change – but the answer to (c) is 0.431. An algorithm to perform this calculation can be constructed using a programming language – or most spreadsheet or statistical software packages. In BASIC, for example, use

```
10 LET P=1
15 PRINT "HOW MANY PEOPLE?"
20 INPUT D
30 FOR K=1 TO D-1
40 LET P=P*(366-K)/366
50 NEXT K
60 PRINT "P(ALL DIFFERENT BIRTHDAYS) IS  ";P
70 END
```

23. This problem can be done by two different methods. In either case, let
    A = getting an ace
    S = getting a spade.
    - Consider the sample space
      The first card could be any of 52 cards, and for each first card there are 51 possible second cards. This makes a total of 52·51 = 2652 equally likely outcomes in the sample space. How many of them are $A_1$ and $S_2$?
      The aces of hearts, diamond and clubs can be paired with any of the 13 spades for a total of 3·13 = 39 favorable possibilities. The ace of spades can only be paired with any of the remaining 12 members of that suit for a total of 12 favorable possibilities. Since there are 39 + 12 = 51 total favorable possibilities among the equally likely outcomes,
      $P(A_1$ and $S_2) = 51/2652 = 0.0192$
    - Use the formulas (and express the event as two mutually exclusive possibilities)
      $P(A_1$ and $S_2) = P([\text{spade}A_1$ and $S_2]$ or $[\text{other}A_1$ and $S_2])$
      $= P(\text{spade}A_1$ and $S_2) + P(\text{other}A_1$ and $S_2)$
      $= P(\text{spade}A_1)·P(S_2|\text{spade}A_1) + P(\text{other}A_1)·P(S_2|\text{other}A_1)$
      $= (1/52)(12/51) + (3/52)(13/51)$
      $= 12/2652 + 39/2652$
      $= 51/2652 = 0.0192$

## 4-5 Multiplication Rule: Complements and Conditional Probability

1. Knowing that "at least one" of ten items is defective does not indicate the exact number of defects – only that it is 1 or greater, and no more than 10.

3. Since about half the population is male, it may be reasonable to estimate that the probability a shopper in a gender-neutral situation is a male is ½. To estimate the probability such a shopper is a male given that a credit card was used for the purchase, one needs information about the proportion of all shoppers that use credit cards and the proportion of males that use credit cards.

5. If it is not true that at least one of them tests positive, then all six test negative – or, equivalently, none of the six tests positive.

7. If it is not true that none of them have the gene, then at least one of the 12 males have the particular X-linked recessive gene.

9. Let F = a credit card is being used fraudulently.
   Let D = a credit card was used in several different countries in one day.
   Answers will vary. One reasonable estimate is $P(F|D) \approx 0.999$.

11. P(at least one girl) = 1 – P(all boys)
    $= 1 - P(B_1$ and $B_2$ and $B_3$ and $B_4)$
    $= 1 - P(B_1)·P(B_2)·P(B_3)·P(B_4)$
    $= 1 - (½)(½)(½)(½)$
    $= 1 - 1/16 = 15/16 = 0.9375$
    Yes, this probability is high enough for the couple to be fairly confident that they will get at least one girl – but since P(all boys) = 1/16 = 0.0625 > 0.05, getting all boys would not be considered an unusual event according to the standard given in the previous chapter.

13. Since each birth is an independent event, $P(G_3|B_1 \text{ and } B_2) = P(G_3) = P(G) = \frac{1}{2}$.
One can also consider the 8-outcome sample space S for 3 births and use the formula for conditional probability as follows.

S = {BBB, BBG, BGB, GBB, BGG, GBG, GGB, GGG}
Let F = the first two children are boys
Let $G_3$ = the third child is a girl
$P(F) = 2/8$
$P(G_3) = 4/8$
$P(F \text{ and } G) = 1/8$

$$P(G_3|F) = \frac{P(G_3 \text{ and } F)}{P(F)} = \frac{1/8}{2/8} = 1/2$$

Note: Using the sample space S also verifies directly that $P(G_3) = 4/8 = 1/2 = P(G)$, independent of what occurred during the first two births.
No. This probability is not the same as $P(BBB) = 1/8$.

15. Make a table like the one at the right.
    $P(\text{neg}|\text{no}) = 154/178 = 0.865$

    This is the probability of correctly declaring that someone does not use marijuana.

|  |  | USE MARIJUANA? | | |
|---|---|---|---|---|
|  |  | yes | no | |
| TEST | pos | 119 | 24 | 143 |
| RESULTS | neg | 3 | 154 | 157 |
|  |  | 122 | 178 | 300 |

17. P(at least one clock works) = 1 − P(all clocks fail)
    $= 1 - P(F_1 \text{ and } F_2 \text{ and } F_3)$
    $= 1 - P(F_1) \cdot P(F_2) \cdot P(F_3)$
    $= 1 - (0.05)(0.05)(0.05)$
    $= 1 - 0.000125 = 0.999875$

    Yes. Her probability of having a functioning alarm goes up from 95% to 99.9875%.

19. P(HIV positive result) = P(at least one person is HIV positive)
    = 1 − P(all persons are HIV negative)
    $= 1 - P(N_1 \text{ and } N_2 \text{ and } N_3)$
    $= 1 - P(N_1) \cdot P(N_2) \cdot P(N_3)$
    $= 1 - (0.9)(0.9)(0.9)$
    $= 1 - 0.729 = 0.271$

    NOTE: This plan is very efficient. Suppose, for example, there were 3000 people to be tested. Only in 0.271 = 27.1% of the groups would a retest need to be done for each of the three individuals. Those (0.271)(1000) = 271 groups would generate (271)(3) = 813 retests. The total number of tests required would be 1813 (the 1000 original + the 813 retests), only 60% of the 3,000 tests that would have been required to test everyone individually.

21. Make a table like the one at the right.
    Let M = a Senator is a male.
    Let F = a Senator is a female.
    Let R = a Senator is a Republican.
    Let D = a Senator is a Democrat.
    Let I = a Senator is an Independent.
    $P(R|M) = 46/86 = 0.535$

|  |  | PARTY | | | |
|---|---|---|---|---|---|
|  |  | Rep | Dem | Ind | |
| GENDER | male | 46 | 39 | 1 | 86 |
|  | female | 5 | 9 | 0 | 14 |
|  |  | 51 | 48 | 1 | 100 |

23. Refer to exercise #21.
    $P(F|I) = 0/1 = 0$

25. For 25 randomly selected people,
    a. P(no two share the same birthday) = 0.432  [or .431, see exercise #21c of section 4-4]
    b. P(at least two share the same birthday) = 1 − P(no two share the same birthday)
    $$= 1 − 0.432 = 0.568$$

27. Let F = getting a seat in the front row.
    $P(F) = 2/24 = 1/12 = 0.0833$
    - formulate the problem in terms of "n" rides
        P(at least one F in n rides) = 1 − P(no F in n rides)
        $$= 1 − (11/12)^n$$
    - try different values for n until P(at least one F) ≥ 0.95
        P(at least one F in 32 rides) = $1 − (11/12)^{32}$
        $$= 1 − 0.0616 = 0.9384$$
        P(at least one F in 33 rides) = $1 − (11/12)^{33}$
        $$= 1 − 0.0566 = 0.9434$$
        P(at least one F in 34 rides) = $1 − (11/12)^{34}$
        $$= 1 − 0.0519 = 0.9481$$
        P(at least one F in 35 rides) = $1 − (11/12)^{35}$
        $$= 1 − 0.0476 = 0.9524$$
    You must ride 35 times in order to have at least a 95% chance of getting a first row seat at least once.

## 4-6 Probabilities Through Simulations

1. A simulation is an approximation of some phenomenon that behaves in the same way as the phenomenon it represents. It often takes the form of a mathematical model for some real-world phenomenon that would be difficult or dangerous to observe first-hand. If a simulation method is used for a probability problem, the result will not necessarily be exactly correct – but it should be close to the true value.

3. No. People have subconscious biases and notions about what is random that prevent them from generating sequences that are truly random.

NOTE: In exercises #5-8 we suggest using random two-digit numbers from 00 to 99 inclusive. For convenience of association with the natural counting numbers, consider 00 to represent 100 so that the conceptual "order" of the digits is 01,02,03,...98,99,00. Even though other schemes using fewer than 100 possibilities could be used, this has the advantage of being able to read through a printed list of random digits and associating them into pairs – without having to discard any values as being outside the range of desired random numbers.

5. Use MATH PRB randInt(0,99,20)→L1 to generate 20 integers as suggested in the above NOTE. Let 01-95 represent a male and 96-00 represent a female.

7. Use MATH PRB randInt(0,99,500)→L1 to generate 500 integers as suggested in the above NOTE. Let 01-02 represent a defective cell phone and 03-00 represent a good cell phone.

9. a. Refer to the table at the right. The simulated percentage of males is 19/20 = 95%, which is exactly the desired value.
   b. The numbers of males in the 10 simulations prepared for this manual were as follows: 19 19 20 20 20 20 20 20 20 20.
   These numbers are very consistent. Based on these results it would be very unusual to find 10 females among 20 randomly selected cyclists.

11. a. The simulation prepared for this manual produced 18/500 = 3.6% defective cell phones, which is fairly close to 2%.
    b. The numbers of defective cell phones in the 5 simulations prepared for this manual were as follows: 18 18 18 11 17.
    Based on these results, it would be very unusual to select 500 such cell phones and find that 0 of them are defective.

| exercise #9 | |
|---|---|
| # | gender |
| 80 | male |
| 86 | male |
| 08 | male |
| 69 | male |
| 81 | male |
| 42 | male |
| 35 | male |
| 96 | female |
| 01 | male |
| 53 | male |
| 08 | male |
| 07 | male |
| 91 | male |
| 45 | male |
| 49 | male |
| 87 | male |
| 30 | male |
| 79 | male |
| 02 | male |
| 52 | male |

13. When the simulation was performed 20 times for this manual, the numbers of smokers who quit (assuming the usual 20% success rate) out of 50 were as follows: 9 14 10 8 13 9 16 13 11 10 13 5 6 13 6 9 14 12 9 6
    Yes. It appears that 12 successes can easily occur by chance when a drug is ineffective and the usual 20% success rate is still in effect. In fact, the estimate is P(12 or more successes with an ineffective drug) $\approx$ 8/20 = 40%.

15. The probability of originally selecting the door with the car is 1/3.
    You can either switch or not switch.
    Let W = winning the car.
    • If you make a selection and do not switch, P(W) = 1/3.
    • If you make a selection and switch, there are only 2 possibilities.
        A = you originally had the winning door and switched to a losing door.
        B = you originally had a losing door and switched to a winning door.
    NOTE: You cannot switch from a losing door to another losing door. Monty Hall always opens a losing door. If you have a losing door and Monty Hall opens a losing door, then the only possible door to switch to is the winning door. Since you win if B occurs, P(W) = P(B) = P(originally had a losing door) = 2/3.
    • Conclusion: the better strategy is to switch.
    While the preceding analysis makes it unnecessary, a simulation of n trails can be done as follows.
    1. Randomly select door 1,2 or 3 to hold the prize.
    2. Randomly select door 1,2 or 3 as your original choice.
    3. Determine m = the number of times the doors match.
    4. Estimated P(winning if you don't switch) = m/n.  [m/n should be about 1/3]
    5. Estimated P(winning if you switch) = (n-m)/n.  [(n-m)/n should be about 2/3]
    NOTE: If you switch, you lose m times and win (n-m) times – if you originally selected a losing door, remember, a switch always wins the car.
    The following TI-83/84 Plus procedure produce the desired simulation, where L1 indicates the winning door and L2 indicates your choice.
    
    MATH PRB randInt(1,3,100) → L1
    randInt(1,3,100) → L2
    L1 – L2 → L3
    STAT SortA(L3)
    
    A zero in L3 indicates a match and that the original choice would win the car. The simulation produced for this manual yielded 37 0's in L3.
        Estimated P(winning if you don't switch) = 37/100 = 37%
        Estimated P(winning if you switch) = 63/100 = 63%

17. No, his reasoning is not correct. No, the proportion of girls will not increase. Each family would consist of zero to several consecutive girls, followed by a boy. The types of possible families and their associated probabilities would be as follows.

$P(B) = ½ = 1/2$                      = 16/32
$P(GB) = (½)(½) = 1/4$             = 8/32
$P(GGB) = (½)(½)(½) = 1/8$        = 4/32
$P(GGGB) = (½)(½)(½)(½) = 1/16$   = 2/32
$P(GGGGB) = (½)(½)(½)(½)(½) = 1/32$ = 1/32
etc.

Each collection of 32 families would be expected to look like this, where * represents one family with 5 or more girls and a boy.

| B | B | B | B | B | B | B | B |
|---|---|---|---|---|---|---|---|
| B | B | B | B | B | B | B | B |
| GB | GB | GB | GB | GB | GB | GB | GB |
| GGB | GGB | GGB | GGB | GGGB | GGGB | GGGGB | * |

A gender count reveals 32 boys and 31 or more girls, an approximately equal distribution of genders. In practice, however, families would likely stop after some maximum number of children – whether they had a boy or not. If that number was 5 for all families, then * = GGGGG and the expected gender count for the 32 families would be an exactly equal distribution of 31 boys and 31 girls.

NOTE: Since the "stopping point" would likely vary from family to family, the problem is actually more complicated. Suppose, for example, that 25% of the families decided in advance to stop at two children regardless of gender. We expect those 1/4 = 8/32 of the families to be distributed approximately evenly among the four rows in the above listing – 2 in each row. The 6 such families in the first 3 rows would be stopped by the king's decree and remain unchanged, but the 2 such families in the last row would both become GG families. Counting the * as GGGGG, this reduces the expected number of boys by 2·7/8 = 1.75 and the expected number of girls by 2·7/8 = 1.75 – and there would still be an equal number of each gender.

## 4-7 Counting

NOTE: In general, the factorial [!], permutations [$_nP_r$], and combination [$_nC_r$] exercises in this section are worked by hand. The solutions may also be obtained directly on the TI-83/84 Plus using the MATH PRB functions #4 [!], #2, [nPr] and #3 [nCr] as noted in the text.

1. The permutations rule applies when different orderings of the same collection of objects represent a different solution. The combinations rule applies when all orderings of the same collection of objects represent the same solution.

3. No. The methods of this section are used when objects are selected at random from various populations. She could find the probability that the letters z, i and p are chosen at random (in any order, or in a particular order) in 3 selections from the alphabet – but this is in no way related to the relative frequency of word usage.

5. $5! = 5·4·3·2·1 = 120$

7. $_{24}C_4 = 24!/(20!4!) = (24·23·22·21·20!)/(20!4!) = (24·23·22·21)/4! = 10,626$
NOTE: This technique of reducing the problem by "canceling out" the 20! from both the numerator and denominator can be used in most combinations and permutations problems. In general, a smaller factorial in the denominator can be completely divided into a larger factorial in the numerator to leave only the "excess" factors not appearing in the denominator.

54   CHAPTER 4   Probability

Furthermore, $_nC_r$ and $_nP_r$ will always be integers.  More generally, a non-integer answer to <u>any counting problem</u> (but not a probability problem) indicates that an error has been made.

9. $_{52}P_2 = 52!/50! = 52 \cdot 51 = 2652$

11. $_{30}C_3 = 30!/(27!3!) = 30 \cdot 29 \cdot 28/3! = 4060$

13. Let W = winning the described lottery with a single selection.
    The total number of possible combinations is $_{35}C_5 = 35!/(30!5!) = 35 \cdot 34 \cdot 33 \cdot 32 \cdot 31/5! = 324,632$.  Since only one combination wins, P(W) = 1/324,632.

15. Let W = winning the described lottery with a single selection.
    The total number of possible combinations is $_{38}C_6 = 38!/(32!6!) = 38 \cdot 37 \cdot \ldots \cdot 34 \cdot 33/6! = 2,760,681$.  Since only one combination wins, P(W) = 1/2,760,681.

17. Let W = winning the described lottery with a single selection.
    The total number of possible permutations is $_{39}P_5 = 39!/34! = 39 \cdot 38 \cdot 37 \cdot 36 \cdot 35 = 69,090,840$.
    Since only one permutation wins, P(W) = 1/69,090,840.

19. Let O = selecting at random and getting the 5 oldest managers.
    The total number of possible combinations is $_{15}C_5 = 15!/(10!5!) = 15 \cdot 14 \cdot 13 \cdot 12 \cdot 11/5! = 3003$.
    Since only one combination includes the oldest 5, P(O) = 1/3003 = 0.000333.
    Yes, that probability is low enough to charge that the dismissals resulted from age discrimination instead of random selection.

21. By the fundamental counting rule, there are $4 \cdot 4 \cdot 4 \cdot 4 \cdot 4 \cdot 4 \cdot 4 \cdot 4 \cdot 4 \cdot 4 = 4^{10} = 1,048,576$ different possible treatment arrangements.

23. a. "Permutations" implies that order makes a difference, but the only distinguishing feature of each pill is whether it is good (G) or defective (D).  The problem reduces to how many ways can you arrange 7G's and 3 D's, which further reduces to how many ways can you pick 3 of the 10 positions to be D's – and the order in which the 3 positions are picked is irrelevant.
    And so the solution is $_{10}C_3 = 10!/(7!3!) = 10 \cdot 9 \cdot 8/3! = 120$.
    b. Since the order in which the pills are picked does not affect the sample, use combinations.
    The total number of possible samples is $_{10}C_3 = 10!/(7!3!) = 10 \cdot 9 \cdot 8/3! = 120$.
    Let A = the sample consists of precisely all 3 defective pills.  Since only one of the possible samples includes all 3 defective pills, P(A) = 1/120.

25. Let S = rolling 5 consecutive 6's.
    By the fundamental counting rule, there are $6 \cdot 6 \cdot 6 \cdot 6 \cdot 6 = 6^5 = 7776$ possible outcomes.
    Since only one of them is 5 consecutive 6's, P(S) = 1/7776 = 0.000129.
    Yes, that probability is low enough to rule out chance as an explanation for Mike's results.

27. a. Since each order is a different slate of officers, use permutations.
    The total number of possible slates is $_{12}P_4 = 12!/8! = 12 \cdot 11 \cdot 10 \cdot 9 = 11,880$.
    b. Since the order does not affect the composition of the committee, use combinations.
    The total number of possible committees is $_{12}C_4 = 12!/(8!4!) = 12 \cdot 11 \cdot 10 \cdot 9/4! = 495$.

29. The first note is given by *.  There are 3 possibilities for each of the next 15 notes.
    By the fundamental counting rule, there are
    $3 \cdot 3 \cdot 3 \cdot 3 \cdot 3 \cdot 3 \cdot 3 \cdot 3 \cdot 3 \cdot 3 \cdot 3 \cdot 3 \cdot 3 \cdot 3 \cdot 3 = 3^{15} = 14,348,907$ possible sequences.
    NOTE: This assumes that each song has at least 16 notes, and it does not guarantee that two different melodies will not have the same representation – if one goes up two steps every time the other goes up one step, for example, they both will show a U in that position.

31. There are 10 digits: 0,1,2,3,4,5,6,7,8,9.
    The first digit could be any of 8 digits: 2,3,4,5,6,7,8,9.
    The middle digit could be any of 2 digits: 0,1.
    The last digit could be any of 9 digits: any digit except the middle one.
    By the fundamental counting rule there were 8·2·9 = 144 possible area codes.

33. a. Since there are 64 teams at the start and only 1 left at the end, 63 teams must be eliminated.
       Since each game eliminates 1 team, it takes 63 games to eliminate 63 teams.
    b. Let W = picking all 63 winners by random guessing.
       Since there are 2 possible outcomes for each game, there are $2^{63}$ = 9.223 x $10^{18}$ possible sets
       of results. Since only one such result gives all the correct winners,
         P(W) = 1/[9.223 x $10^{18}$] = 1.084 x $10^{-19}$.
       One could also reason: since each guess has a 50% chance of being correct,
         P(W) = $(0.5)^{63}$ = 1.084 x $10^{-19}$.
    c. Let E = picking all 63 winners by expert guessing.
       Since each guess has a 70% chance of being correct, P(W) = $(0.7)^{63}$ = 1.743 x $10^{-10}$.
       Since 1.743 x $10^{-10}$ = 1/5,738,831,575, the expert has a 1 in 5,738,831,575 chance [or about
       a 1 in 5.7 billion chance] of selecting all 63 winners.

35. There are $_{47}C_5$ = 47!/(42!5!) = 1,533,939 ways to select the five regular numbers.
    There are 27 ways to select the special number.
    By the fundamental counting rule, there are 1,533,939·27 = 41,416,353 total possible entry
    selections. Since only one of them is a winner, the probability of winning the jackpot with a
    single entry selection is 1/41,416,353 = 0.0000000241.

37. There are 26 possible first characters, and 36 possible characters for the other positions.
    Find the number of possible names using 1,2,3,...,8 characters and then add to get the total.

    | characters | possible names | | |
    |---|---|---|---|
    | 1 | 26 | = | 26 |
    | 2 | 26·36 | = | 936 |
    | 3 | 26·36·36 | = | 33,696 |
    | 4 | 26·36·36·36 | = | 1,213,056 |
    | 5 | 26·36·36·36·36 | = | 43,670,016 |
    | 6 | 26·36·36·36·36·36 | = | 1,572,120,576 |
    | 7 | 26·36·36·36·36·36·36 | = | 56,596,340,736 |
    | 8 | 26·36·36·36·36·36·36·36 | = | 2,037,468,266,496 |
    | | | total = | 2,095,681,645,538 |

39. a. The calculator factorial key gives 50! = 3.04140932 x $10^{64}$
       Using the approximation, K = (50.5)·log(50) + 0.39908993 – 0.43429448(50)
         = 85.79798522 + 0.39908993 – 21.71472400
         = 64.48235225
       and then 50! = $10^K$ = $10^{64.48235225}$ = 3.036345215 x $10^{64}$
       NOTE: The two answers differ by 0.0051 x $10^{64}$ – i.e., 5.1 x $10^{61}$, or 51 followed by 60
       zeros. While such an error of "zillions and zillions" may seem quite large, it is only an error
       of (5.1 x $10^{61}$)/(3.04 x $10^{64}$) = 1.7%.
    b. The number of possible routes is 300!
       Using the approximation, K = (300.5)·log(300) + 0.39908993 – 0.43429448(300)
         = 744.374937 + 0.39908993 – 130.288344
         = 614.485683
       and then 300! = $10^K$ = $10^{614.485683}$
       Since the number of digits in $10^x$ is the next whole number above x, 300! has 615 digits.

## 56  CHAPTER 4  Probability

41. This problem cannot be solved directly using permutations, combinations or other techniques presented in this chapter. It is best solved by listing all the possible solutions in an orderly fashion and then counting the number of solutions. Often this is the most reasonable approach to a counting problem.

   While you are encouraged to develop your own systematic approach to the problem, the following table represents one way to organize the solution. The table is organized by rows, according to the numbers of pennies in each way that change can be made. The numbers in each row give the numbers of the other coins as explained in the footnotes below the table.

   The three numbers in the 80 row, for example, indicate that there are three ways to make change using 80 pennies. The 4 in the "only 5¢" column represents one way (80 pennies, 4 nickels). The 2 in the 10¢ column of the "nickels and one other coin" columns represents another way (80 pennies, 2 dimes). The 1 in the 10¢ column of the "nickels and one other coin" columns represents a third way (80 pennies, 1 dime, and [by default] 2 nickels).

   The bold-face **3** in the 20 row and the 10¢/25¢ column of the "nickels and two other coins" columns represents 20 pennies, 3 dimes, one quarter, and [by default] 5 nickels.

   Using the following table, or the system of your own design, you should be able to
   (1) take a table entry and determine what way to make change it represents and
   (2) take a known way to make change and find its representation in the table.

| | only | nickels and one other coin[a] | | | nickels and two other coins[b] | | | | all[c] |
|---|---|---|---|---|---|---|---|---|---|
| | | | | | 10¢ | 10¢ | 25¢ | | 10¢ |
| | | | | | 25¢ | 50¢ | 50¢ | | 25¢ 50¢ |
| 1¢ | 5¢ | 10¢ | 25¢ | 50¢ | | | | | |
| 100 | 0 | . | . | . | . | . | . | | . . |
| 95  | 1 | . | . | . | . | . | . | | . |
| 90  | 2 | 1 | . | . | . | . | . | | . |
| 85  | 3 | 1 | . | . | . | . | . | | . |
| 80  | 4 | 2,1 | . | . | . | . | . | | . |
| 75  | 5 | 2,1 | 1 | . | . | . | . | | . |
| 70  | 6 | 3,2,1 | 1 | . | . | . | . | | . |
| 65  | 7 | 3,2,1 | 1 | . | 1 | . | . | | . |
| 60  | 8 | 4,3,2,1 | 1 | . | 1 | . | . | | . |
| 55  | 9 | 4,3,2,1 | 1 | . | 2,1 | . | . | | . |
| 50  | 10 | 5,4,3,2,1 | 2,1 | 1 | 2,1 | . | . | | . |
| 45  | 11 | 5,4,3,2,1 | 2,1 | 1 | 3,2,1 | . | . | | . |
| 40  | 12 | 6,5,4,3,2,1 | 2,1 | 1 | 3,2,1 | 1 | . | | . |
| 35  | 13 | 6,5,4,3,2,1 | 2,1 | 1 | 4,3,2,1 | 1 | . | | . |
| 30  | 14 | 7,6,5,4,3,2,1 | 2,1 | 1 | 4,3,2,1 | 2,1 | . | | . |
| 25  | 15 | 7,6,5,4,3,2,1 | 3,2,1 | 1 | 5,4,3,2,1 | 2,1 | 1 | | . |
| 20  | 16 | 8,7,6,5,4,3,2,1 | 3,2,1 | 1 | 5,4,**3**,2,1 | 3,2,1 | 1 | | . |
| 15  | 17 | 8,7,6,5,4,3,2,1 | 3,2,1 | 1 | 6,5,4,3,2,1 | 3,2,1 | 1 | | 1 |
| 10  | 18 | 9,8,7,6,5,4,3,2,1 | 3,2,1 | 1 | 6,5,4,3,2,1 | 4,3,2,1 | 1 | | 1 |
| 5   | 19 | 9,8,7,6,5,4,3,2,1 | 3,2,1 | 1 | 7,6,5,4,3,2,1 | 4,3,2,1 | 1 | | 2,1 |
| 0   | 20 | 10,9,8,7,6,5,4,3,2,1 | 4,3,2,1 | 2,1 | 7,6,5,4,3,2,1 | 5,4,3,2,1 | 2,1 | | 2,1 |
| ways | 21 | 100 | 34 | 12 | 56 | 25[d] | 7 | | 6[e] |

[a] The possible numbers of the non-nickel coin are given. The numbers of nickels are found by default.

[b] This is for a single occurrence of the second coin listed. The possible numbers of the leading non-nickel coin are given. The numbers of nickels are found by default.

[c] The is for a single occurrence of the second and third coin listed. The possible numbers of the leading non-nickel coin are given. The numbers of nickels are found by default.

[d] There are another 25 ways when the 50¢ is in the form of 2 quarters.

[e] There are another 6 ways when the 50¢ is in the form of 2 quarters.

The total number of ways identified is 21+100+34+12+56+25+25+7+6+6 = 292.
If a one-dollar coin is also considered as change for a dollar, there are 293 ways.

## Statistical Literacy and Critical Thinking

1. A probability must be a value between 0 and 1 inclusive. The term "50-50" is not a valid probability. The correct statement is that the probability a fair coin turns up heads is 0.5.

2. No, chance cannot be ruled out as a reasonable explanation. The probability of an event occurring by chance alone must be less than 0.05 in order for that event to be considered unusual enough to attribute it to factors other than chance.

3. No, his reasoning is not correct. Using the classical approach to calculate probabilities requires that all the outcomes are equally likely. These two outcomes (life exists, life does not exist) are not equally likely.

4. When two events are disjoint, they cannot both occur at the same time. When two events are independent, the occurrence of one does not affect the occurrence of the other. Disjoint events cannot be independent – since they cannot both occur at the same time, the occurrence of one definitely affects (i.e., it actually prevents) the occurrence of the other.

## Review Exercises

1. Make a chart like the one on the right.
   Let H = a person had a headache
   N = a person had no headache
   D = a person used the drug
   P = a person used the placebo
   P(H) = 80/100 = 0.80

|  |  | TREATMENT | |  |
|---|---|---|---|---|
|  |  | drug | placebo |  |
| RESULT | headache | 15 | 65 | 80 |
|  | no headache | 17 | 3 | 20 |
|  |  | 32 | 68 | 100 |

2. Refer to exercise #1.
   P(D) = 32/100 = 0.32

3. Refer to exercise #1.
   P(H or H) = P(H) + P(D) – P(H and D)
   = 80/100 + 32/1000 – 15/100
   = 97/100 = 0.97

4. Refer to exercise #1.
   P(P or N) = P(P) + P(N) – P(P and N)
   = 68/100 + 20/100 – 3/100
   = 85/100 = 0.85

5. Refer to exercise #1.
   $P(P_1$ and $P_2) = P(P_1) \cdot P(P_2|P_1)$
   = (68/100)(67/99)
   = 1139/2475 = 0.460

6. Refer to exercise #1.
   $P(H_1$ and $H_2) = P(H_1) \cdot P(H_2|H_1)$
   = (80/100)(79/99)
   = 316/495 = 0.638

7. Refer to exercise #1.
   P(H|D) = 15/32 = 0.469

8. Refer to exercise #1.
   P(D|H) = 15/80 = 0.1875

9. a. P(birthday of randomly selected person is October 18) = 1/365.
   b. P(birthday of randomly selected person is in October) = 31/365.
   c. Answers will vary, but a reasonable answer for America might be
      P(randomly selected adult knows October 18 is NSD in Japan) ≈ 1/1,000,000.
   d. Yes. Since 1/1,000,000 = 0.000001 < 0.05, such a selection would be unusual.

10. There were 132 + 880 = 1012 people in the survey.
    Let D = the randomly selected person thinks it should be used as a doorstop.
    P(D) = 132/1012 = 0.130

11. P($G_1$ and $G_2$ and…and $G_{12}$) = P($G_1$)·P($G_2$)·…·P($G_{12}$)
                                    = (½)(½)…(½)
                                    = (½)$^{12}$ = 0.000244
    Yes. The probability of getting 12 girls in 12 births by chance alone is so small that the results appear to support the company's claim.

12. Let S = a 27-year-old man survives for the year.
    P(S) = 0.9982, for each such man
    P($S_1$ and $S_2$ and…and $S_{12}$) = P($S_1$)·P($S_2$)·…·P($S_{12}$)
                                    = (0.9982)(0.9982)…(0.9982)
                                    = (0.9982)$^{12}$ = 0.9786

13. Let L = selecting the two wires that are live
    There are $_5C_2$ = 5!/(3!2!) = 10 ways to select two of the five wires.
    Since only one of those selections is the two wires that are live, P(L) = 1/10.

14. Let G = selecting a good aspirin.
    Since the population is so large, assume all probabilities remain constant as tablets are selected for examination.
    P(G) = 0.98, for each aspirin selected.
    P(at least one defective) = 1 − P(all are good)
                             = 1 − P($G_1$ and $G_2$ and $G_3$ and $G_4$)
                             = 1 − P($G_1$)·P($G_2$)·P($G_3$)·P($G_4$)
                             = 1 − (0.98)$^4$
                             = 1 − 0.9224 = 0.0776
    NOTE: The exact answer, not assuming the probabilities remain constant, is 0.0777.
    P(at least one defective) = 1 − P(all are good)
                             = 1 − P($G_1$ and $G_2$ and $G_3$ and $G_4$)
                             = 1 − P($G_1$)·P($G_2$)·P($G_3$)·P($G_4$)
                             = 1 − (2450/2500)(2449/1499)(2448/2498)(2447/2497)
                             = 1 − 0.9223 = 0.0777

15. Let C = a person has chlamydia.
    Since the population is so large, assume all probabilities remain constant as persons are selected.
    a. P(C) = 278.32/100,000 = 0.0027832, for each person selected.
    b. P($C_1$ and $C_2$) = P($C_1$)·P($C_2$)
                      = (0.0027832)(0.0027832)
                      = 0.00000775
    c. P($\overline{C}$) = 1 − P(C) = 1 − 0.0027832 = 0.9972168, for each person selected.
       P($\overline{C}_1$ and $\overline{C}_2$) = P($\overline{C}_1$)·P($\overline{C}_2$)
                      = (0.9972168)(0.9972168)
                      = 0.9944413

16. There are 10 possibilities (0,1,2,3,4,5,6,7,8,9) for each digit.
    By the fundamental counting rule, the number of possibilities is now
        10·10·…·10 = 10$^{13}$ = 10,000,000,000,000 = 10 trillion

Cumulative Review Exercises     59

**Cumulative Review Exercises**

1. The values in order are: 0 0 0 2 3 3 3 3 4 4 4 5 5 5 5 6 6 6 6 7 7
   The summary statistics are: $n = 21$    $\Sigma x = 84$    $\Sigma x^2 = 430$

   a. $\bar{x} = (\Sigma x)/n = 84/21 = 4.0$

   b. $\tilde{x} = 4$

   c. $s^2 = [n(\Sigma x^2) - (\Sigma x)^2]/[n(n-1)]$
      $= [21(430) - (84)^2]/[21(20)]$
      $= 1974/420 = 4.70$
      $s = 2.2$

   d. $s^2 = 4.7$ [from part (c) above]

   e. Yes. Since there were no negative scores at all, it appears that the treatment was effective.

   f. Since 18 of the 21 values are positive,
      P(selecting a positive value at random) $= 18/21 = 6/7 = 0.857$.

   g. $P(P_1 \text{ and } P_2) = P(P_1) \cdot P(P_2|P_1)$
      $= (18/21)(17/20)$
      $= 51/70 = 0.729$

   h. $P(P_1 \text{ and } P_2 \text{ and}\ldots\text{and } P_{18}) = P(P_1) \cdot P(P_2) \cdot \ldots \cdot P(P_{18})$
      $= (½)(½)\ldots(½)$
      $= (½)^{18} = 0.00000381$
      Yes. The probability of getting 18 positive values in 18 subjects by chance alone is so small as to justify rejecting the assumption that the treatment is ineffective and to conclude that the treatment appears to be effective.

2. From the given five-number summary,
   minimum = 62    25th percentile = 76    median = 76    75th percentile = 80    maximum = 85

   a. $\tilde{x} = 76$ °F

   b. $P(72 \leq t \leq 76) = 0.50 - 0.25 = 0.25$

   c. $P(t < 72 \text{ or } t > 76) = 1 - P(72 \leq t \leq 76)$
      $= 1 - 0.25$
      $= 0.75$

   d. Let B = selecting a day for which $72 \leq t \leq 76$
      Since the population is relatively large compared to the sample, assume all probabilities remain constant as subsequent days are selected.
      $P(B) = 0.25$, for each day
      $P(B_1 \text{ and } B_2) = P(B_1) \cdot P(B_2)$
      $= (0.25)(0.25) = 0.0625$

   e. No. Hot and cold days tend to occur in cycles. If the temperature is above 80°F one day, it is more likely than usual to be above 80°F on the next day.

# Chapter 5

## Discrete Probability Distributions

### 5-2 Random Variables

1. The requested probability distribution is given below.

    | x | P(x) |
    |---|------|
    | 1 | 1/6 |
    | 2 | 1/6 |
    | 3 | 1/6 |
    | 4 | 1/6 |
    | 5 | 1/6 |
    | 6 | 1/6 |
    |   | 6/6 = 1 |

NOTE: Officially a probability distribution for a random variable is a list or other representation of the possible values for the variable and their associated probabilities. Just as we included the sum of the frequencies as an integral part of a frequency distribution in Chapter 2, we now choose to include the sum of the probabilities as an integral part of a probability distribution. This sum should always be 1.00 and serves as a check to the set-up of the problem.

3. The "probability distribution" described would be as follows.

    | x | P(x) |
    |---|------|
    | 1 | 0.1 |
    | 2 | 0.2 |
    | 3 | 0.3 |
    | 4 | 0.4 |
    | 5 | 0.5 |
    | 6 | 0.6 |
    |   | 2.1 |

    No, the gambler cannot load the die as described.
    Since $\Sigma P(x) = 2.1 \neq 1.0$, this is not a possible probability distribution.

5. a. Continuous, since height can be any value on a continuum.
   b. Discrete, since the number of bald eagles must be an integer.
   c. Continuous, since time can be any value on a continuum.
   d. Discrete, since the number of authors so occupied must be an integer.
   e. Discrete, since the number of students so occupied must be an integer.

NOTE: If one of the conditions for a probability distribution does not hold, the formulas do not apply – and they produce numbers that have no meaning. In general, this manual will find the mean and standard deviation of a probability distribution using the 1-VAR STATS function on the TI-83/84 Plus, with the x values in L1 and the P(x) values in L2. When only the mean is required, or when the exercise requires presenting a table, it is often just as convenient to use the given formulas and work the problem by hand. When working with probability distributions and the given formulas by hand, always keep these important facts in mind.
- $\Sigma P(x)$ must always equal 1.000
- $\Sigma[x \cdot P(x)]$ gives the mean of the x values and must be a number between the highest and lowest x values.

Random Variables  SECTION 5-2   61

- $\Sigma[x^2 \cdot P(x)]$ gives the mean of the $x^2$ values and must be a number between the highest and lowest $x^2$ values.
- $\Sigma x$ and $\Sigma x^2$ have no meaning and should not be calculated
- The quantity "$\Sigma[x^2 \cdot P(x)] - \mu^2$" cannot possibly be negative – if it is, there is a mistake.
- Always be careful to use the unrounded mean in the calculation of the variance, and to take the square root of the unrounded variance to find the standard deviation.

NOTE: In exercises #7-#12, we use the decimal values for P(x) given in the text. In truth these are decimal approximations for fractions or other exact mathematical values. This sometimes produces slight round-off errors. In exercise #11, for example, exact calculations would produce $\Sigma P(x) = 1$ exactly. In general, this manual ignores such discrepancies. The relevant TI-83/84 Plus information and screen displays for exercises 7-12 are as follows.

#7
L1 = 0, 1, ...
L2 = 0.4219, 0.4219, ...
1-Var Stats
x̄=.7499
Σx=.7499
Σx²=1.1247
Sx=
σx=.7498999867
↓n=1

#11
L1 = 4, 5, ...
L2 = 0.1818, 0.2121, ...
1-Var Stats
x̄=5.797979798
Σx=5.7974
Σx²=34.8854
Sx=
σx=1.127971255
↓n=.9999

7. The given table is a probability distribution since $0 \le P(x) \le 1$ for each x and $\Sigma P(x)=1$.
From the screen display above, $\mu = 0.7499$, rounded to 0.7
$\sigma = 0.7499$, rounded to 0.7

9. The given table is not a probability distribution since $\Sigma P(x) = 1.16 \ne 1$.

11. The given table is a probability distribution since $0 \le P(x) \le 1$ for each x and $\Sigma P(x)=1$.
From the screen display above, $\mu = 5.7974$, rounded to 5.8
$\sigma = 1.12797$, rounded to 1.1

13. a. P(x=5) = 0.003
b. P(x≤5) = 0⁺ + 0⁺ + 0⁺ + 0⁺ +0.001 + 0.003 = 0.004
c. The probability in part (b) is relevant. When there are many possible outcomes, the probability that any one of them occurs will typically be small – whether or not that outcome is really unusual. Determining whether an outcome is rare should consider that outcome and "more unusual" outcomes, as this places the outcome in a proper context.
d. Since 0.004 ≤ 0.05, selecting five or fewer Mexican-Americans by chance alone would be an unusual event. Either an unusual event has occurred or factors other than chance are at work in the selection process. Yes, conclude that there is discrimination.

15. a. P(x≤8) = 0⁺ + 0⁺ + 0⁺ + 0⁺ + 0.001 + 0.003 + 0.016 + 0.053 + 0.133 = 0.206
b. Since 0.206 > 0.05, selecting 8 or fewer Mexican-Americans by chance alone would not be an unusual event. No, there is no evidence of discrimination.

17. Consider the following tables, where x represents the net financial result.

a. betting on 7

| x | P(x) | x·P(x) |
|---|------|--------|
| -5 | 37/38 | -185/38 |
| 175 | 1/38 | 175/38 |
|  | 38/38 | -10/38 |

b. betting on an odd number

| x | P(x) | x·P(x) |
|---|------|--------|
| -5 | 20/38 | -100/38 |
| 5 | 18/38 | 90/38 |
|  | 38/38 | -10/38 |

c. not betting

| x | P(x) | x·P(x) |
|---|------|--------|
| 0 | 1.00 | 0 |
|  | 1.00 | 0 |

(continued on the next page)

(continued from the previous page)

a. betting on 7
$E = \Sigma[x \cdot P(x)]$
$= -10/38$
$= -26.3¢$

b. betting on an odd number
$E = \Sigma[x \cdot P(x)]$
$= -10/38$
$= -26.3¢$

c. not betting
$E = \Sigma[x \cdot P(x)]$
$= 0$
$= 0¢$

The expected value for (a) and (b) is -26.3¢. This means that on the average you lose 26.3 cents for every time you play. The expected value for (c) is 0. This means you expect to break even if you don't play – which makes perfect sense. The best option is (c).

19. a. He "wins" $100,000 – $250 = $99,750 if he dies.
    He "loses" $250 if he lives.
    b. The following table describes the situation.

    | x | P(x) | x·P(x) |
    |---|------|--------|
    | -250 | 0.9985 | -249.625 |
    | 99750 | 0.0015 | 149.625 |
    |  | 1.0000 | -100.000 |

    $E = \Sigma[x \cdot P(x)] = -\$100.000$. The expected value is a loss of $100.
    c. Since the company is making a $100 profit at this price, it would break even if it sold the policy for $100 less – i.e., for $250 - $100 = $150. NOTE: This oversimplified analysis ignores the cost of doing business. If the costs (printing, salaries, advertising, etc.) associated with offering the policy are $25, for example, then the company's profit is only $75 and selling the policy for $75 less (i.e., for $175) would represent the break-even point.
    d. Buying life insurance is similar to purchasing other services. You let someone make a profit by shoveling the snow in your driveway, for example, because the cost is worth the frustration you save. You let someone make a profit by selling you life insurance because the cost is worth the peace you have from providing for the financial security of your heirs.

21. There are eight equally likely possible outcomes: GGG GGB GBG BGG GBB BGB BBG BBB. The following table describes the situation, where x is the number of girls per family of 3.

    | x | P(x) | x·P(x) | $x^2$ | $x^2 \cdot P(x)$ |
    |---|------|--------|-------|------------------|
    | 0 | 0.125 | 0 | 0 | 0 |
    | 1 | 0.375 | 0.375 | 1 | 0.375 |
    | 2 | 0.375 | 0.750 | 4 | 1.500 |
    | 3 | 0.125 | 0.375 | 9 | 1.125 |
    |   | 1.000 | 1.500 |   | 3.000 |

    $\mu = \Sigma[x \cdot P(x)]$
    $= 1.500$, rounded to 1.5
    $\sigma^2 = \Sigma[x^2 \cdot P(x)] - \mu^2$
    $= 3.000 - (1.500)^2$
    $= 0.75$
    $\sigma = 0.8660$, rounded to 0.9

    No. Since P(x=3) = 0.125 > 0.05, it is not unusual for a family of 3 children to have all girls.

23. The following table describes the situation.

    | x | P(x) | x·P(x) | $x^2$ | $x^2 \cdot P(x)$ |
    |---|------|--------|-------|------------------|
    | 0 | 0.1 | 0 | 0 | 0 |
    | 1 | 0.1 | 0.1 | 1 | 0.1 |
    | 2 | 0.1 | 0.2 | 4 | 0.4 |
    | 3 | 0.1 | 0.3 | 9 | 0.9 |
    | 4 | 0.1 | 0.4 | 16 | 1.6 |
    | 5 | 0.1 | 0.5 | 25 | 2.5 |
    | 6 | 0.1 | 0.6 | 36 | 3.6 |
    | 7 | 0.1 | 0.7 | 49 | 4.9 |
    | 8 | 0.1 | 0.8 | 64 | 6.4 |
    | 9 | 0.1 | 0.9 | 81 | 8.1 |
    |   | 1.0 | 4.5 |   | 28.5 |

    $\mu = \Sigma[x \cdot P(x)]$
    $= 4.5$
    $\sigma^2 = \Sigma[x^2 \cdot P(x)] - \mu^2$
    $= 28.5 - (4.5)^2$
    $= 8.25$
    $\sigma = 2.8723$, rounded to 2.9

    The probability histogram for this distribution would be flat, with each bar having the same height.

25. Both distributions are tables with two columns – the first column stating the possible outcomes, and the second column giving some information about the outcomes. In a frequency distribution, the second column consists of whole numbers that give the actual observed counts. In a probability distribution, the second column consists of values between 0 and 1 that give the theoretical proportion of times the outcomes are expected to occur in the long run.

27. Let C and N represent correctly and not correctly calibrated altimeters respectively. For 8 C's and 2 N's, there are 7 possible samples of size n=3. These are given, along with their associated probabilities, in the box at the right. Letting x be the number of N's selected yields the following table.

$P(CCC) = (8/10)(7/9)(6/8) = 336/720$
$P(CCN) = (8/10)(7/9)(2/8) = 112/720$
$P(CNC) = (8/10)(2/9)(7/8) = 112/720$
$P(NCC) = (2/10)(8/9)(7/8) = 112/720$
$P(CNN) = (8/10)(2/9)(1/8) = 16/720$
$P(NCN) = (2/10)(8/9)(1/8) = 16/720$
$P(NNC) = (2/10)(1/9)(8/8) = 16/720$
                                       720/720

| x | P(x) | x·P(x) | $x^2$ | $x^2$·P(x) |
|---|------|--------|-------|------------|
| 0 | 336/720 | 0 | 0 | 0 |
| 1 | 336/720 | 336/720 | 1 | 336/720 |
| 2 | 48/720 | 96/720 | 4 | 192/720 |
|   | 720/720 | 432/720 |   | 528/720 |

$\mu = \Sigma[x \cdot P(x)]$
$\quad = 432/720 = 0.600$, rounded to 0.6
$\sigma^2 = \Sigma[x^2 \cdot P(x)] - \mu^2$
$\quad = (528/720) - (432/720)^2$
$\quad = 0.3733$
$\sigma = 0.611$, rounded to 0.6

## 5-3 Binomial Probability Distributions

1. The symbols p and x are defined so as to refer to the same event. If p is the probability of getting a correct answer, then x must be the number of correct answers.

3. Table A-1 is limited because it includes only selected values of p and stops at n=15.

NOTE: For exercises #5-#12, the four requirements for a binomial experiment are
 (1) There is a fixed number of trials.
 (2) The trials are independent.
 (3) Each trial has exactly two possible named outcomes.
 (4) The probabilities remain constant for each trial.

5. No. Requirement (3) is not met. There are more than two possible outcomes.

7. No. Requirement (3) is not met. There are more than two possible outcomes.

9. Yes. All four requirements are met.

11. Yes. All four requirements are met.

13. a. P(WWC) = P(W)·P(W)·P(C) = (4/5)(4/5)(1/5) = 16/125 = 0.128
    b. There are three possible arrangements: WWC, WCW, CWW
         P(WWC) = P(W)·P(W)·P(C) = (4/5)(4/5)(1/5) = 16/125
         P(WCW) = P(W)·P(C)·P(W) = (4/5)(1/5)(4/5) = 16/125
         P(CWW) = P(C)·P(W)·P(W) = (1/5)(4/5)(4/5) = 16/125
    c. P(exactly one correct) = P(WWC or WCW or CWW)
                              = P(WWC) + P(WCW) + P(CWW)
                              = 16/125 + 16/125 + 16/125 = 48/125 = 0.384

15. From Table A-1 in the .05 column and the 3-0 row, P(x=0) = 0.857.

17. From Table A-1 in the .05 column and the 8-4 row, P(x=4) = 0⁺ [rounds to 0 to 3 decimals].

19. From Table A-1 in the .30 column and the 14-2 row, P(x=2) = 0.113.

21. $P(x) = \{n!/[(n-x)!x!]\}p^x q^{n-x}$
    $P(x=2) = [5!/(3!2!)](0.25)^2(0.75)^3$
    $= [10](0.0625)(0.421875) = 0.264$

NOTE: The intermediate values 10, 0.0625, and 0.421875 are given in exercise #21 to help those with an incorrect answer to identify the portion of the problem in which the mistake was made. In the future, only the value n!/[(n-x)!x!] will be given separately. In practice, all calculations can be done in one step on a calculator. You may choose (or be asked to) write down intermediate values for you own (or the instructor's) benefit, but…
- Never round off in the middle of a problem.
- Do not write the values down on paper and then re-enter them in the calculator.
- Use the memory to let the calculator remember with complete accuracy any intermediate values that will be used in subsequent calculations.

In addition, always make certain that n!/[(n-x)!x!] is a whole number and that the final answer is between 0 and 1.

23. $P(x) = \{n!/[(n-x)!x!]\}p^x q^{n-x}$
    $P(x=3) = [9!/(6!3!)](0.25)^3(0.75)^6$
    $= [84](0.25)^3(0.75)^6$
    $= 0.234$

25. $P(x \geq 5) = P(x=5) + P(x=6)$
    $= 0.00065 + 0.000022$
    $= 0.00067$

    Yes. Since 0.00067 ≤ 0.05, it is unusual for at least 5 subjects to experience headaches.
    NOTE: Since the 6th decimal place is not given for P(x=5), it is really not possible to given an answer of .000652.

27. $P(x>1) = 1 - P(x \leq 1)$
    $= 1 - P(x=0) + P(x=1)$
    $= 1 - [0.33410 + 0.40188]$
    $= 1 - 0.73598$
    $= 0.26402$

    No. Since 0.26402 > 0.05, it is not unusual for more than 1 subject to experience headaches.

29. Let x = the number of households tuned to *60 Minutes*.
    binomial problem: n=10 and p=0.20, use Table A-1
    a. P(x=0) = 0.107
    b. $P(x \geq 1) = 1 - P(x=0)$
        $= 1 - 0.107$
        $= 0.893$
    c. $P(x \leq 1) = P(x=0) + P(x=1)$
        $= 0.107 + 0.268$
        $= 0.375$
    d. No. Since 0.375 > 0.05, it would not be unusual to find at most one household tuned to *60 Minutes* when that show had a 20% share of the market.

31. Let x = the number of defects found in the sample.
    binomial problem: n=24 and p=0.04, use the TI-83/84 Plus DISTR function binompdf
    P(shipment accepted) = P(x≤1)
    $$= P(x=0) + P(x=1)$$
    $$= \text{binompdf}(24,0.04,0) + \text{binompdf}(24,0.04,1)$$
    $$= 0.3754 + 0.3754$$
    $$= 0.7508, \text{ rounded to } 0.751$$

33. Let x = the number of booked passengers that actually arrive.
    binomial problem: n=15 and p=0.85, use the TI-83/84 Plus DISTR function binompdf
    P(x=15) = binompdf(15,0.85,15) = 0.0874
    No. Since 0.0874 > 0.05, not having enough seats is not an unusual event. The probability is not low enough to be of no concern.

35. Let x = the number of women hired.
    binomial problem: n=20 and p=0.5, use the TI-83/84 Plus DISTR function binompdf
    P(x≤2) = P(x=0) + P(x=1) + P(x=2)
    $$= \text{binompdf}(20,0.5,0) + \text{binompdf}(20,0.5,1) + \text{binompdf}(20,0.5,2)$$
    $$= 0.000000954 + 0.000019073 + 0.000181198$$
    $$= 0.0002012, \text{ rounded to } 0.000201$$
    Yes. Since 0.0002012 ≤ 0.05, it would be unusual to have only 2 or fewer female hires by chance alone. All other factors being equal, the probability does support a charge of gender discrimination.

37. Let x = the number of components tested to find the first defect.
    geometric problem: p=0.2, use the geometric formula $P(x) = p(1-p)^{x-1}$
    $P(x=7) = (0.2)(0.8)^6 = 0.0524$

39. Extending the patter to cover six types of outcomes, where Σx = n and Σp = 1,
    n=20 and $p_1=p_2=p_3=p_4=p_5=p_6=1/6$ and $x_1=5, x_2=4, x_3=3, x_4=2, x_5=3, x_6=3$
    use the multinomial formula
    $$P(x_1,x_2,x_3,x_4,x_5,x_6) = [n!/(x_1!x_2!x_3!x_4!x_5!x_6!)](p_1)^{x_1}(p_2)^{x_2}(p_3)^{x_3}(p_4)^{x_4}(p_5)^{x_5}(p_6)^{x_6}$$
    $$= [20!/(5!4!3!2!3!3!)](1/6)^5(1/6)^4(1/6)^3(1/6)^2(1/6)^3(1/6)^3$$
    $$= [1,955,457,504,030] \cdot (1/6)^{20}$$
    $$= 0.000535$$

## 5-4 Mean, Variance, and Standard Deviation for the Binomial Distribution

1. Yes. Since 70 is more than two standard deviations from the mean, it would be unusual to get 70 girls in 100 births. Any number higher than 50 + 2(5) = 60 would be considered unusual.

3. If the standard deviation is 1.2 females, the variance is the square of that – i.e., 1.44 females$^2$.

5. μ = np = (200)(0.4) = 80.0
   $\sigma^2$ = npq = (200)(0.4)(0.6) = 48; σ = 6.928, rounded to 6.9
   minimum usual value = μ − 2σ = 80 − 2(6.928) = 66.1
   maximum usual value = μ + 2σ = 80 + 2(6.928) = 93.9

7. μ = np = (1492)(0.25) = 373.0
   $\sigma^2$ = npq = (1492)(0.25)(0.75) = 279.75; σ = 16.726, rounded to 16.7
   minimum usual value = μ − 2σ = 373.0 − 2(16.726) = 339.5
   maximum usual value = μ + 2σ = 373.0 + 2(16.726) = 406.5

9. Let x = the number of correct answers.
   binomial problem: n=16 and p=0.5
   a. $\mu = np = (16)(0.5) = 8.0$
   $\sigma^2 = npq = (16)(0.5)(0.5) = 4$
   $\sigma = 2.0$
   b. Unusual values are those outside $\mu \pm 2\sigma$
   $$8.0 \pm 2(2.0)$$
   $$8.0 \pm 4.0$$
   $$4.0 \text{ to } 12.0$$
   No. Since 10 is within the above limits, it would not be unusual for a student to pass by getting at least 10 correct answers.

11. Let x = the number of orange M&M's.
    binomial problem: n=100 and p=0.2
    a. $\mu = np = (100)(0.2) = 20.0$
    $\sigma^2 = npq = (100)(0.2)(0.8) = 16$
    $\sigma = 4.0$
    b. Unusual values are those outside $\mu \pm 2\sigma$
    $$20.0 \pm 2(4.0)$$
    $$20.0 \pm 8.0$$
    $$12.0 \text{ to } 28.0$$
    No. Since 25 is within the above limits, it would not be unusual to find 25 orange M&M's. There is no reason to conclude that the 20% rate is wrong.

13. Let x = the number of girl births.
    binomial problem: n=325 and p=0.5
    a. $\mu = np = (325)(0.5) = 162.5$
    $\sigma^2 = npq = (325)(0.5)(0.5) = 81.25$
    $\sigma = 9.0$
    b. Unusual values are those outside $\mu \pm 2\sigma$
    $$162.5 \pm 2(9.0)$$
    $$162.5 \pm 18.0$$
    $$144.5 \text{ to } 180.5$$
    Yes. Since 295 is not within the above limits, it would be an unusual number of girl births. The gender-selection method appears to be effective.

15. Let x = the number of letter r's per page.
    binomial problem: n=2600 and p=0.077
    a. $\mu = np = (2600)(0.077) = 200.2$
    $\sigma^2 = npq = (2600)(0.077)(0.923) = 184.78$
    $\sigma = 13.6$
    b. Unusual values are those outside $\mu \pm 2\sigma$
    $$200.2 \pm 2(13.6)$$
    $$200.2 \pm 27.2$$
    $$173.0 \text{ to } 227.4$$
    No. Since 175 is within the above limits, it would not be an unusual result.

# Mean, Variance, and Standard Deviation for the Binomial Distribution  SECTION 5-4

17. Let x = the number of persons who actually voted.
   binomial problem: n=1002 and p=0.61
   a. $\mu = np = (1002)(0.61) = 611.2$
   $\sigma^2 = npq = (1002)(0.61)(0.39) = 238.3758$
   $\sigma = 15.439$, rounded to 15.4
   b. Unusual values are those outside $\mu \pm 2\sigma$
   $$611.2 \pm 2(15.439)$$
   $$611.2 \pm 30.9$$
   $$580.3 \text{ to } 642.1 \text{ [using rounded } \sigma, 580.4 \text{ to } 642.0]$$
   No. Since 701 is not within the above limits, it is not consistent with the actual voter turnout.
   c. No. Based on these results, it appears that accurate voting data cannot be obtained by asking possible voters how they acted.

19. Let x = the number of persons receiving the drug who experience flu symptoms.
   binomial problem: n=863 and p=0.019
   a. $\mu = np = (863)(0.019) = 16.4$
   $\sigma^2 = npq = (863)(0.019)(0.981) = 16.09$
   $\sigma = 4.0$
   b. Unusual values are those outside $\mu \pm 2\sigma$
   $$16.4 \pm 2(4.0)$$
   $$16.4 \pm 8.0$$
   $$8.4 \text{ to } 24.4$$
   No. Since 19 is within the above limits, it is not an unusual result.
   c. No. These results do not suggest that persons receiving the drug experience flu symptoms any differently from persons not receiving the drug.

21. Let x = the number of girl births.
   binomial problem: n=100 and p=0.5
   a. $\mu = np = (100)(0.5) = 50.0$
   $\sigma^2 = npq = (100)(0.5)(0.5) = 25.0$
   $\sigma = 5.0$
   minimum usual value = $\mu - 2\sigma = 50.0 - 2(5.0) = 40.0$
   maximum usual value = $\mu + 2\sigma = 50.0 + 2(5.0) = 60.0$
   b. Yes, the probability distribution is bell-shaped. Values near 50.0 are most likely – and the farther a value is from 50.0 in either direction, the less likely it is to occur.
   c. According to the empirical rule for bell-shaped distributions, the number of girls will fall between 40 and 60 (i.e., within $2\sigma$ of the mean) about 95% of the time.

## 5-5 Poisson Probability Distributions

1. The Poisson distribution describes the numbers of occurrences, over some defined interval, of independent random events that are uniformly distributed over that interval.

3. The formula to generate the entire distribution needs only the value of $\mu$, the mean. The standard deviation, variance and shape of the distribution are all determined by the mean.

5. Poisson problem: $P(x) = \mu^x e^{-\mu}/x!$ with $\mu=5$
   $P(x=4) = (5)^4(e^{-5})/4!$
   $= (625)(0.0067)/24 = 0.175$

68    CHAPTER 5   Discrete Probability Distributions

NOTE: This problem is worked by hand and the intermediate values of 625, 0.0067 and 24 are given to help students who have difficulty understanding the formula and need to identify where they are going wrong. In practice, the answer can be obtained on the TI-83/84 Plus calculator in one step without writing down intermediate values. Use the poissonpdf [i.e., "B"] option from the DISTR [i.e., 2nd VARS] list. For this problem,
Poisson problem: $\mu=5$
$P(x=4) = \text{poissonpdf}(5,4) = 0.175$

7. Poisson problem: $\mu=0.5$

   $P(x=3) = \text{poissonpdf}(0.5,3)$
   $= 0.0126$

9. Let x = the number of dandelions per square meter.
   Poisson problem: $\mu = 7.0$
   a. $P(x=0) = \text{poissonpdf}(7.0,0)$
      $= 0.000912$
   b. $P(x \geq 1) = 1 - P(x=0)$
      $= 1 - 0.000912$
      $= 0.999088$, rounded to 0.9991
   c. $P(x \leq 2) = P(x=0) + P(x=1) + P(x=2)$
      $= \text{poissonpdf}(7.0,0) + \text{poissonpdf}(7.0,1) + \text{poissonpdf}(7.0,2)$
      $= 0.0009 + 0.0064 + 0.0223$
      $= 0.0296$

IMPORTANT NOTE: In the problems that follow, store the unrounded value for $\mu$ for use in all Poisson calculations. One way to do this is as follows:
   With the calculated value for $\mu$ on the display, press "STO→" then "ALPHA ÷" then "ENTER".
   The following will appear on the display: "Ans→M", followed by the value now stored as "M".
   To use the value M as needed, press "ALPHA ÷" then "ENTER".

11. Let x = the number of atoms lost per day.
    Poisson problem: $\mu = (1,000,000 - 997,287)/365 = 62.2274$  STO→ M
    a. The number of atoms lost is $1,000,000 - 997,287 = 22,713$.
       $\mu = 22,713/365 = 62.2$
    b. $P(x=50) = \text{poissonpdf}(M,50) = 0.0155$

NOTE: The following exercises compare actual results to predictions by the Poisson probability distribution. The comparison can be made either by changing the actual frequencies to relative frequencies (by dividing by $n = \Sigma f$) and comparing them to the predicted Poisson probabilities, or by changing the Poisson probabilities to predicted frequencies (by multiplying by $n = \Sigma f$) and comparing them to the actual frequencies. We use the first approach, but the choice is arbitrary.

13. Let x = the number of homicides per day.
    Poisson problem: $\mu = 116/365 = 0.3178$  STO→ M
    a. $P(x=0) = \text{poissonpdf}(M,0) = 0.7277$
    b. $P(x=1) = \text{poissonpdf}(M,1) = 0.2313$
    c. $P(x=2) = \text{poissonpdf}(M,2) = 0.0368$
    d. $P(x=3) = \text{poissonpdf}(M,3) = 0.00389$
    e. $P(x=4) = \text{poissonpdf}(M,4) = 0.000309$

The following table compares the actual relative frequencies to the Poisson probabilities.

| x | f | r.f. | P(x) |
|---|---|------|------|
| 0 | 268 | 0.7342 | 0.7277 |
| 1 | 79 | 0.2164 | 0.2313 |
| 2 | 17 | 0.0466 | 0.0368 |
| 3 | 1 | 0.0027 | 0.0038 |
| 4 or more | 0 | 0.0000 | 0.0004 (by subtraction) |
|  | 365 | 1.0000 | 1.0000 |

NOTE: r.f. = f/Σf = f/365

The agreement is very good.

15. Let x = the number of successes in 100 trials.
    binomial problem: n=100 and p=0.1
    Poisson approximation appropriate since n = 100 ≥ 100 and np = (100)(0.1) = 10 ≤ 10.
     Poisson problem: μ = np = (100)(0.1) = 10  STO→ M
    P(x=101) = poissonpdf(M,101)
             = 4.82E-64 = 4.82 x $10^{-64}$
    This value is so small that it can be considered 0, agreeing with the fact that 101 is impossible in such a binomial distribution.

## Statistical Literacy and Critical Thinking

1. A probability distribution is a statement (as a graph, table, or formula) of the values a random variable can assume and the probability associated with each of those values.

2. The two main requirements of a probability distribution are as follows.
   (1) Each individual probability must be between 0 and 1 inclusive.
   (2) The sum of all the individual probabilities must be 1.000.

3. A discrete probability distribution is one for which all the individual possible values of the random variable can be listed in some logical order so that each one can be identified by a counting number – even if there is an infinite number of them. There are also continuous probability distributions – for which there are an infinite number of possible values for the random variable that occur over some continuum.

4. No. The discrete probability distribution below, for example, describing the possible outcomes for tossing a single die, is neither a binominal probability distribution nor a Poisson probability distribution.

| x | P(x) |
|---|------|
| 1 | 1/6 |
| 2 | 1/6 |
| 3 | 1/6 |
| 4 | 1/6 |
| 5 | 1/6 |
| 6 | 1/6 |
|   | 6/6 |

# CHAPTER 5  Discrete Probability Distributions

## Review Exercises

1. The following table summarizes the calculations for this exercise.

| x | P(x) | x·P(x) | $x^2$ | $x^2$·P(x) | |
|---|---|---|---|---|---|
| 0 | 0.107 | 0 | 0 | 0 | $\mu = \Sigma[x \cdot P(x)]$ |
| 1 | 0.268 | 0.268 | 1 | 0.268 | $\quad = 2.000$, rounded to 2.0 |
| 2 | 0.302 | 0.604 | 4 | 1.208 | $\sigma^2 = \Sigma[x^2 \cdot P(x)] - \mu^2$ |
| 3 | 0.201 | 0.603 | 9 | 1.089 | $\quad = 5.608 - (2.000)^2$ |
| 4 | 0.088 | 0.352 | 16 | 1.408 | $\quad = 1.608$ |
| 5 | 0.026 | 0.130 | 25 | 0.650 | $\sigma = 1.268$, rounded to 1.3 |
| 6 | 0.006 | 0.036 | 36 | 0.216 | |
| 7 | 0.001 | 0.007 | 49 | 0.049 | |
| 8 | $0.000^+$ | $0.000^+$ | 64 | $0.000^+$ | |
| 9 | $0.000^+$ | $0.000^+$ | 81 | $0.000^+$ | |
| 10 | $0.000^+$ | $0.000^+$ | 100 | $0.000^+$ | |
| | 0.999 | 2.000 | | 5.608 | |

a. The table satisfies the requirements for a probability distribution since
   (1) $0 \leq P(x) \leq 1$ for each x
   (2) $\Sigma P(x) = 1$
   NOTE: The "missing" 0.001 in the table is distributed among the three $0.000^+$ categories.

b. Since this is a binomial distribution with n=10 and p=0.2, $\mu = np = (10)(0.2) = 2$. This agrees with the 2.000 in the above table.

c. Since this is a binomial distribution with n=10 and p=0.2, $\sigma^2 = npq = (10)(0.2)(0.8) = 1.6$ and $\sigma = 1.265$. This agrees (within round-off error) with the 1.268 in the table above.

d. $P(x \geq 5) = P(x=5) + P(x=6) + P(x=7) + P(x=8) + P(x=9) + P(x=10)$
   $= 0.026 + 0.006 + 0.001 + 0.000^+ + 0.000^+ + 0.000^+ = 0.033$

e. $E = \Sigma[x \cdot P(x)] = 2.000$

f. $P(x \geq 1) = 1 - P(x=0)$
   $\quad = 1 - 0.107$
   $\quad = 0.893$

g. No. Since $0.893 > 0.05$, getting at least one answer correct by chance alone is not unusual.

2. Let x = the number of TV sets tuned to *Cold Case*.
   binomial problem: n=12 and p=0.15

a. $E = \mu = np = (12)(0.15) = 1.8$

b. $\mu = np = (12)(0.15) = 1.8$

c. $\sigma^2 = npq = (12)(0.15)(0.85) = 1.53$
   $\sigma = 1.237$, rounded to 1.2.

d. $P(x=3) = \text{binomialpdf}(12, 0.15, 3)$
   $\quad = 0.1720$

e. $P(x=0) = \text{binomialpdf}(12, 0.15, 0)$
   $\quad = 0.1422$
   No. Since $0.1422 > 0.05$, it would not be unusual to find in a sample of 12 sets that 0 sets were tuned to *Cold Case*.

3. Let x = the number of workers fired for inability to get along with others.
   binomial problem: n=5 and p=0.17

   a. P(x≥4) = P(x=4) + P(x=5)
            = binomialpdf(5,0.17,4) + binomialpdf(5,0.17,5)
            = 0.00347 + 0.00014
            = 0.00361

   b. Yes. Since 0.00361 ≤ 0.05, having at least 4 of the 5 firings due to an inability to get along with others would be unusual for a company with the standard rate of 17%.

4. Let x = the number of deaths per year.
   Poisson problem: μ = 0.01918 STO→ M

   a. μ = 7/365 = 0.019

   b. P(x=0) = poissonpdf(M,0)
             = 0.9810

   c. P(x=1) = poissonpdf(M,1)
             = 0.0188

   d. P(x≥1) = 1 – [P(x=0) + P(x=1)]
             = 1 – [.9810 + 0.0188]
             = 1 – 0.9998
             = 0.0002

   e. No. Since 0.0002 ≤ 0.05 by such a wide margin, having more than one death on a single day is such a rare event that no such contingency plans are necessary.
   NOTE: This exercise specifically stated that the information concerned the deaths of village residents, not deaths occurring within the village limits or the nearby area. If there are other circumstances to consider (e.g., a nearby highway prone to fatal accidents), then having such contingency plans would be advisable.

## Cumulative Review Exercises

1. Refer to the table given in the text.

   a. 1-VAR STATS L1, L2 generates the following display,
      with L1 = 0, 1, 2, ...
           L2 = 7, 14, 5, ...
   ```
   1-Var Stats
   x̄=4.3625
   Σx=349
   Σx²=2251
   Sx=3.036669353
   σx=3.017630486
   ↓n=80
   ```
   From the above display, $\bar{x}$ = 4.4
                            s = 3.0

b. The relative frequency distribution is given at the right.
Since n = Σf = 80, each relative frequency is f/80.

c. The probability distribution is given at the right.
1-VAR STATS L1,L2 generates the following display,
with L1 = 0, 1, 2, ...
L2 = 0.1, 0.1, 0.1, ...

```
1-Var Stats
x̄=4.5
Σx=4.5
Σx²=28.5
Sx=
σx=2.872281323
↓n=1
```

From the above display, μ = 4.5
σ = 2.87

| relative frequency distribution | | theoretical distribution | |
|---|---|---|---|
| x | r.f. | x | P(x) |
| 0 | 0.0987 | 0 | 0.1 |
| 1 | 0.1750 | 1 | 0.1 |
| 2 | 0.0625 | 2 | 0.1 |
| 3 | 0.1375 | 3 | 0.1 |
| 4 | 0.1000 | 4 | 0.1 |
| 5 | 0.0500 | 5 | 0.1 |
| 6 | 0.0625 | 6 | 0.1 |
| 7 | 0.0750 | 7 | 0.1 |
| 8 | 0.1500 | 8 | 0.1 |
| 9 | 0.1000 | 9 | 0.1 |
|  | 1.0000 |  | 1.0 |

d. Yes. It appears that the given last digits roughly agree with the distribution we expect with random selection.

2. Let x = the number of HIV cases.
binomial problem: n=150 and p=0.10

a. μ = np = (150)(0.10) = 15.0
$\sigma^2$ = npq = (150)(0.10)(0.90) = 13.5
σ = 3.674, rounded to 3.7

b. minimum usual value = μ – 2σ = 15.0 – 2(3.67) = 7.7    [using rounded σ, 7.6]
maximum usual value = μ + 2σ = 15.0 + 2(3.67) = 22.3    [using rounded σ, 22.4]
No. Since 12 is within the above limits, it is not an unusually low result. There is not sufficient evidence to suggest that the program is effective in lowering the 10% rate.

3. The table at the right summarizes the calculations.

a. Yes. Since 0 ≤ P(x) ≤ 1 for each x, and ΣP(x) = 1, it is a valid probability distribution.

b. The population represented is all of the student's friends. No, the population is not all credit card holders in the United States.

c. Yes. Conclusions from the data cannot be generalized to a general population.

| x | P(x) | x·P(x) | x²·P(x) |
|---|---|---|---|
| 0 | 0.16 | 0 | 0 |
| 1 | 0.24 | 0.24 | 0.24 |
| 2 | 0.40 | 0.80 | 1.60 |
| 3 | 0.16 | 0.48 | 1.44 |
| 4 | 0.04 | 0.16 | 0.64 |
|  | 1.00 | 1.68 | 3.92 |

d. μ = Σ[x·P(x)] = 1.68, rounded to 1.7

e. $\sigma^2$ = Σ[x²·P(x)] – μ² = 3.92 – (1.68)² = 1.0976
σ = 1.048, rounded to 1.0

# Chapter 6

# Normal Probability Distributions

## 6-2 The Standard Normal Distribution

1. The word "normal" as used when referring to a normal distribution does carry with it some of the meaning the word has in ordinary language. Normal distributions occur in nature and describe the normal, or natural, state of many common phenomena. But in statistics the term "normal" has a specific and well-defined meaning in addition to its generic connotations of being "typical" – it refers to a specific bell-shaped distribution generated by a particular mathematical formula.

3. A normal distribution can be centered about any value and have any level of spread. A *standard* normal distribution has a center (as measured by the mean) of 0 and has a spread (as measured by the standard deviation) of 1.

5. The height of the rectangle is 0.5. Probability corresponds to area, and the area of a rectangle is (width)·(height).
$P(x<51.5)$ = (width)·(height)
   $= (51.5 - 50.0)(0.5)$
   $= (1.5)(0.5)$
   $= 0.75$

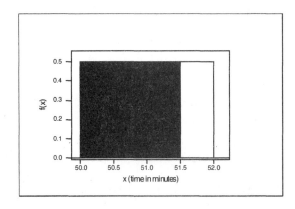

7. The height of the rectangle is 0.5. Probability corresponds to area, and the area of a rectangle is (width)·(height).
$P(50.5<x<51.5)$ = (width)·(height)
   $= (51.5 - 50.5)(0.5)$
   $= (1.0)(0.5)$
   $= 0.50$

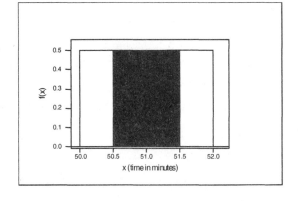

NOTE: The sketch is the key to exercises #9-#28. It tells whether to subtract two Table A-2 probabilities, to subtract a Table A-2 probability from 1, etc. For the remainder of chapter 6, THE ACCOMPANYING SKETCHES ARE NOT TO SCALE and are intended only as aids to help the reader understand how to use the table values to answer the questions. In addition, the probability of any single point in a continuous distribution is zero – i.e., $P(x=a) = 0$ for any single point a. For normal distributions, therefore, this manual ignores $P(x=a)$ and uses $P(x>a) = 1 - P(x<a)$. Each problem is worked first using Table A-2, and then using the appropriate function from the DISTR [i.e., 2nd VARS] menu of the TI-83/84 Plus.

9. P(z<-1.00) = 0.1587
   normalcdf(-999,-1.00) = 0.1587

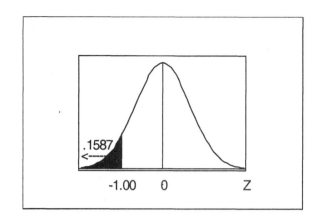

11. P(z<1.00) = 0.8413
    normalcdf(-999,1.00) = 0.8413

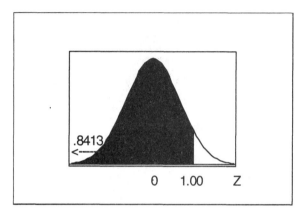

13. P(z>1.25)
    = 1 − 0.8944
    = 0.1056
    normalcdf(1.25,999) = 0.1056

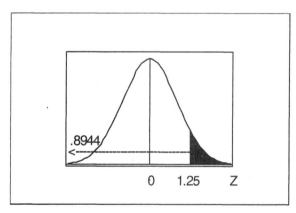

15. P(z>-1.75)
    = 1 − 0.0401
    = 0.9599
    normalcdf(-1.75,999) = 0.9599

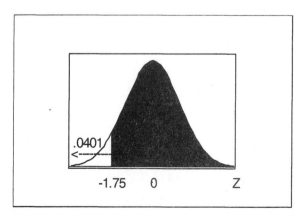

17. P(1.00<z<2.00)
    = 0.9772 − 0.8413
    = 0.1359
    normalcdf(1.00,2.00) = 0.1359

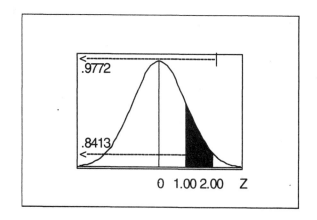

19. P(−2.45<z<−2.00)
    = 0.0228 − 0.0071
    = 0.0157
    normalcdf(−2.45,−2.00) = 0.0156

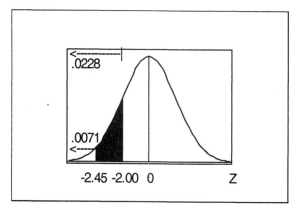

21. P(−2.11<z<1.55)
    = 0.9394 − 0.0174
    = 0.9220
    normalcdf(−2.11,1.55) = 0.9220

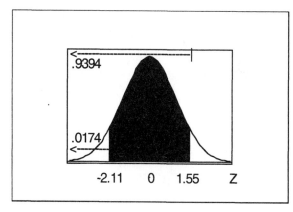

23. P(−1.00<z<4.00)
    = 0.9999 − 0.1587
    = 0.8412
    normalcdf(−1.00,4.00) = 0.8413

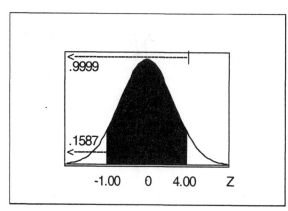

25. P(z>3.52)
    = 1 − 0.9999
    = 0.0001
    normalcdf(3.52,999) = 2.158E-4 = 0.0002

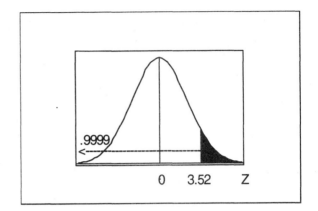

27. P(z>0.00)
    = 1 − 0.5000
    = 0.5000
    normalcdf(0.00,999) = 0.5000

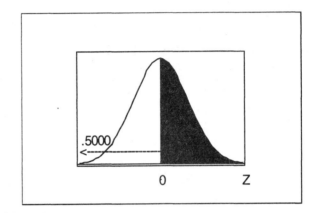

29. P(−1.00<z<1.00)
    = P(z<1.00) − P(z<−1.00)
    = 0.8413 − 0.1587
    = 0.6826 or 68.26%
    normalcdf(−1.00,1.00) = 0.6827 = 68.27%

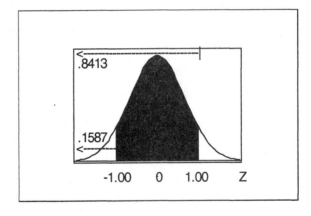

31. P(−3.00<z<3.00)
    = P(z<3.00) − P(z<−3.00)
    = 0.9987 − 0.0013
    = 0.9974 or 99.74%
    normalcdf(−3.00,3.00) = 0.9973 = 99.73%

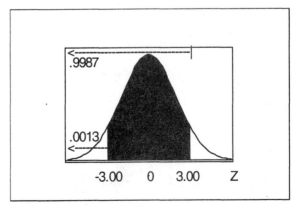

33. P(-1.96<z<1.96)
    = P(z<1.96) − P(z<-1.96)
    = 0.9750 − 0.0250
    = 0.9500
   normalcdf(-1.96,1.96) = 0.9500

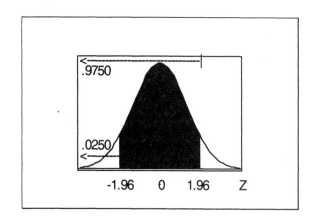

35. P(z<-2.575 or z>2.575)
    = P(z<-2.575) + P(z>2.575)
    = 0.0050 + [1 − 0.9950]
    = 0.0050 + 0.0050
    = 0.0100
   normalcdf(-999,-2.575)
       + normalcdf(2.575,999) =        0.0100

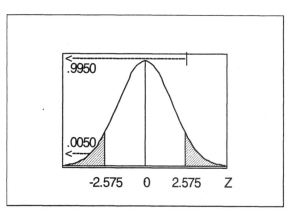

NOTE: The sketch is the key to exercises #37-#40. It tells what probability (i.e., cumulative area) to look up when reading Table A-2 "backwards." It also provides a check against gross errors by indicating at a glance whether a z score is above or below 0. Each problem is worked first using Table A-2, and then using the appropriate function from the DISTR [i.e., 2nd VARS] menu of the TI-83/84 Plus.

37. For $P_{90}$, the cumulative area is 0.9000.
    The closest entry is 0.8997,
    for which z = 1.28.
    invNorm(0.900) = 1.28

39. For the lowest 2.5%, the cumulative
    area is 0.0250 (exact entry in the table),
    indicated by z = -1.96. By symmetry,
    the highest 2.5% [=1−0.9750] are above
    z = 1.96.
    invNorm(0.025) = -1.96
    invNorm(1−0.025) = 1.96

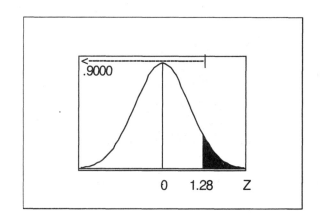
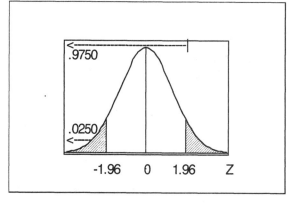

41. Rewrite each statement in terms of z, recalling that z is the number of standard deviations a score is from the mean.

a. P(-1.00<z<1.00)
   = P(z<1.00) − P(z<-1.00)
   = 0.8413 − 0.1587
   = 0.6826 or 68.26%
   normcdf(-1.00,1.00) = .6827 = 68.27%

b. P(-1.96<z<1.96)
   = P(z<1.96) − P(z<-1.96)
   = 0.9750 − 0.0250
   = 0.9500 or 95.00%
   normcdf(-1.96,1.96) = .9500 = 95.00%

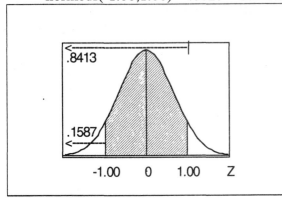

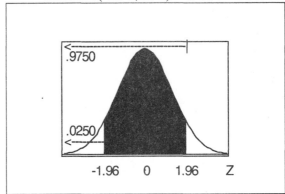

c. P(-3.00<z<3.00)
   = P(z<3.00) − P(z<-3.00)
   = 0.9987 − 0.0013
   = 0.9974 or 99.74%
   normcdf(-3.00,3.00) = .9973 = 99.73%

d. P(-1.00<z<2.00)
   = P(z<2.00) − P(z<-1.00)
   = 0.9772 − 0.1587
   = 0.8185 or 81.85%
   normcdf(-1.00,2.00) = .8186 = 81.86%

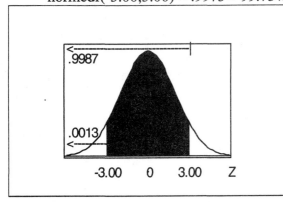

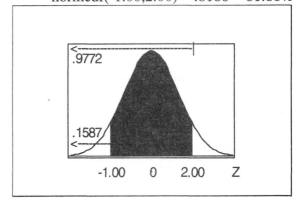

e. P(z<-2.00 or z>2.00)
   = P(z<-2.00) + P(z>2.00)
   = P(z<-2.00) + [1 − P(z<2.00)]
   = 0.0228 + [1 − 0.9772]
   = 0.0228 + 0.0228
   = 0.0456 or 4.56%
   normcdf(-999,-2.00) + normcdf(2.00,999)
   = 0.0455 = 4.55%

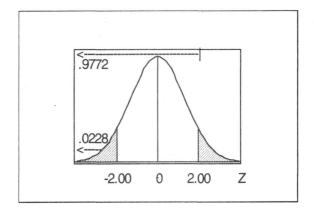

## The Standard Normal Distribution SECTION 6-2   79

43. The sketches are the key. They tell what probability (i.e., cumulative area) to look up when reading Table A-2 "backwards." They also provide a check against gross errors by indicating whether a score is above or below zero.

a. $P(0<z<a) = P(z<a) - P(z<0)$
$0.3907 = P(z<a) - 0.5000$
$0.8907 = P(z<a)$
$a = 1.23$
invNorm$(0.5000 + 0.3907) = 1.23$

b. Since $P(z<-b) = P(z>b)$ by symmetry & $\Sigma P(z) = 1$,
$P(z<-b) + P(-b<z<b) + P(z>b) = 1$
$P(z<-b) + 0.8664 + P(z<-b) = 1$
$2 \cdot P(z<-b) + 0.8664 = 1$
$2 \cdot P(z<-b) = 0.1336$
$P(z<-b) = 0.0668$
$-b = -1.50$, and so $b = 1.50$
invNorm$((1-0.8664)/2) = -1.50$, and so $b = 1.50$

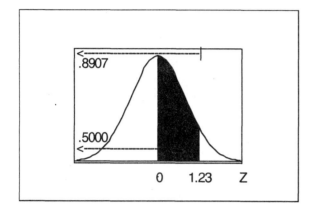
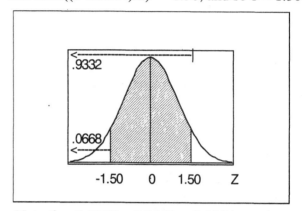

Note that $0.9332 - 0.0668 = 0.8664$, as given.

c. $P(z>c) = 0.0643)$
$P(z<c) = 1 - 0.0643 = 0.9357$
$c = 1.52$
invNorm$(1-0.0643) = 1.52$

d. $P(z>d) = 0.9922$
$P(z<d) = 1 - 0.9922 = 0.0078$
$d = -2.42$
invNorm$(1-0.9922) = -2.42$

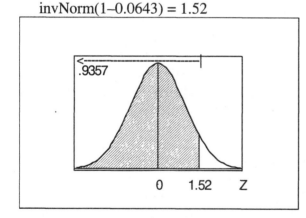
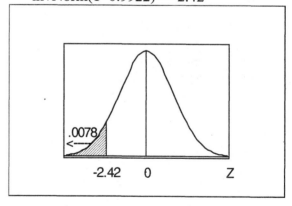

e. $P(z<e) = 0.4500$ [closest entry is 0.4483]
$e = -0.13$
invNorm$(0.4500) = -0.13$

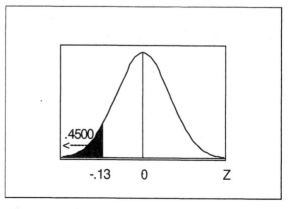

80 CHAPTER 6 Normal Probability Distributions

45. a. Moving across the x values from the minimum to the maximum will accumulate probability at a constant rate from 0 to 1. The result will be a straight line (i.e., with a constant slope) rom (minimum x, 0) to (maximum x, 1). The sketch is given at the right.

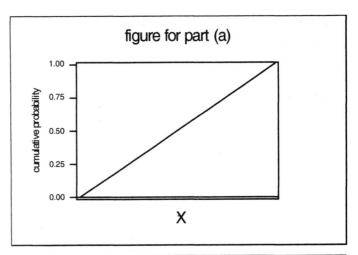

b. Moving across the x values from the minimum to the maximum will accumulate probability slowly at first, at a larger rate near the middle, and more slowly again at the upper end. The result will be a curve that has a variable slope from (minimum x, 0) to (maximum x, 1) – starting with a slope near 0 at the lower end, gradually increasing until reaching a maximum slope at μ, and then gradually decreasing again toward a slope near 0 at the upper end. The sketch is given at the right.

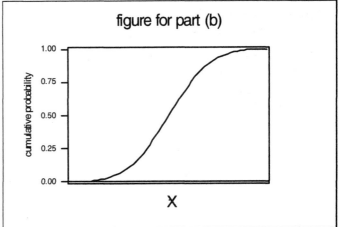

## 6-3 Applications of Normal Distributions

1. A normal distribution can have any mean and any non-negative standard deviation. A standard normal distribution has mean 0 and standard deviation 1 – and it follows a "nice" bell-shaped curve. Non-standard normal distributions can follow bell-shaped curves that are tall and thin, or short and fat.

3. No. Since the distribution of random digits does not follow a normal distribution, the methods of this section are not relevant. Because random digits follow a discrete uniform distribution, in which each digit is equally likely, probability questions can be answered using the methods of chapter 4. When selecting a random digit, there are 5 possibilities [0,1,2,3,4] less than 5 and P(x<5) = 5/10 = 0.5.

NOTE: In each nonstandard normal distribution, x scores are converted to z scores using the formula $z = (x-\mu)/\sigma$ and rounded to two decimal places. For "backwards" problems, solving the formula for x yields $x = \mu + z\sigma$. The area (cumulative probability) given in Table A-2 will be designated by A. As in the previous section, drawing and labeling the sketch is the key to successful completion of the exercises.

5. normal distribution: $\mu = 100$ and $\sigma = 15$
   $P(x<130)$
   $\quad = P(z<2.00)$
   $\quad = 0.9772$
   normalcdf(-999,130,100,15) = 0.9772

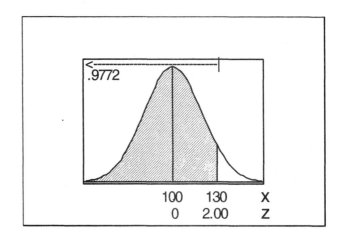

7. normal distribution: $\mu = 100$ and $\sigma = 15$
   $P(90<x<110)$
   $\quad = P(-0.67<z<0.67)$
   $\quad = P(z<0.67) - P(z<-0.67)$
   $\quad = 0.7486 - 0.2514$
   $\quad = 0.4972$
   normalcdf(90,110,100,15) = 0.4950

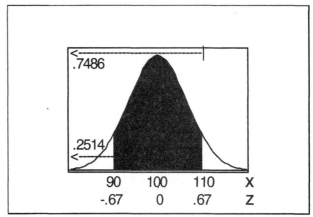

9. normal distribution: $\mu = 100$ and $\sigma = 15$
   For $P_{10}$, A = 0.1000 [0.1003]
   and z = -1.28.
   $x = \mu + z\sigma$
   $\quad = 100 + (-1.28)(15)$
   $\quad = 100 - 19.2$
   $\quad = 80.8$
   invNorm(0.1000,100,15) = 80.8

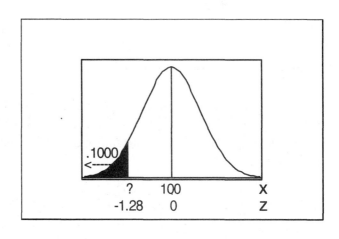

11. normal distribution: $\mu = 100$ and $\sigma = 15$
    For the top 35%,
    $\quad$ A = 0.6500 [0.6517] and z = 0.39.
    $x = \mu + z\sigma$
    $\quad = 100 + (0.39)(15)$
    $\quad = 100 + 5.85$
    $\quad = 105.85$
    invNorm(0.6500,100,15) = 105.8

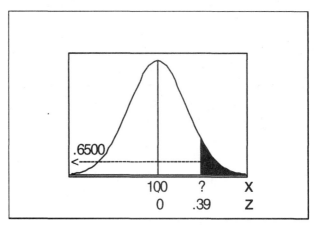

13. normal distribution: μ = 63.6 and σ = 2.5
    P(x>70)
        = 1 − P(x<70)
        = 1 − P(z<2.56)
        = 1 − 0.9948
        = 0.0052 or 0.52%
    normalcdf(70,999,63.6,2.5) = 0.0052
                                = 0.52%

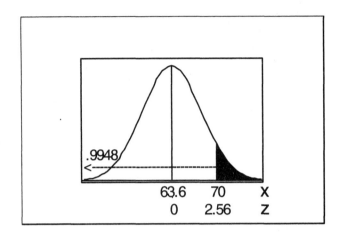

15. normal distributions
    a. men: μ = 69.0 and σ = 2.8         women: μ = 63.6 and σ = 2.5
       P(x>80)                                  P(x>80)
         = 1 − P(x<80)                      = 1 − P(x<80)
         = 1 − P(z<3.93)                  = 1 − P(z<6.44)
         = 1 − 0.9999 = 0.0001           = 1 − 0.9999 = 0.0001
    normalcdf(80,999,69.0,2.8) = 0.00004   normalcdf(80,999,63.6,2.5) = 2.70E-11 ≈ 0

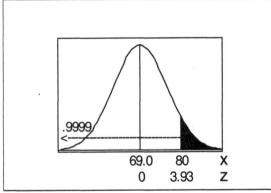 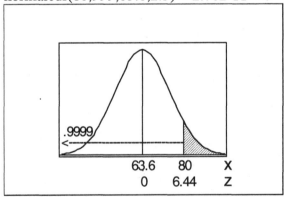

Yes. It appears that the current doorway design is OK.

b. normal distribution: μ = 69.0 and σ = 2.8
    For the tallest 5%,
        A = 0.9500 and z = 1.645.
    x = μ + zσ
       = 69.0 + (1.645)(2.8)
       = 69.0 + 4.6
       = 73.6 in
    invNorm(1−0.0500,69.0,2.8) = 73.6 in

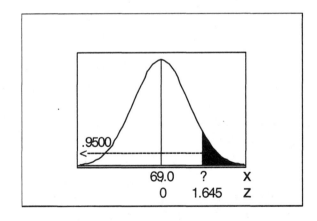

17. normal distribution: $\mu = 3420$ and $\sigma = 495$
For the lightest 2%,
   $A = 0.0200\ [0.0202]$ and $z = -2.05$.
$x = \mu + z\sigma$
   $= 3420 + (-2.05)(495)$
   $= 3420 - 1015$
   $= 2405$ g
invNorm(0.02,3420,495) = 2403 g

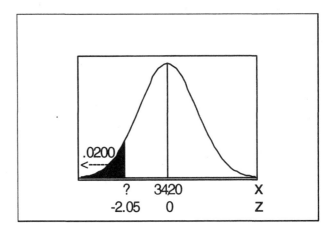

19. normal distribution: $\mu = 184.0$ and $\sigma = 55.0$
   $P(x>230)$
      $= 1 - P(x<230)$
      $= 1 - P(z<0.84)$
      $= 1 - 0.7995$
      $= 0.2005$
   normalcdf(230,999,184.0,55.0) = 0.2015

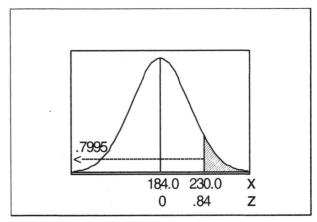

21. normal distribution: $\mu = 268$ and $\sigma = 15$
   a. $P(x>308)$
      $= 1 - P(x<308)$
      $= 1 - P(z<2.67)$
      $= 1 - 0.9962$
      $= 0.0038$
   normalcdf(308,999,268,15) = 0.0038
   The result suggests that an unusual event has occurred – but certainly not an impossible one, as about 38 of every 10,000 pregnancies can be expected to last as long.

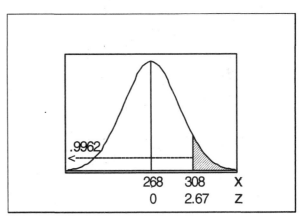

   b. For the lowest 4%,
      $A = 0.0400\ [0.0401]$ and $z = -1.75$.
   $x = \mu + z\sigma$
      $= 268 + (-1.75)(15)$
      $= 268 - 26$
      $= 242$ days
   invNorm(0.04,268,15) = 242 days

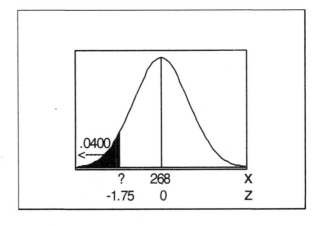

84    CHAPTER 6    Normal Probability Distributions

23. normal distribution: $\mu = 6.0$ and $\sigma = 1.0$
For the smallest 2.5%,
    A = 0.0250 and z = -1.96.
For the largest 2.5%,
    A = 0.9750 and z = 1.96
$x_S = \mu + z\sigma = 6.0 + (-1.96)(1.0)$
        $= 6.0 - 1.96$
        $= 4.04$, rounded to 4.0 in
invNorm(0.025,6.0,1.0) = 4.0 in
$x_L = \mu + z\sigma = 6.0 + (1.96)(1.0)$
        $= 6.0 + 1.96$
        $= 7.96$, rounded to 8.0 in
invNorm(1–0.025,6.0,1.0) = 8.0 in

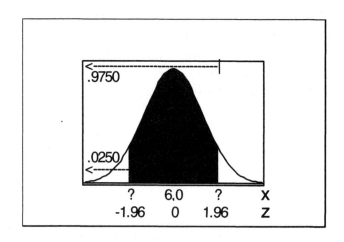

25. a. 1-VAR STATS with the list MSYS gives screen #1 below.
STAT PLOT with the settings in screen #2 and ZOOM 9 give screen #3 below.
The histogram produced by the default settings is approximately normal.
The settings in screen #4 and GRAPH give the more normal-looking histogram in screen #5.

screen #1          screen #2          screen #3          screen #4          screen #5

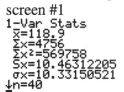  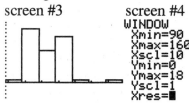 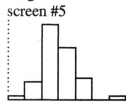

b. normal distribution:
    $\mu = 118.9$ and $\sigma = 10.46$
    For $P_5$, A = 0.0500 and z = -1.645.
    For $P_{95}$, A = 0.9500 and z = 1.645.
    $x_5 = \mu + z\sigma$
        $= 118.9 + (-1.645)(10.46)$
        $= 118.9 - 17.2$
        $= 101.7$ mm  [using rounded $\sigma$, 101.6]
    invNorm(0.05,118.9,10.46) = 101.7 mm
    $x_{95} = \mu + z\sigma$
        $= 118.9 + (1.645)(10.46)$
        $= 118.9 + 17.2$
        $= 136.1$ mm  [using rounded $\sigma$, 136.2]
    invNorm(0.95,118.9,10.46) = 136.1 mm

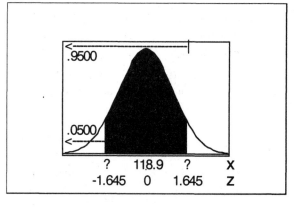

27. a. The z scores are always unit free. Because the numerator and the denominator of the fraction $z = (x-\mu)/\sigma$ have the same units, the units will divide out.
    b. For a population of size N, $\mu = \Sigma x/N$ and $\sigma^2 = \Sigma(x-\mu)^2/N$.
        As shown below, $\mu = 0$ and $\sigma = 1$ will be true for *any* set of z scores – regardless of the shape of the original distribution.
        $\Sigma z = \Sigma(x-\mu)/\sigma = (1/\sigma)[\Sigma(x-\mu)] = (1/\sigma)[\Sigma x - \Sigma \mu] = (1/\sigma)[N\mu - N\mu] = (1/\sigma)[0] = 0$
        $\Sigma z^2 = \Sigma[(x-\mu)/\sigma]^2 = (1/\sigma)^2[\Sigma(x-\mu)^2] = (1/\sigma^2)[N\sigma^2] = N$
        $\mu_z = (\Sigma z)/N = 0/N = 0$
        $\sigma^2_z = \Sigma(z-\mu_z)^2/N = \Sigma(z-0)^2/N = \Sigma z^2/N = N/N = 1$ and $\sigma_z = 1$
        The re-scaling from x scores to z scores will not affect the basic shape of the distribution – and so in this case the z scores will be normal, as was the original distribution.

Applications of Normal Distributions  SECTION 6-3  85

29. normal distribution: $\mu = 25$ and $\sigma = 5$

a. For a population of size N, $\mu = \Sigma x/N$ and $\sigma^2 = \Sigma(x-\mu)^2/N$.
As shown below, adding a constant to each score increases the mean by that amount but does not affect the standard deviation. In non-statistical terms, shifting everything by k units does not affect the spread of the scores. This is true for *any* set of scores – regardless of the shape of the original distribution. Let $y = x + k$.

$\mu_Y = [\Sigma(x+k)]/N$
$= [\Sigma x + \Sigma k]/N$
$= [\Sigma x + Nk]/N$
$= (\Sigma x)/N + (Nk)/N$
$= \mu_X + k$

$\sigma^2_Y = \Sigma[y - \mu_Y]^2/N$
$= \Sigma[(x+k) - (\mu_X + k)]^2/N$
$= \Sigma[x - \mu_X]^2/N$
$= \sigma^2_X$ and so $\sigma_Y = \sigma_X$

If the teacher adds 50 to each grade,
new mean = 25 + 50 = 75
new standard deviation = 5 (same as before).

b. No. Curving should consider the variation. Had the test been more appropriately constructed, it is not likely that every student would score exactly 50 points higher. If the typical student score increased by 50, we would expect the better students to increase by more than 50 and the poorer students to increase by less than 50. This would make the scores more spread out and would increase the standard deviation.

c. For the top 10%, invNorm(1–0.10,25,5) = 31.4
For the bottom 70%, invNorm(0.70,25,5) = 27.6
For the bottom 30%, invNorm(0.30,25,5) = 22.4
For the bottom 10%, invNorm(0.10,25,5) = 18.6
This produces the grading scheme given at the right.

```
A: higher than 31.4
B: 27.6 to 31.4
C: 22.4 4o 27.6
D: 18.6 to 22.4
E: less than 18.6
```

d. The curving scheme in part (c) is fairer because it takes into account the variation as discussed in part (b). Assuming the usual 90-80-70-60 letter grade cut-offs, for example, the percentage of A's under the scheme in part (a) with $\mu = 75$ and $\sigma = 5$ is

$P(x>90) = 1 - P(x<90)$
$= 1 - P(z<3.00)$
$= 1 - 0.9987$
$= 0.0013$ or 0.13%   normalcdf(90,999,75,5) = 0.0013 = 0.13%

This is considerably less than the 10% A's under the scheme in part (c) and reflects the fact that the variation in part (a) is unrealistically small.

## 6-4 Sampling Distributions and Estimators

1. Given a population distribution of scores, one can take a sample of size n and calculate any of several statistics. A sampling distribution is the distribution of all possible values of such a particular statistic.

3. Table 6-7 indicates that the sample median, range and standard deviation are not necessarily unbiased estimators of the population median, range and standard deviation. Table 6-7 indicates that the sample mean, variance and proportion were unbiased estimators of the population mean, variance and proportion for the given example – and the text states that such is always the case.

5. No. The sample proportion is an unbiased estimator of the population proportion. This means that it gives the correct value on the average, but not necessarily for any one particular sample.

NOTE: Section 5-2 defined the mean of a probability distribution of x's as $\mu_x = \Sigma[x \cdot P(x)]$. If the variable is designated by the symbol y, then the mean of a probability distribution of y's is $\mu_y = \Sigma[y \cdot P(y)]$. In this section, the variables are statistics – like $\bar{x}$ and $\hat{p}$. In such cases, the formula for the mean may be adjusted – to $\mu_{\bar{x}} = \Sigma[\bar{x} \cdot P(\bar{x})]$ and $\mu_{\hat{p}} = \Sigma[\hat{p} \cdot P(\hat{p})]$. In a similar manner, the formula for the variance of a probability distribution may also be adjusted to match the variable being considered.

7. For the original population,
$\mu = \Sigma x/N$
$= (10+6+5)/3$
$= 21/3 = 7.0$ calls

| sample | $\bar{x}$ | $\bar{x}$ | $P(\bar{x})$ | $\bar{x} \cdot P(\bar{x})$ |
|---|---|---|---|---|
| 5,5 | 5.0 | 5.0 | 1/9 | 5.0/9 |
| 5,6 | 5.5 | 5.5 | 2/9 | 11.0/9 |
| 5,10 | 7.5 | 6.0 | 1/9 | 6.0/9 |
| 6,5 | 5.5 | 7.5 | 2/9 | 15.0/9 |
| 6,6 | 6.0 | 8.0 | 2/9 | 16.0/9 |
| 6,10 | 8.0 | 10.0 | 1/9 | 10.0/9 |
| 10,5 | 7.5 | | 9/9 | 63.0/9 |
| 10,6 | 8.0 | | | |
| 10,10 | 10.0 | | | |
| | 63.0 | | | |

a. The 9 equally likely samples and their means are given at the immediate right.

b. Each sample has a probability of 1/9. The sampling distribution of the means, a list of each possible sample mean and its total probability, is given at the far right. The distribution appears to "bunch up" in the middle and taper off at the ends.

c. $\mu_{\bar{x}} = \Sigma[\bar{x} \cdot P(\bar{x})] = 63.0/9 = 7.0$ calls

d. Yes. The sample mean is always an unbiased estimator of the population mean.

NOTE: One can also find $\mu_{\bar{x}}$ as follows: put the $\bar{x}$ values in list L1 and the $P(\bar{x})$ values in L2, and then use STAT CALC 1-Var-Stat L1,L2. The resulting $\bar{x}$ is the desired mean of the sampling distribution.

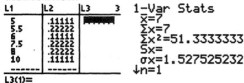

9. For the original population,
   $\mu = \Sigma x/N$
   $= (47+43+21+20+20)/5$
   $= 151/5 = 30.2$ billion $

   a. The 25 equally likely samples and their means are given at the immediate right.

   b. Each sample has a probability of 1/25. The sampling distribution of the means, a list of each possible sample mean and its total probability, is given at the far right. The distribution appears to "bunch up" toward the lower end.

   c. $\mu_{\bar{x}} = \Sigma[\bar{x} \cdot P(\bar{x})] = 755.0/25 = 30.2$ billion $

   d. Yes. The sample mean is always an unbiased estimator of the population mean.

   NOTE: Using the technique outlined for exercise #7,
   1-Var Stats
   $\bar{x}=30.2$
   $\Sigma x=30.2$
   $\Sigma x^2=985.92$
   $Sx=$
   $\sigma x=8.595347579$
   $\downarrow n=1$

| sample | $\bar{x}$ | $\bar{x}$ | $P(\bar{x})$ | $\bar{x} \cdot P(\bar{x})$ |
|---|---|---|---|---|
| 20,20 | 20.0 | 20.0 | 4/25 | 80.0/25 |
| 20,20 | 20.0 | 20.5 | 4/25 | 82.0/25 |
| 20,21 | 20.5 | 21.0 | 1/25 | 21.0/25 |
| 20,43 | 31.5 | 31.5 | 4/25 | 126.0/25 |
| 20,47 | 33.5 | 32.0 | 2/25 | 64.0/25 |
| 20,20 | 20.0 | 33.5 | 4/25 | 134.0/25 |
| 20,20 | 20.0 | 34.0 | 2/25 | 68.0/25 |
| 20,21 | 20.5 | 43.0 | 1/25 | 43.0/25 |
| 20,43 | 31.5 | 45.0 | 2/25 | 90.0/25 |
| 20,47 | 33.5 | 47.0 | 1/25 | 47.0/25 |
| 21,20 | 20.5 | | 25/25 | 755.0/25 |
| 21,20 | 20.5 | | | |
| 21,21 | 21.0 | | | |
| 21,43 | 32.0 | | | |
| 21,47 | 34.0 | | | |
| 43,20 | 31.5 | | | |
| 43,20 | 31.5 | | | |
| 43,21 | 32.0 | | | |
| 43,43 | 43.0 | | | |
| 43,47 | 45.0 | | | |
| 47,20 | 33.5 | | | |
| 47,20 | 33.5 | | | |
| 47,21 | 34.0 | | | |
| 47,43 | 45.0 | | | |
| 47,47 | 47.0 | | | |
| | 755.0 | | | |

11. For the original population, $p = 3/4 = 0.75$.

   a. The 16 equally likely samples and their proportions are given at the immediate right. Each sample has a probability of 1/16. The sampling distribution of the proportions, a list of each possible sample proportion and its total probability, is given at the far right. The distribution appears to "bunch up" toward the upper end.

   b. $\mu_{\hat{p}} = \Sigma[\hat{p} \cdot P(\hat{p})] = 12.0/16 = 0.75$

   c. Yes. The sample proportion is always an unbiased estimator of the population proportion..

   NOTE: Using the technique outlined for exercise #7,
   1-Var Stats
   $\bar{x}=.75$
   $\Sigma x=.75$
   $\Sigma x^2=.65625$
   $Sx=$
   $\sigma x=.3061862178$
   $\downarrow n=1$

| sample | $\hat{p}$ | $\hat{p}$ | $P(\hat{p})$ | $\hat{p} \cdot P(\hat{p})$ |
|---|---|---|---|---|
| MM | 0.0 | 0.0 | 1/16 | 0.0/16 |
| MA | 0.5 | 0.5 | 6/16 | 3.0/16 |
| MB | 0.5 | 1.0 | 9/16 | 9.0/16 |
| MC | 0.5 | | 16/16 | 12.0/16 |
| AM | 0.5 | | | |
| AA | 1.0 | | | |
| AB | 1.0 | | | |
| AC | 1.0 | | | |
| BM | 0.5 | | | |
| BA | 1.0 | | | |
| BB | 1.0 | | | |
| BC | 1.0 | | | |
| CM | 0.5 | | | |
| CA | 1.0 | | | |
| CB | 1.0 | | | |
| CC | 1.0 | | | |
| | 12.0 | | | |

13. For the original population,
    μ = Σx/N = (3+9+23)/3 = 35/3 = 11.7

    a. The 9 equally likely samples and their means are given at the immediate right. Each sample has a probability of 1/9. The sampling distribution of the means, a list of each possible sample mean and its total probability, is given at the far right. The distribution appears to "bunch up" toward the middle and taper off at the ends. As expected,
    $\mu_{\bar{x}} = \Sigma[\bar{x} \cdot P(\bar{x})] = 105.0/9 = 11.7$

| sample | $\bar{x}$ | $\bar{x}$ | $P(\bar{x})$ | $\bar{x} \cdot P(\bar{x})$ |
|---|---|---|---|---|
| 3,3 | 3.0 | 3.0 | 1/9 | 3.0/9 |
| 3,9 | 6.0 | 6.0 | 2/9 | 12.0/9 |
| 3,23 | 13.0 | 9.0 | 1/9 | 9.0/9 |
| 9,3 | 6.0 | 13.0 | 2/9 | 26.0/9 |
| 9,9 | 9.0 | 16.0 | 2/9 | 32.0/9 |
| 9,23 | 16.0 | 23.0 | 1/9 | 23.0/9 |
| 23,3 | 13.0 | | 9/9 | 105.0/9 |
| 23,9 | 16.0 | | | |
| 23,23 | 23.0 | | | |
|  | 105.0 | | | |

b. Rank data is at the ordinal level – and the difference between ranks 1 and 2 is not necessarily the same, in actual performance, as the difference between ranks 25 and 26. And since the values of the ranks are predetermined (1,2,…,n), the data are not independent. But one can still average ranks to make valid comparisons and make sample selections from a population of ranks. In that sense ranks are just as meaningful as other scores and it makes sense to identify sampling distributions for statistics calculated using ranks.  NOTE: Using the technique outlined for exercise #7,

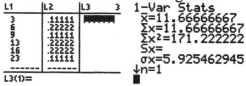

15. Use of the formula for the given values is shown below. The resulting distribution agrees with the sampling distribution for the proportion of girls in a family of size 2.
    P(x) = 1/[2(2-2x)!(2x)!]
    P(x=0) = 1/[2(2)!(0)!] = 1/[2·2·1] = 1/4
    P(x=0.5) = 1/[2(1)!(1)!] = 1/[2·1·1] = 1/2
    P(x=1) = 1/[2(0)!(2)!] = 1/[2·1·2] = 1/4

## 6-5 The Central Limit Theorem

1. The standard error of a statistic is the standard deviation of its sampling distribution. For any given sample size n, the standard error of the mean is the standard deviation of the distribution of all possible sample means of size n. Its symbol and numerical value are $\sigma_{\bar{x}} = \sigma/\sqrt{n}$.

3. The subscript $\bar{x}$ is used to distinguish the mean and standard deviation of the sample means from the mean and standard deviation of the original population. This produces the notation
    $\mu_{\bar{x}} = \mu$    $\sigma_{\bar{x}} = \sigma/\sqrt{n}$

5. a. normal distribution
    μ = 63.6
    σ = 2.5
    P(x<64)
       = P(z<0.16)
       = 0.5636
    normalcdf(-999,64,63.6,2.5) = 0.5636

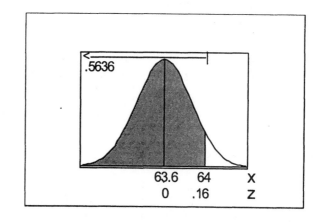

b. normal distribution,
   since the original distribution is so
   $\mu_{\bar{x}} = \mu = 63.6$
   $\sigma_{\bar{x}} = \sigma/\sqrt{n} = 2.5/\sqrt{36} = 0.417$
   $P(\bar{x}<64)$
   $= P(z<0.96)$
   $= 0.8315$
   normalcdf(-999,64,63.6,2.5/√36)
   $= 0.8315$

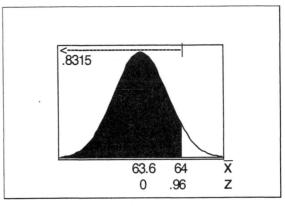

NOTE: This manual uses the above notation, even though the √ function on the TI-83/84 Plus automatically inserts a parenthesis so that the screen display shows "√(". Using ENTER after the "36" above automatically closes all the operations. If there is more to follow, insert the appropriate number of parentheses.

7. a. normal distribution
   $\mu = 63.6$
   $\sigma = 2.5$
   $P(63.5<x<64.5)$
   $= P(-0.04<z<0.36)$
   $= P(z<0.36) - P(z<-0.04)$
   $= 0.6406 - 0.4840$
   $= 0.1566$
   normalcdf(63.5,64.5,63.6,2.5) = 0.1565

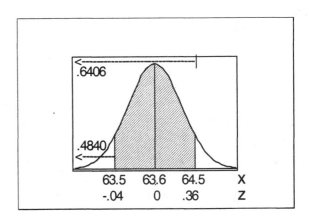

b. normal distribution,
   since the original distribution is so
   $\mu_{\bar{x}} = \mu = 63.6$
   $\sigma_{\bar{x}} = \sigma/\sqrt{n} = 2.5/\sqrt{9} = 0.833$
   $P(63.5<\bar{x}<64.5)$
   $= P(-0.12<z<0.1.08)$
   $= 0.8599 - 0.4522$
   $= 0.4077$
   normalcdf(63.5,64.5,63.6,2.5/√9)
   $= 0.4077$

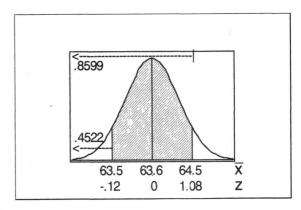

c. Since the original distribution is normal, the Central Limit Theorem can be used in part (b) even though the sample size does not exceed 30.

9. a. normal distribution
   $\mu = 172$
   $\sigma = 29$
   $P(x > 167)$
   $= 1 - P(x < 167)$
   $= 1 - P(z < -0.17)$
   $= 1 - 0.4325$
   $= 0.5675$
   normalcdf(167,999,172,29) = 0.5684

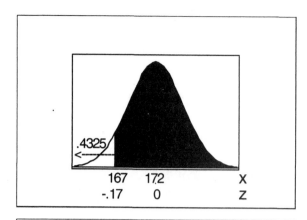

b. normal distribution,
   since the original distribution is so
   $\mu_{\bar{x}} = \mu = 172$
   $\sigma_{\bar{x}} = \sigma/\sqrt{n} = 29/\sqrt{12} = 8.372$
   $P(\bar{x} > 167)$
   $= 1 - P(\bar{x} < 167)$
   $= 1 - P(z < -0.60)$
   $= 1 - 0.2743$
   $= 0.7257$
   normalcdf(167,999,172,29/√12)
   $= 0.7248$

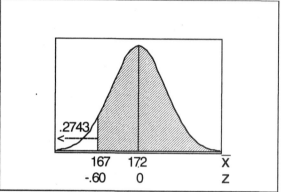

c. No. It appears that the 12 person capacity could easily exceed the 2004 lbs – especially when the weight of clothes and other equipment is considered. On the other hand, skiers may not be representative of the general population – they may be lighter than the general population, as skiing may not be an activity that attracts heavier persons.

11. a. normal distribution,
    by the Central Limit Theorem
    $\mu_{\bar{x}} = \mu = 12.00$
    $\sigma_{\bar{x}} = \sigma/\sqrt{n} = 0.11/\sqrt{36} = 0.0183$
    $P(\bar{x} > 12.19)$
    $= 1 - P(\bar{x} < 12.19)$
    $= 1 - P(z < 10.36)$
    $= 1 - 0.9999$
    $= 0.0001$
    normalcdf(12.19,999,12.00,0.11/√36)
    $= 1.85\text{E-}25 \approx 0$

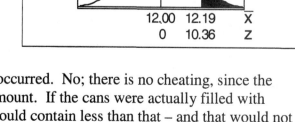

b. No; if $\mu = 12.00$, then a very rare event has occurred. No; there is no cheating, since the actual amount is more than the advertised amount. If the cans were actually filled with $\mu = 12.00$, then about one-half of the cans would contain less than that – and that would not be acceptable.

13. a. normal distribution
   $\mu = 3.00$
   $\sigma = 0.40$
   $P(x>4.00)$
   $= 1 - P(x<4.00)$
   $= 1 - P(z<2.50)$
   $= 1 - 0.9938$
   $= 0.0062$
   normalcdf(4.00,999,3.00,0.40) = 0.0062

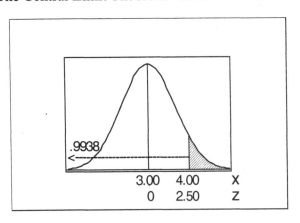

b. normal distribution,
      since the original distribution is so
   $\mu_{\bar{x}} = \mu = 4.00$
   $\sigma_{\bar{x}} = \sigma/\sqrt{n} = 0.40/\sqrt{60} = 0.0516$
   $P(\bar{x}>4.00)$
   $= 1 - P(\bar{x}<4.00)$
   $= 1 - P(z<19.36)$
   $= 1 - 0.9999$
   $= 0.0001$
   normalcdf(4.00,999,3.00,0.40/√60)
   $= 8.21\text{E-}84 \approx 0$

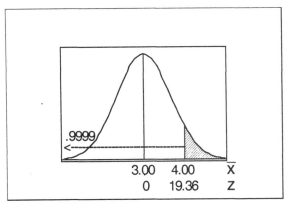

c. Part (a). It is the individual flashes that guide pilots, and not the mean of several flashes.

15. a. normal distribution
   $\mu = 114.8$
   $\sigma = 13.1$
   $P(x>140)$
   $= 1 - P(x<140)$
   $= 1 - P(z<1.92)$
   $= 1 - 0.9726$
   $= 0.0274$
   normalcdf(140,999,114.8,13.1) = 0.272

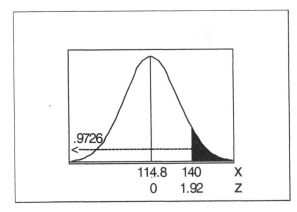

b. normal distribution,
      since the original distribution is so
   $\mu_{\bar{x}} = \mu = 114.8$
   $\sigma_{\bar{x}} = \sigma/\sqrt{n} = 13.1/\sqrt{4} = 6.55$
   $P(\bar{x}>140)$
   $= 1 - P(\bar{x}<140)$
   $= 1 - P(z<3.85)$
   $= 1 - 0.9999$
   $= 0.0001$
   normalcdf(140,999,114.8,13.1/√4)
   $= 0.0000597$, rounded to 0.0001

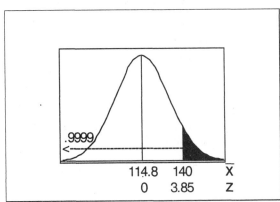

92   CHAPTER 6   Normal Probability Distributions

c. Since the original distribution is normal, the Central Limit Theorem can be used in part (b) even though the sample size does not exceed 30.

d. No. The mean can be less than 140 when one or more of the values is greater than 140.

17. a. normal distribution
$\mu = 143$
$\sigma = 29$
$P(140<x<211)$
  $= P(-0.10<z<2.34)$
  $= P(z<2.34) - P(z<-0.10)$
  $= 0.9904 - 0.4602$
  $= 0.5302$
normalcdf(140,211,143,29) = 0.5317

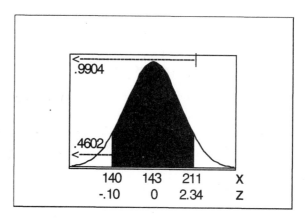

b. normal distribution,
   since the original distribution is so
$\mu_{\bar{x}} = \mu = 143$
$\sigma_{\bar{x}} = \sigma/\sqrt{n} = 29/\sqrt{36} = 4.833$
$P(140<\bar{x}<211)$
  $= P(-0.62<z<14.07)$
  $= P(z<14.07) - P(z<-0.62)$
  $= 0.9999 - 0.2676$
  $= 0.7323$
normalcdf(140,211,143,29/√36)
  $= 0.7326$

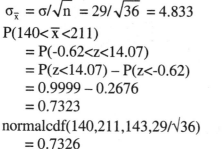

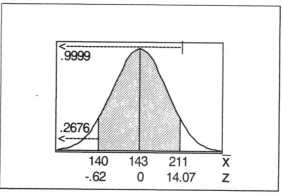

c. The information from part (a) is more relevant, since the seats will be occupied by one woman at a time.

19. a. normal distribution
$\mu = 5.670$
$\sigma = 0.062$
$P(5.550<x<5.790)$
  $= P(-1.94<z<1.94)$
  $= P(z<1.94) - P(z<-1.94)$
  $= 0.9738 - 0.0262$
  $= 0.9476$
normalcdf(5.550,5.790,5.670,0.062)
  $= 0.9471$

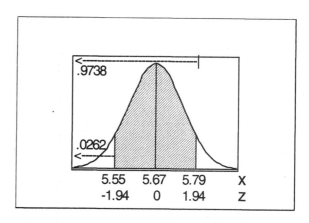

If 0.9476 of the quarters are accepted, then 1 − 0.9471 = 0.0529 of the quarters are rejected. For 280 quarters, we expect (0.0529)(280) = 14.8 of them to be rejected.

b. normal distribution,
   since the original distribution is so
   $\mu_{\bar{x}} = \mu = 5.670$
   $\sigma_{\bar{x}} = \sigma/\sqrt{n} = 0.062/\sqrt{280} = 0.00371$
   $P(5.550 < \bar{x} < 5.790)$
   $= P(-32.39 < z < 32.39)$
   $= P(z < 32.39) - P(z < -32.39)$
   $= 0.9999 - 0.0001$
   $= 0.9998$
   normalcdf(5.550,5.790,5.670,0.062/√280)
   $= 1.0000$

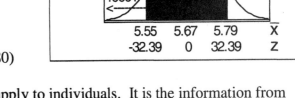

c. Probabilities concerning means do not apply to individuals. It is the information from part (a) that is relevant, since the vending machine deals with quarters one at a time.

21. a. normal distribution,
    since the original distribution is so
    $\mu_{\bar{x}} = \mu = 14.4$
    $\sigma_{\bar{x}} = \sigma/\sqrt{n} = 1.0/\sqrt{18} = 0.236$
    This is a "backwards" normal problem,
    With A = 0.9750 and z = 1.96.
    $\bar{x} = \mu_{\bar{x}} + z\sigma_{\bar{x}} = 14.4 + (1.96)(0.236)$
    $= 14.4 + 0.46$
    $= 14.86$
    invNorm(0.975,14.4,1.0/√18) = 14.86
    For 18 men, the total width is (18)(14.86) = 267.5 in.

    b. The 14.4 inch figure applies to the general male population, not to football players (who would tend to be larger than normal anyway) wearing pads. Furthermore, the calculations in part (a) do not consider any spacing between the players.

23. a. The box below gives the calculations to determine μ=8 and σ=5.385 for the original population of N=6 values.

    b. The 15 equally like samples of size n=2 are listed in the first column in the box below.

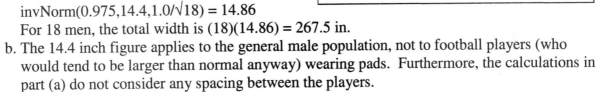

| x | x-μ | (x-μ)² |
|---|---|---|
| 2 | -6 | 36 |
| 3 | -5 | 25 |
| 6 | -2 | 4 |
| 8 | 0 | 0 |
| 11 | 3 | 9 |
| 18 | 10 | 100 |
| 48 | 0 | 174 |

$\mu = \Sigma x/N = 48/6 = 8.0$

$\sigma^2 = \Sigma(x-\mu)^2/N = 174/6 = 29$

$\sigma = 5.385$

| sample | $\bar{x}$ | $\bar{x} - \mu_{\bar{x}}$ | $(\bar{x} - \mu_{\bar{x}})^2$ |
|---|---|---|---|
| 2,3 | 2.5 | -5.5 | 30.25 |
| 2,6 | 4.0 | -4.0 | 16.00 |
| 2,8 | 5.0 | -3.0 | 9.00 |
| 2,11 | 6.5 | -1.5 | 2.25 |
| 2,18 | 10.0 | 2.0 | 4.00 |
| 3,6 | 4.5 | -3.5 | 12.25 |
| 3,8 | 5.5 | -2.5 | 6.25 |
| 3,11 | 7.0 | -1.0 | 1.00 |
| 3,18 | 10.5 | 2.5 | 6.25 |
| 6,8 | 7.0 | -1.0 | 1.00 |
| 6,11 | 8.5 | 0.5 | 0.25 |
| 6,18 | 12.0 | 4.0 | 16.00 |
| 8,11 | 9.5 | 1.5 | 2.25 |
| 8,18 | 13.0 | 5.0 | 25.00 |
| 11,18 | 14.5 | 6.5 | 42.25 |
|  | 120.0 | 0.0 | 174.00 |

c. The values of $\bar{x}$ are listed in the second column in the box below.

d. Since there are 15 equally likely samples, the desired values can be determined using the box above on the right as follows.
$\mu_{\bar{x}} = \Sigma\bar{x}/15 = 120/15 = 8.0$
$\sigma_{\bar{x}}^2 = \Sigma(\bar{x}-\mu_{\bar{x}})^2/15 = 174/15 = 11.6$
$\sigma_{\bar{x}} = \sqrt{11.6} = 3.406$

e. $\mu = 8 = \mu_{\bar{x}}$
$\dfrac{\sigma}{\sqrt{n}}\sqrt{\dfrac{N-n}{N-1}} = \dfrac{5.385}{\sqrt{2}}\sqrt{\dfrac{6-2}{6-1}} = 3.406 = \sigma_{\bar{x}}$

NOTE: The mean $\mu$ and standard deviation $\sigma$ in part (a) can also be found by placing the x values in L1 and using 1-VAR STATS. The same is true, using the $\bar{x}$ values, for part (d).

## 6-6 Normal as Approximation to Binomial

1. The histogram will be bell-shaped – with the highest frequencies near 0.50, and tapering off evenly toward each end.

3. No. With n=6 and p=0.001, the requirements that np≥5 and nq≥5 are not met.

5. The area to the right of 15.5.    In symbols, $P(x>15) = P_C(x>15.5)$.

7. The area to the left of 11.5.    In symbols, $P(x<12) = P_C(x<11.5)$.

9. The area to the left of 4.5.    In symbols $P(x\le 4) = P_C(x<4.5)$.

11. The area between 7.5 and 10.5.    In symbols, $P(8\le x\le 10) = P_C(7.5<x<10.5)$.

13. binomial: n=12 and p=0.6
    a. from Table A-1, $P(x=7) = 0.227$
    b. normal approximation not appropriate since nq = 12(0.4) = 4.8 < 5

15. binomial: n = 11 and p = 0.5
    a. from Table A-1,
    $P(x\ge 4) = P(x=4) + P(x=5) +\ldots+ P(x=11)$
    $= 0.161 + 0.226 + 0.226 + 0.161 + 0.081 + 0.027 + 0.005 + 0^+ = 0.887$
    b. normal approximation appropriate since
    $np = 11(0.5) = 5.5 \ge 5$
    $nq = 11(0.5) = 5.5 \ge 5$
    $\mu = np = 11(0.5) = 5.5$
    $\sigma = \sqrt{npq} = \sqrt{11(0.5)(0.5)} = 1.658$
    $P(x\ge 4) = P_C(x>3.5)$
    $= 1 - P(x<3.5)$
    $= 1 - P(z<-1.21)$
    $= 1 - 0.1131 = 0.8869$
    normalcdf(3.5,999,5.5,√(11×0.5×0.5))
    $= 0.8861$

    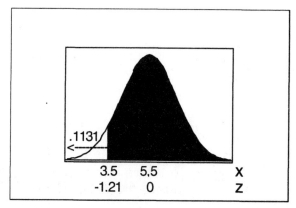

    NOTE: 1 – binomcdf(11,0.5,3) = 0.8867 gives a more accurate answer.

17. Let x = the number of girl births.
    binomial: n=64 and p=0.5
    normal approximation appropriate since
    $\quad np = 64(0.5) = 32 \geq 5$
    $\quad nq = 64(0.5) = 32 \geq 5$
    $\mu = np = 64(0.5) = 32$
    $\sigma = \sqrt{npq} = \sqrt{64(0.5)(0.5)} = 4$
    $P(x>36) = P_C(x>36.5)$
    $\quad\quad\quad\quad = 1 - P(x<36.5)$
    $\quad\quad\quad\quad = 1 - P(z<1.125)$
    $\quad\quad\quad\quad = 1 - 0.8697$
    $\quad\quad\quad\quad = 0.1303 \quad [0.8697 = (0.8686+0.8708)/2]$

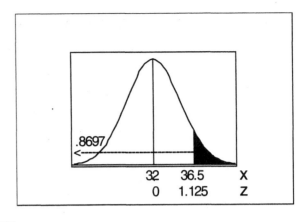

    normalcdf(36.5,999,32,√(64×0.5×0.5)) = 0.1303
    No. Since 0.1303 > 0.05, getting more than 36 girls is not an unusual occurrence.
    NOTE: 1 – binomcdf(64,0.5,36) = 0.1302 gives a more accurate answer.

19. Let x = the number that actually voted.
    binomial: n=1002 and p=0.61
    normal approximation appropriate since
    $\quad np = 1002(0.61) = 611.22 \geq 5$
    $\quad nq = 1002(0.39) = 390.78 \geq 5$
    $\mu = np = 1002(0.61) = 611.22$
    $\sigma = \sqrt{npq} = \sqrt{1002(0.61)(0.39)} = 15.439$
    $P(x \geq 701) = P_C(x>700.5)$
    $\quad\quad\quad\quad = 1 - P(x<700.5)$
    $\quad\quad\quad\quad = 1 - P(z<5.78)$
    $\quad\quad\quad\quad = 1 - 0.9999 = 0.0001$

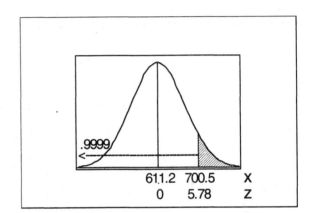

    normalcdf(700.5,9999,611.22,√(1002×0.61×0.39)) = 3.69E-9 ≈ 0
    Under the stated conditions, getting 701 persons that actually voted would be a very rare occurrence. It appears that the people in the survey were not telling the truth.
    NOTE: 1 – binomcdf(1002,0.61,700) = 2.05E-9 ≈ 0 gives a more accurate answer.

21. Let x = the number with yellow pods.
    binomial: n=580 and p=0.25
    normal approximation appropriate since
    $\quad np = 580(0.25) = 145 \geq 5$
    $\quad nq = 580(0.75) = 435 \geq 5$
    $\mu = np = 580(0.25) = 145$
    $\sigma = \sqrt{npq} = \sqrt{580(0.25)(0.75)} = 10.428$
    $P(x \geq 152) = P_C(x>151.5)$
    $\quad\quad\quad\quad = 1 - P(x<151.5)$
    $\quad\quad\quad\quad = 1 - P(z<-0.62)$
    $\quad\quad\quad\quad = 1 - 0.7324 = 0.2676$

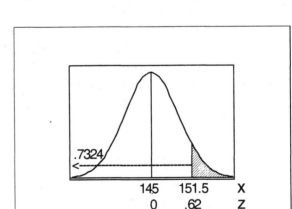

    normalcdf(151.5,999,145,√(580×0.25×0.75)) = 0.2665
    No. There is no evidence to suggest that Mendel's 25% rate is incorrect.
    NOTE: 1 – binomcdf(580,0.25,151) = 0.2650 gives a more accurate answer.

23. Let x = the number with such cancer.
  binomial: n=420,095 and p=0.00034
  normal approximation appropriate since
    np = 420,095(0.00034) = 142.8 ≥5
    nq = 420,095(0.99966) = 419857.2 ≥5
  μ = np = 420,095(0.00034) = 142.8
  σ = $\sqrt{npq}$ = $\sqrt{420095(0.00034)(0.99966)}$
    = 11.949
  P(x≤135) = $P_C$(x<135.5)
    = P(z<-0.61)
    = 0.2709

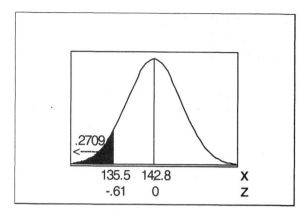

normalcdf(-999,135.5,142.8,√(420095×0.00034×0.99966)) = 0.2706
Finding "135 or fewer cancer cases" is not an unusual event, and so the cancer rate among cell phone users appears not to differ significantly from that of the general population.
NOTE: To test the media claim that cell phone usage is associated with a higher rate, one should calculate the probability of finding "135 or more cancer cases."
NOTE: binomcdf(420095,0.00034,135) = 0.2726 gives a more accurate answer.

25. Let x = the number females hired.
  binomial: n=62 and p=0.5
  normal approximation appropriate since
    np = 62(0.5) = 31 ≥5
    nq = 62(0.5) = 31 ≥5
  μ = np = 62(0.5) = 31
  σ = $\sqrt{npq}$ = $\sqrt{62(0.5)(0.5)}$ = 3.937
  P(x≤21) = $P_C$(x<21.5)
    = P(z<-2.41)
    = 0.0080

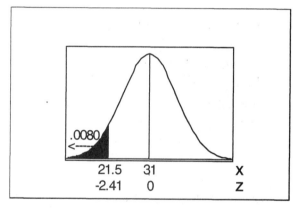

normalcdf(-999,21.5,31,√(62×0.5×0.5)) = 0.0079
Hiring 21 or fewer females by chance alone would be an unusual event. It appears that factors other than chance are at work, including the possibility of gender discrimination.
NOTE: binomcdf(62,0.5,21) = 0.0076 gives a more accurate answer.

27. Let x = the number with Group O blood.
  binomial: n=400 and p=0.45
  normal approximation appropriate since
    np = 400(0.45) = 180 ≥5
    nq = 400(0.55) = 220 ≥5
  μ = np = 400(0.45) = 180
  σ = $\sqrt{npq}$ = $\sqrt{400(0.45)(0.55)}$ = 9.950
  P(x≥177) = $P_C$(x>176.5)
    = 1 – P(x<176.5)
    = 1 – P(z<-0.35)
    = 1 – 0.3632 = 0.6368

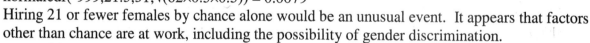

normalcdf(176.5,999,180,√(400×0.45×0.55)) = 0.6375
Yes. A pool of 400 is likely to be sufficient, but it might be prudent to try for a larger pool.
NOTE: 1 – binomcdf(400,0.45,176) = 0.6369 gives a more accurate answer.

Normal as Approximation to Binomial    SECTION 6-6   97

29. Let x = the number opposed to cloning.
    binomial: n=1012 and p=0.50
    normal approximation appropriate since
        np = 1012(0.50) = 506 ≥ 5
        nq = 1012(0.5) = 506 ≥ 5
    μ = np = 1012(0.50) = 506
    σ = $\sqrt{npq}$ = $\sqrt{1012(0.50)(0.50)}$ = 15.906
    P(x≥901) = $P_C$(x>900.5)
             = 1 – P(x<900.5)
             = 1 – P(z<24.80)
             = 1 – 0.9999 = 0.0001
    normalcdf(900.5,9999,506,√(1012×0.50×0.50)) = 0.0000
    Yes. If only 50% of the people oppose cloning, a very rare event has occurred. The result is strong evidence in favor of the claim that more than 50% of the people are opposed to cloning.
    NOTE: 1 – binomcdf(1012,0.50,900) = 0.0000 gives a more accurate answer.

31. Let x = the number of charges over $100.
    binomial: n=30 and p=0.358
    normal approximation appropriate since
        np = 30(0.358) = 10.74 ≥ 5
        nq = 30(0.642) = 19.26 ≥ 5
    μ = np = 30(0.358) = 10.74
    σ = $\sqrt{npq}$ = $\sqrt{30(0.358)(0.642)}$ = 2.626
    P(x≥18) = $P_C$(x>17.5)
            = 1 – P(x<17.5)
            = 1 – P(z<2.57)
            = 1 – 0.9949 = 0.0051
    normalcdf(17.5,999,10.74,√(30×0.358×0.642)) = 0.0050
    Yes. If the usual usage pattern still applies, then there has been an unusual occurrence. Verification of correct usage of the card is warranted.
    NOTE: 1 – binomcdf(30,0.358,17) = 0.0059 gives a more accurate answer.

33. Let x = the number of times Marc wins $35.
    binomial: n=200 and p=1/38
    normal approximation appropriate since
        np = 200(1/38) = 5.26 ≥ 5
        nq = 200(37/38) = 194.76 ≥ 5
    μ = np = 784(0.079) = 61.936
    σ = $\sqrt{npq}$ = $\sqrt{200(1/38)(37/38)}$ = 2.264
    Marc needs at least 6 $35 wins for a profit.
    P(x≥6) = $P_C$(x>5.5)
           = 1 – P(x<5.5)
           = 1 – P(z<0.10)
           = 1 – 0.5398 = 0.4602
    normalcdf(5.5,999,200/38,√(200×37/38^2)) = 0.4583
    NOTE: 1 – binomcdf(200,1/38,5) = 0.4307 gives a more accurate answer.

35. a. binomial: n=4 and p=0.350
   $P(x \geq 1) = 1 - P(x=0)$
   $= 1 - [4!/4!0!](0.350)^0(0.650)^4$
   $= 1 - 0.1785$
   $= 0.8215$

   b. binomial: n=56(4)=224 and p=0.350
   normal approximation appropriate since
   np = 224(0.350) = 78.4 ≥ 5
   nq = 224(0.650) = 145.6 ≥ 5
   µ = np = 224(0.350) = 78.4
   $\sigma = \sqrt{npq} = \sqrt{224(0.350)(0.650)} = 7.139$
   $P(x \geq 56) = P_C(x>55.5)$
   $= 1 - P(x<55.5)$
   $= 1 - P(z<-3.20)$
   $= 1 - 0.0007 = 0.9993$
   normalcdf(55.5,999,78.4,√(224×0.350×0.650)) = 0.9993
   NOTE: 1 – binomcdf(224,0.350,55) = 0.9995 gives a more accurate answer.

   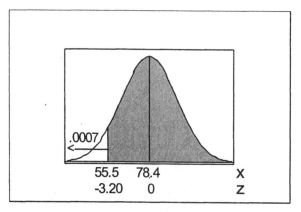

   c. Let H = getting at least one hit in 4 at bats.
   P(H) = 0.8215 [from part (a)]
   For 56 consecutive games, $[P(H)]^{56} = [0.8125]^{56} = 0.0000165$

   d. The solution below employs the techniques and notation of parts (a) and (c).
   for $[P(H)]^{56} > 0.10$
   $P(H) > (0.10)^{1/56}$
   $P(H) > 0.9597$
   for $P(H) = P(x \geq 1) > 0.9597$
   $1 - P(x=0) > 0.9597$
   $0.0403 > P(x=0)$
   $0.0403 > [4!/(4!0!)]p^0(1-p)^4$
   $0.0403 > (1-p)^4$
   $(0.0403)^{1/4} > 1 - p$
   $p > 1 - (0.0403)^{1/4}$
   $p > 1 - 0.448$
   $p > 0.552$

37. Most calculators and software will not display final answers less than 1.00E-100 or greater than 1.00E100. Older ones could not deal with such intermediate values. Since 70! Is greater than 1.00E100, some calculators and software could not complete binomial calculations involving factorials larger than 70 – and the normal approximation would have to be used. Present calculators, including the TI=83/84 Plus, use essentially the same numerical approximations for extremes in both the binomial and the normal distributions – and so binomial answers usually exist whenever normal answers exist, and vice-versa. Any situation for which the TI-83/84 Plus gives a normal approximation answer but not an exact binomial answer would be a numerical anomaly, and not of significance.

## 6-7 Assessing Normality

1. A normal quantile plot can be used to determine whether sample data come form a normal distribution. In theory, it compares the z scores for the sample data with the z scores for normally distributed data with the same cumulative relative frequencies as the sample data. In practice, it uses the sample data directly – since converting to z scores is a linear transformation that re-labels the scores but does not change their distribution.

3. The 100 points would approximate a straight line and show no systematic pattern.

5. Not normal. There is a systematic pattern.

7. Normal.

9. Yes. The frequency distribution (from the list on the following page) and the histogram (from STATPLOT using MBI and the ZOOM 9 default settings) indicate an approximately normal distribution. The frequencies taper off from the modal class relatively symmetrically in both directions.

| MBMI value | frequency |
|---|---|
| 18.75 – 21.25 | 3 |
| 21.25 – 23.75 | 7 |
| 23.75 – 26.25 | 11 |
| 26.25 – 28.75 | 13 |
| 28.75 – 31.25 | 1 |
| 31.25 – 33.75 | 5 |
| | 40 |

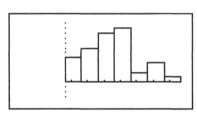

11. No. The frequency distribution (from the list on the following page) and the histogram (from STATPLOT using PREC and the ZOOM 9 default settings) do not indicate an approximately normal distribution. he scores bunch up at the lower end.

| precipitation | frequency |
|---|---|
| -0.375 – 0.375 | 30 |
| 0.375 – 1.125 | 2 |
| 1.125 – 1.875 | 1 |
| 1.875 – 2.625 | 1 |
| 2.625 – 3.375 | 1 |
| | 35 |

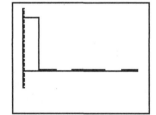

NOTE: The normal quantile plots for exercises #13-#16 may be constructed using the appropriate columns from the table on the following page. *Exercise #16 uses the exercise #12 columns and most clearly illustrates the process. Each of the 10 scores is 10% of the data set. The first score covers 0.00 to 0.10, the midpoint of which is 0.05 (the cp value).* Plot the x values on the horizontal axis and the corresponding normal z values (determined by the cp value) on the vertical axis. Plotting the z scores "stretches out" the evenly spaced cp values to agree with the cumulative probabilities of the normal distribution. As with most exercises in this section, the judgment as to whether the points approximate a straight line is a subjective one. The figures given for these exercises were generated from STATPLOT and the ZOOM 9 default settings as directed in the text.

## CHAPTER 6  Normal Probability Distributions

NOTE: This table gives the information for working exercises #9-#12 and #13-#16 by hand. While these exercises can be answered using computer software, seeing the actual procedure and calculations involved promotes a better understanding of the concepts and processes of this section. For each problem,

   x = the original scores arranged in order
   cp = the cumulative probability values 1/2n, 3/2n, 5/2n,..., (2n-1)/2n
   z = the normal z scores that have the cp to their left.

| # | 9. MBMI | | | 10. CPWHT | | | 11. PREC | | | 12. ADT | | |
|---|---|---|---|---|---|---|---|---|---|---|---|---|
|   | x | cp | z | x | cp | z | x | cp | z | x | cp | z |
| 1  | 19.6 | 0.013 | -2.24 | 3.0050 | .104 | -2.19 | 0    | 0.014 | -2.19 | 26 | 0.05 | -1.64 |
| 2  | 19.9 | 0.038 | -1.78 | 3.0054 | .043 | -1.72 | 0    | 0.043 | -1.72 | 33 | 0.15 | -1.04 |
| 3  | 20.7 | 0.063 | -1.53 | 3.0064 | .071 | -1.47 | 0    | 0.071 | -1.47 | 34 | 0.25 | -0.67 |
| 4  | 21.5 | 0.088 | -1.36 | 3.0131 | .100 | -1.28 | 0    | 0.100 | -1.28 | 39 | 0.35 | -0.39 |
| 5. | 21.6 | 0.113 | -1.21 | 3.0185 | .129 | -1.13 | 0    | 0.129 | -1.13 | 48 | 0.45 | -0.13 |
| 6  | 22.7 | 0.138 | -1.09 | 3.0290 | .157 | -1.01 | 0    | 0.157 | -1.01 | 58 | 0.55 | 0.13 |
| 7  | 23.2 | 0.163 | -0.98 | 3.0357 | .186 | -0.89 | 0    | 0.186 | -0.89 | 66 | 0.65 | 0.39 |
| 8  | 23.3 | 0.188 | -0.89 | 3.0377 | .214 | -0.79 | 0    | 0.214 | -0.79 | 67 | 0.75 | 0.67 |
| 9  | 23.4 | 0.213 | -0.80 | 3.0408 | .243 | -0.70 | 0    | 0.243 | -0.70 | 71 | 0.85 | 1.04 |
| 10 | 23.5 | 0.238 | -0.71 | 3.0476 | .271 | -0.61 | 0    | 0.271 | -0.61 | 72 | 0.95 | 1.64 |
| 11 | 23.8 | 0.263 | -0.64 | 3.0480 | .300 | -0.52 | 0    | 0.300 | -0.52 |    |      |      |
| 12 | 23.8 | 0.288 | -0.56 | 3.0561 | .329 | -0.44 | 0    | 0.329 | -0.44 |    |      |      |
| 13 | 24.2 | 0.313 | -0.49 | 3.0570 | .357 | -0.37 | 0    | 0.357 | -0.37 |    |      |      |
| 14 | 24.5 | 0.338 | -0.42 | 3.0755 | .386 | -0.29 | 0    | 0.386 | -0.29 |    |      |      |
| 15 | 24.6 | 0.363 | -0.35 | 3.0765 | .414 | -0.22 | 0    | 0.414 | -0.22 |    |      |      |
| 16 | 24.6 | 0.388 | -0.29 | 3.0786 | .443 | -0.14 | 0    | 0.443 | -0.14 |    |      |      |
| 17 | 25.2 | 0.413 | -0.22 | 3.0816 | .471 | -0.07 | 0    | 0.471 | -0.07 |    |      |      |
| 18 | 25.5 | 0.438 | -0.16 | 3.0862 | .500 | 0     | 0    | 0.500 | 0     |    |      |      |
| 19 | 25.6 | 0.463 | -0.09 | 3.0936 | .529 | 0.07  | 0    | 0.529 | 0.07  |    |      |      |
| 20 | 26.2 | 0.488 | -0.03 | 3.0965 | .557 | 0.14  | 0.01 | 0.557 | 0.14  |    |      |      |
| 21 | 26.2 | 0.513 | 0.03  | 3.0976 | .586 | 0.22  | 0.01 | 0.586 | 0.22  |    |      |      |
| 22 | 26.3 | 0.539 | 0.09  | 3.0994 | .614 | 0.29  | 0.01 | 0.614 | 0.29  |    |      |      |
| 23 | 26.4 | 0.563 | 0.16  | 3.1029 | .643 | 0.37  | 0.01 | 0.643 | 0.37  |    |      |      |
| 24 | 26.4 | 0.588 | 0.22  | 3.1031 | .671 | 0.44  | 0.01 | 0.671 | 0.44  |    |      |      |
| 25 | 26.6 | 0.613 | 0.29  | 3.1038 | .700 | 0.52  | 0.01 | 0.700 | 0.52  |    |      |      |
| 26 | 26.7 | 0.638 | 0.35  | 3.1083 | .729 | 0.61  | 0.02 | 0.729 | 0.61  |    |      |      |
| 27 | 26.9 | 0.663 | 0.42  | 3.1114 | .757 | 0.70  | 0.05 | 0.757 | 0.70  |    |      |      |
| 28 | 27.0 | 0.688 | 0.49  | 3.1141 | .786 | 0.79  | 0.06 | 0.786 | 0.79  |    |      |      |
| 29 | 27.1 | 0.713 | 0.56  | 3.1267 | .814 | 0.89  | 0.07 | 0.814 | 0.89  |    |      |      |
| 30 | 27.4 | 0.738 | 0.64  | 3.1366 | .843 | 1.01  | 0.21 | 0.843 | 1.01  |    |      |      |
| 31 | 27.8 | 0.763 | 0.71  | 3.1461 | .871 | 1.13  | 0.47 | 0.871 | 1.13  |    |      |      |
| 32 | 28.1 | 0.788 | 0.80  | 3.1524 | .900 | 1.28  | 0.67 | 0.900 | 1.28  |    |      |      |
| 33 | 28.3 | 0.813 | 0.89  | 3.1535 | .929 | 1.47  | 1.59 | 0.929 | 1.47  |    |      |      |
| 34 | 28.7 | 0.838 | 0.98  | 3.1692 | .957 | 1.72  | 1.99 | 0.957 | 1.72  |    |      |      |
| 35 | 30.9 | 0.863 | 1.09  | 3.1934 | .986 | 2.19  | 2.85 | 0.986 | 2.19  |    |      |      |
| 36 | 31.4 | 0.888 | 1.21  |        |      |       |      |       |       |    |      |      |
| 37 | 31.9 | 0.913 | 1.35  |        |      |       |      |       |       |    |      |      |
| 38 | 32.1 | 0.938 | 1.53  |        |      |       |      |       |       |    |      |      |
| 39 | 33.1 | 0.963 | 1.78  |        |      |       |      |       |       |    |      |      |
| 40 | 33.2 | 0.988 | 2.24  |        |      |       |      |       |       |    |      |      |

13. Yes. Since the points lie close to a straight line and there is no other obvious pattern, conclude that the population distribution is approximately normal.

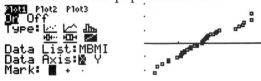

15. No. Since the points do not lie close to a straight line, conclude that the population distribution is not approximately normal.

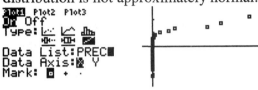

17. The two histograms are shown below. The heights (on the left) appear to be approximately normally distributed, while the cholesterol levels (on the right) appear to be positively skewed. Many natural phenomena are normally distributed. Height is a natural phenomenon unaffected by human input, but cholesterol levels are humanly influenced (by diet, exercise, medication, etc.) in ways that might alter any naturally occurring distribution.

Height                                   Cholesterol Level

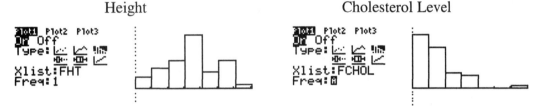

19. The corresponding z scores in the table below were determined as follows.
    (1) Arrange the n scores in order and place them in the x column.
    (2) For each $x_i$, calculate the cumulative probability using $cp_i = (2i-1)/2n$ for $i = 1,2,\ldots,n$.
    (3) For each $cp_i$, find the $z_i$ for which $P(z<z_i) = cp_i$ for $i = 1,2,\ldots,n$.
    The normal quantile plot may be obtained by hand by graphing x (on the x axis) vs. z (on the y axis), or by placing the x values in L1 and using STAT PLOT as shown. The figure indicates that the data appear to come from a population with a normal distribution.

    | i | x  | cp   | z     |
    |---|----|------|-------|
    | 1 | 73 | 0.10 | -1.28 |
    | 2 | 78 | 0.30 | -0.52 |
    | 3 | 79 | 0.50 | 0     |
    | 4 | 82 | 0.70 | 0.52  |
    | 5 | 85 | 0.90 | 1.28  |

21. No. The z scores from the cumulative probabilities must be used and not the z scores from the raw data. The z formula is a linear transformation, $z = (x-\mu)/\sigma = (1/\sigma)x - (\mu/\sigma) = ax + b$. The (x,z) pairs all lie exactly on the straight line $z = (1/\sigma)x - (\mu/\sigma)$ regardless of the distribution.

## Statistical Literacy and Critical Thinking

1. A normal distribution is one that is symmetric and bell-shaped – more technically, it is one that can be described by the following formula, where μ and σ are specified values,

$$f(x) = \frac{e^{-\frac{1}{2\sigma^2}(x-\mu)^2}}{\sigma\sqrt{2\pi}}.$$

A standard normal distribution is one that has a mean of 0 and a standard deviation of 1 – more technically, it is the distribution that results when μ = 0 and σ = 1 in the above formula.

2. Regardless of the distribution of grip strengths in the original population, the sample means will be normally distributed around the mean of the grip strengths in the original population.

3. No. The normal distribution formed by his large sample is not necessarily the same normal distribution formed by the population. The normal distribution formed from his friends and relatives could, for example, have a different mean – i.e., it could have the same shape as the population distribution but be centered around a higher or lower value.

4. The Central Limit Theorem tells us that as we take many large independent random samples of size n from <u>any</u> population (i.e., regardless of its shape) with mean μ and standard deviation σ, the resulting sample means will be normally distributed with mean μ and standard deviation $\sigma/\sqrt{n}$.

## Review Exercises

1. a. normal distribution
   μ = 0
   σ = 1
   P(-0.5<x<0.5)
   = P(-0.50<z<0.50)
   = P(x<0.50) – P(z<-0.50)
   = 0.6915 – 0.3085
   = 0.3830
   normalcdf(-0.50,0.50) = 0.3829

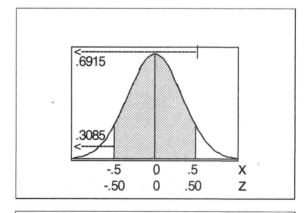

   b. normal distribution,
      since the original distribution is so
   $\mu_{\bar{x}} = \mu = 0$
   $\sigma_{\bar{x}} = \sigma/\sqrt{n} = 1/\sqrt{16} = 0.250$
   P(-0.5<$\bar{x}$<0.5)
   = P(-2.00<z<2.00)
   = P(z<2.00) – P(z<-2.00)
   = 0.9772 – 0.0228
   = 0.9544
   normalcdf(-0.50,0.50,0,1/√16) = 0.9545

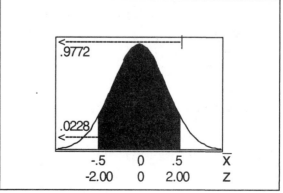

c. For $P_{90}$, A = 0.9000 [0.8997] and z = 1.28
$$x = \mu + z\sigma$$
$$= 0 + (1.28)(1)$$
$$= 0 + 1.28$$
$$= 1.28$$
invNorm(0.90,0,1) = 1.28

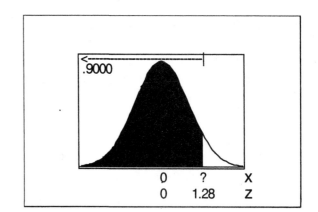

2. a. normal distribution
$\mu = 69.0$
$\sigma = 2.8$
$P(x>74.0)$
$\quad = 1 - P(x<74.0)$
$\quad = 1 - P(z<1.79)$
$\quad = 1 - 0.9633$
$\quad = 0.0367$ or 3.67%
normalcdf(74.0,999,69.0,2,8)
$\quad = 0.0371$ or 3.71%

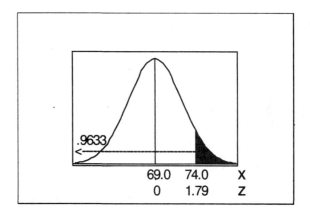

b. normal distribution,
since the original distribution is so
$\mu_{\bar{x}} = \mu = 69.0$
$\sigma_{\bar{x}} = \sigma/\sqrt{n} = 2.8/\sqrt{4} = 1.4$
$P(\bar{x}>74.0)$
$\quad = 1 - P(\bar{x}<74.0)$
$\quad = 1 - P(z<3.57)$
$\quad = 1 - 0.9999$
$\quad = 0.0001$
normalcdf(74.0,999,69.0,2.8/$\sqrt{4}$)
$\quad = 1.78$E-05, rounded to 0.0002

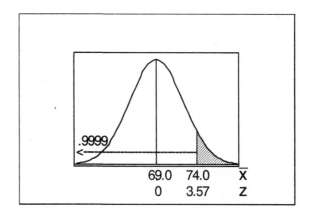

c. For the tallest 10%,
A = 0.9000 [0.8997] and z = 1.28
$$x = \mu + z\sigma$$
$$= 69.0 + (1.28)(2.8)$$
$$= 69.0 + 3.6$$
$$= 72.6 \text{ in}$$
invNorm(1−0.10,69.0,2.8) = 72.6 in

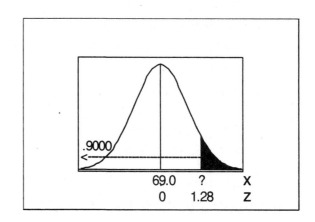

3. a. normal distribution
   $\mu = 178.1$
   $\sigma = 40.7$
   $P(x>260)$
   $= 1 - P(x<260)$
   $= 1 - P(z<2.01)$
   $= 1 - 0.9778$
   $= 0.0222$
   normalcdf(260,999,178.1,40.7)
   $= 0.0221$

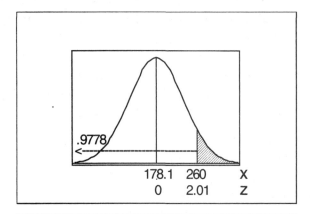

b. normal distribution
   $\mu = 178.1$
   $\sigma = 40.7$
   $P(170<x<200)$
   $= P(-0.20<z<0.54)$
   $= P(z<0.54) - P(z<-0.20)$
   $= 0.7054 - 0.4207$
   $= 0.2847$
   normalcdf(170,200,178.1,40.7)
   $= 0.2836$

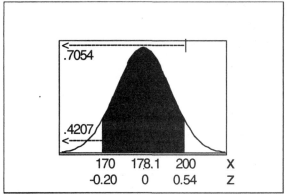

c. normal distribution,
   since the original distribution is so
   $\mu_{\bar{x}} = \mu = 178.1$
   $\sigma_{\bar{x}} = \sigma/\sqrt{n} = 40.7/\sqrt{9} = 13.567$
   $P(170<\bar{x}<200)$
   $= P(-0.60<z<1.61)$
   $= P(z<1.61) - P(z<-0.60)$
   $= 0.9463 - 0.2743$
   $= 0.6720$
   normalcdf(170,200,178.1,40.7/$\sqrt{9}$)
   $= 0.6715$

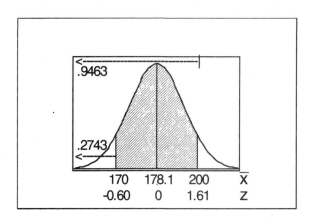

d. For the top 3%,
   $A = 0.9700\ [0.9699]$ and $z = 1.88$
   $x = \mu + z\sigma$
   $= 178.1 + (1.88)(40.7)$
   $= 178.1 + 76.5$
   $= 254.6$ mg/100 mL
   invNorm(1−0.03,178.1,40.7) =
   $= 254.6$ mg/100 mL

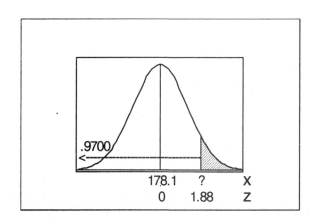

4. Let x = the number incorrect responders who are female.
   binomial problem: n=20 and p=0.5, use the binomial formula $P(x) = \{n!/[(n-x)!x!]\}p^x q^{n-x}$
   $P(x \geq 18) = P(x=18) + P(x=19) + P(x=20)$
   $= [20!/(2!18!)](0.5)^{18}(0.5)^2 + [20!/(1!19!)](0.5)^{19}(0.5)^1 + [20!/(0!20!)](0.5)^{20}(0.5)^0$
   $= [190](0.5)^{18}(0.25) + [20](0.5)^{19}(0.5) + [1](0.5)^{20}(1)$
   $= 0.000181 + 0.000019 + 0.000001 = 0.000201$
   $1 - \text{binomcdf}(20, 0.5, 17) = 2.01\text{E-}04 = 0.000201$

   normal approximation appropriate since
   $np = 20(0.5) = 10 \geq 5$
   $nq = 20(0.5) = 10 \geq 5$
   $\mu = np = 20(0.5) = 10.0$
   $\sigma = \sqrt{np(1-p)} = \sqrt{20(0.5)(0.5)} = 2.236$
   $P(x \geq 18)$
   $= P_C(x>17.5)$
   $= 1 - P(x<17.5)$
   $= 1 - P(z<3.35)$
   $= 1 - 0.9996$
   $= 0.0004$
   normalcdf$(17.5, 999.5, \sqrt{(20 \times 0.5 \times 0.5)}) = 0.0004$

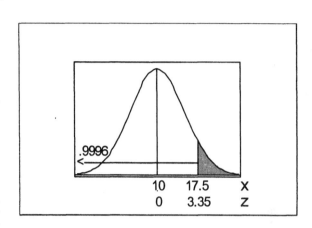

Yes. By either the exact binomial or the normal approximation, having 18 or more females among the 20 incorrect responders would be a very unusual event. This is evidence that the question is more often answered correctly by men than by women, and in that sense the question favors men.

5. Let x = the number of females hired.
   binomial problem: n=20 and p=0.3, use the binomial formula $P(x) = \{n!/[(n-x)!x!]\}p^x q^{n-x}$
   $P(x \leq 3) = P(x=0) + P(x=1) + P(x=2) + P(x=3)$
   $= [20!/(20!0!)](0.3)^0(0.7)^{20} + [20!/(19!1!)](0.3)^1(0.7)^{19} + [20!/(18!2!)](0.3)^2(0.7)^{18} + [20!/(17!3!)](0.3)^3(0.7)^{17}$
   $= [1](1)(0.7)^{20} + [20](0.3)(0.7)^{19} + [190](0.09)(0.7)^{18} + [1140](0.027)(0.7)^{17}$
   $= 0.000798 + 0.006839 + 0.027846 + 0.071604 = 0.107087$
   binomcdf$(20, 0.3, 3) = 0.1071$

   normal approximation appropriate since
   $np = 20(0.3) = 6 \geq 5$
   $nq = 20(0.7) = 14 \geq 5$
   $\mu = np = 20(0.5) = 10.0$
   $\sigma = \sqrt{np(1-p)} = \sqrt{20(0.3)(0.7)} = 2.049$
   $P(x \leq 3)$
   $= P_C(x<3.5)$
   $= P(z<-1.22)$
   $= 0.1112$
   normalcdf$(-999, 3.5, 6, \sqrt{(20 \times 0.3 \times 0.7)})$
   $= 0.1113$

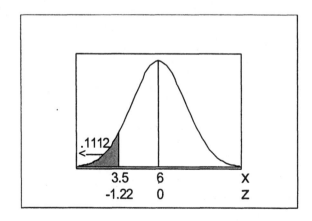

No. By either the exact binomial or the normal approximation, having 3 or fewer females among the 20 persons hired would not be an unusual event. There is not evidence that the company is discriminating based on gender.

# 106 CHAPTER 6 Normal Probability Distributions

6. Arranged in order the 19 scores are:
   1 4 4 4 10 13 18 19 34 44 45 48 64 67 68 103 124 125 125.
   Two reasonable frequency distributions for the data are given below.

   | days | frequency |
   |---|---|
   | 1 – 24 | 8 |
   | 25 – 49 | 4 |
   | 50 – 74 | 3 |
   | 75 – 99 | 0 |
   | 100 –124 | 2 |
   | 125 –149 | 2 |
   | | 19 |

   | days | frequency |
   |---|---|
   | 1 – 29 | 8 |
   | 30 – 59 | 4 |
   | 60 – 89 | 3 |
   | 90 –129 | 4 |
   | | 19 |

   No. The lengths of time do not appear to come from a population with a normal distribution. It appears that the most frequently occurring values are at the lower end, and not near the middle of the distribution.

## Cumulative Review Exercises

1. The n=8 ordered scores are: 0 1 3 7 10 12 24 27.
   summary statistics: $\Sigma x = 84$  $\Sigma x^2 = 1608$

   a. $\bar{x} = \Sigma x/n = 84/8 = 10.5$ g

   b. $\tilde{x} = (7+10)/2 = 8.5$ g

   c. $s^2 = [n(\Sigma x^2) - (\Sigma x)^2]/[n(n-1)]$
      $= [8(1608) - (84)^2]/[8(7)]$
      $= 5808/56 = 103.714$
      $s = 10.184$, rounded to 10.2 g

   d. $s^2 = 103.71$ g$^2$ [from part (c)]

   e. $z = (x - \bar{x})/s$
      $z_3 = (3 - 10.5)/10.184 = -0.74$

   f. The actual percentage of scores that exceeds 3 is 5/8 = 62.5%.

   g. Assuming a normal distribution and using the sample values for $\bar{x}$ and s to estimate $\mu$ and $\sigma$, the percentage of the population that exceeds 3 is
      $P(x>3) = P(z>-0.74)$
      $= 1 - P(z<-0.74)$
      $= 1 - 0.2296$
      $= 0.7704$
      normalcdf(3,999,10.5,10.184) = 0.7693

   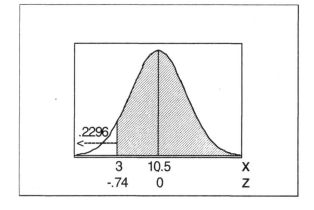

   h. Ratio, since differences are meaningful and there is a natural zero.

   i. Continuous, since the exact amounts can be any value on a continuum.

   j. No. The items were selected for analysis because they represent an interesting variety, not because they are proportionately representative of human consumption.

2. a. Let L = a person is left-handed.
P(L) = 0.10, for each random selection
P($L_1$ and $L_2$ and $L_3$) = P($L_1$)·P($L_2$)·P($L_3$)
= (0.10)(0.10)(0.10)
= 0.001

b. Let N = a person is not left-handed.
P(N) = 0.90, for each random selection
P(at least one left-hander) = 1 − P(no left-handers)
= 1 − P($N_1$ and $N_2$ and $N_3$)
= 1 − P($N_1$)·P($N_2$)·P($N_3$)
= 1 − (0.90)(0.90)(0.90)
= 1 − 0.729
= 0.271

c. binomial: n=3 and p=0.10
normal approximation not appropriate since
np = 3(0.10) = 3 < 5

d. binomial: n=50 and p=0.10
μ = np = 50(0.10) = 5

e. binomial problem: n=50 and p=0.10
σ = $\sqrt{npq}$ = $\sqrt{50(0.10)(0.90)}$ = 2.121

f. There are two previous approaches that may be used to answer this question.
(1) An unusual score is one that is more than two standard deviations from the mean. Use the values for μ and σ from parts (d) and (e).
z = (x − μ)/σ
$z_8$ = (8 − 5)/2.121
= 1.41
Since 8 is 1.41<2 standard deviations from the mean, it would not be an unusual result.
(2) A score is unusual if the probability of getting that result or a more extreme result is less than or equal to 0.05.
binomial: n = 50 and p = 0.1
normal approximation appropriate since
np = 50(0.1) = 5 ≥ 5
nq = 50(0.9) = 45 ≥ 5
Use the values for μ and σ
from parts (d) and (e).
P(x≥8) = $P_C$(x>7.5)
= 1 − P(x<7.5)
= 1 − P(z<1.18)
= 1 − 0.8810
= 0.1190
normalcdf(7.5,999,5,√(50×0.1×0.9))
= 0.1193

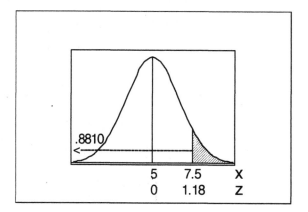

Since 0.1193>0.05, getting 8 left-handers in a group of 50 is not an unusual event.
NOTE: 1 − binomcdf(50,0.1,7) = 0.1221

# Chapter 7

# Estimates and Sample Sizes

## 7-2 Estimating a Population Proportion

1. A critical value for a normal distribution is the z score that separates, at some prescribed level, the more extreme (i.e., less likely) high values from the more common (i.e., more likely) ones. The level is indicated by a subscript, and so if $\alpha = 0.10$ and $\alpha/2 = 0.05$, then $z_{\alpha/2} = 1.645$ is the critical value that separates the highest $0.05 = 5\%$ of the scores from the others. Since the standard normal distribution is symmetric around 0, the critical value for the lower tail of the distribution is the negative of the critical value for the upper tail – and so if $\alpha = 0.10$ and $\alpha/2 = 0.05$, then $-z_{\alpha/2} = -1.645$ is the critical value that separates the lowest $0.05 = 5\%$ of the scores from the others.

NOTE: It is also acceptable to use the actual numerical value as the subscript – and so if $\alpha = 0.10$ and $\alpha/2 = 0.05$, then one may indicate that $z_{\alpha/2} = z_{0.05} = 1.645$.

3. By including a statement of the maximum likely error, a confidence interval provides information about the accuracy of an estimate.

5. For 99% confidence, $\alpha = 1 - 0.99 = 0.01$ and $\alpha/2 = 0.01/2 = 0.005$.
   For the upper 0.005, $A = 0.9950$ and $z = 2.575$.
   $z_{\alpha/2} = 2.575$          invNorm(0.995) = 2.576

7. For 98% confidence, $\alpha = 1 - 0.98 = 0.02$ and $\alpha/2 = 0.02/2 = 0.01$.
   For the upper 0.01, $A = 0.9900$ [0.9901] and $z = 2.33$.
   $z_{\alpha/2} = 2.33$          invNorm(0.99) = 2.326

9. Let L = the lower confidence limit; U = the upper confidence limit.
   $\hat{p} = (L+U)/2 = (0.222+0.444)/2 = 0.666/2 = 0.333$
   $E = (U-L)/2 = (0.444-0.222)/2 = 0.222/2 = 0.111$
   The interval can be expressed as $0.333 \pm 0.111$.

11. Let L = the lower confidence limit; U = the upper confidence limit.
    $\hat{p} = (L+U)/2 = (0.206+0.286)/2 = 0.492/2 = 0.246$
    $E = (U-L)/2 = (0.286-0.206)/2 = 0.080/2 = 0.040$
    The interval can be expressed as $0.246 \pm 0.040$.

13. Let L = the lower confidence limit; U = the upper confidence limit.
    $\hat{p} = (L+U)/2 = (0.868+0.890)/2 = 1.758/2 = 0.879$
    $E = (U-L)/2 = (0.890-0.868)/2 = 0.022/2 = 0.011$

15. Let L = the lower confidence limit; U = the upper confidence limit.
    $\hat{p} = (L+U)/2 = (0.607+0.713)/2 = 1.320/2 = 0.660$
    $E = (U-L)/2 = (0.713-0.607)/2 = 0.106/2 = 0.053$

NOTE: When calculating $E = z_{\alpha/2}\sqrt{\hat{p}\hat{q}/n}$ do not round off in the middle of the problem. If necessary, calculate $\hat{p} = x/n$ and STO→ the value. Inputting the fractions x/n and (n-x)/n whenever $\hat{p}$ does not "come out even" eliminates the need for storing the decimal value for $\hat{p}$. The exercises in this section are worked using Table A-2 with the usual notation as given in the text AND using only the TI-83/84 Plus.

17. $\alpha = 0.05$ and $\hat{p} = x/n = 200/500 = 0.40$
    $E = z_{\alpha/2}\sqrt{\hat{p}\hat{q}/n} = 1.96\sqrt{(0.40)(0.60)/500} = 0.0429$
    $E = z_{\alpha/2}\sqrt{\hat{p}\hat{q}/n} = \text{invNorm}(0.975)*\sqrt{(0.40*0.60/500)} = 0.0429$

19. $\alpha = 0.02$ and $\hat{p} = x/n = [267]/1068 = 0.25$
    $E = z_{\alpha/2}\sqrt{\hat{p}\hat{q}/n} = 2.33\sqrt{(0.25)(0.75)/1068} = 0.0309$
    $E = z_{\alpha/2}\sqrt{\hat{p}\hat{q}/n} = \text{invNorm}(0.99)*\sqrt{(0.25*0.75/1068)} = 0.0308$
    NOTE: The value x=[267] was not given. In truth, any $262 \leq x \leq 272$ rounds to the given $\hat{p} = x/1068 = 25\%$. For want of a more precise value, $\hat{p} = 0.25$ is used in the calculation of E.

NOTE: The exercises in this section are worked using Table A-2 with the usual notation as given in the text AND using only the TI-83/84 Plus, employing the STAT TESTS 1-PropZInt options as presented in the text.

21. $\alpha = 0.05$ and $\hat{p} = x/n = 200/500 = 0.4000$
    $\hat{p} \pm z_{\alpha/2}\sqrt{\hat{p}\hat{q}/n}$
    $0.4000 \pm 1.96\sqrt{(0.4000)(0.6000)/500}$
    $0.4000 \pm 0.0429$
    $0.357 < p < 0.443$

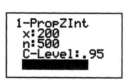 

$0.357 < p < 0.443$

23. $\alpha = 0.02$ and $\hat{p} = x/n = 267/1068 = 0.2500$
    $\hat{p} \pm z_{\alpha/2}\sqrt{\hat{p}\hat{q}/n}$
    $0.2500 \pm 2.33\sqrt{(0.2500)(0.7500)/1068}$
    $0.2500 \pm 0.0309$
    $0.219 < p < 0.281$

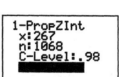

$0.219 < p < 0.281$

NOTE: For exercises #25-#28 $z_{\alpha/2}$ was found using [unrounded] invNorm(1−α/2).

25. $\alpha = 0.05$ and $E = 0.020$; $\hat{p}$ unknown, use $\hat{p} = 0.5$
    $n = [(z_{\alpha/2})^2 \hat{p}\hat{q}]/E^2$
    $= [(1.960)^2(0.5)(0.5)]/(0.020)^2 = 2400.91$, rounded up to 2401

27. $\alpha = 0.05$ and $E = 0.03$; $\hat{p}$ estimated to be 0.27
    $n = [(z_{\alpha/2})^2 \hat{p}\hat{q}]/E^2$
    $= [(1.960)^2(0.27)(0.73)]/(0.03)^2 = 841.28$, rounded up to 842

29. Let x = the number of girls born using the method.
    Use STAT TESTS 1-PropZINT.
    x : 295
    n : 325
    C-level: 0.99
    $0.866 < p < 0.949$

    ```
    1-PropZInt
    (.86633,.94905)
    p̂=.9076923077
    n=325
    ```

    Yes. Since 0.5 is not within the confidence interval, and below the interval, we can be 99% certain that the method is effective.

31. Let x = the number of deaths in the week before Thanksgiving.
    Use STAT TESTS 1-PropZINT.
    x : 6062
    n : 12000
    C-level: 0.95
    $0.496 < p < 0.514$

    ```
    1-PropZInt
    (.49622,.51411)
    p̂=.5051666667
    n=12000
    ```

    No. Since 0.5 is within the confidence interval, it is a reasonable possibility for the true population value.

33. Let x = the number with yellow pods. There are 428 + 152 = 580 total trials.
    a. Use STAT TESTS 1-PropZINT.
       x : 152
       n : 580
       C-level: 0.95
       $0.226 < p < 0.298$

       ```
       1-PropZInt
       (.22628,.29786)
       p̂=.2620689655
       n=580
       ```

    b. No. Since 0.25 is within the confidence interval, it is a reasonable possibility for the true population value. The results do not contradict the theory.

35. Let x = the number that develop those types of cancer.
    a. Use STAT TESTS 1-PropZINT.
       x : 135
       n : 420095
       C-level: 0.95
       $0.00027 < p < 0.00038$

       ```
       1-PropZInt
       (2.7E-4,3.8E-4)
       p̂=3.2135588E-4
       n=420095
       ```

       $0.000267 < p < 0.000376$ [using $\hat{p} \pm z_{\alpha/2}\sqrt{\hat{p}\hat{q}/n}$ by hand to keep 3 significant digits]

    b. No. Since 0.0340% = 0.000340 is within the confidence interval, it is a reasonable possibility for the true population value. The results do not provide evidence that cell phone users have a different cancer rate than the general population.

37. NOTE: Since the actual value of x is not given, this problem is limited to the two decimal accuracy reported in 39%. Any $335 \leq x \leq 343$ rounds to $\hat{p} = x/n = x/870 = 0.39$. All that can be said is that $0.38506 \leq \hat{p} \leq 0.39425$, and 3 decimal confidence interval accuracy is not possible.
    Let x = the number of Mexican-Americans selected for grand jury duty.
    $\alpha = 0.01$ and $\hat{p} = x/n = x/870 = 0.39$

    $\hat{p} \pm z_{\alpha/2}\sqrt{\hat{p}\hat{q}/n}$
    $0.39 \pm 2.575\sqrt{(0.39)(0.61)/870}$
    $0.39 \pm 0.04$
    $0.35 < p < 0.43$

Using STAT TESTS 1-PropZINT with x = (0.39)(870) = 339,
x : 339
n : 870
C-level: 0.99
0.347 < p < 0.432

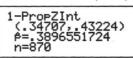

Yes. Since 0.791 is not in the confidence interval, we can be 99% certain that the selection process does not reflect the general population.

39. NOTE: Since the actual value of x is not given, this problem is limited to the two decimal accuracy reported in 94%. Any $3999 \leq x \leq 4040$ rounds to $\hat{p} = x/n = x/4276 = 0.94$. All that can be said is that $0.93522 \leq \hat{p} \leq 0.94480$, and 3 decimal confidence interval accuracy is not possible.
Let x = the number of households with at least one telephone.
$\alpha = 0.01$ and $\hat{p} = x/n = x/4276 = 0.94$

$\hat{p} \pm z_{\alpha/2} \sqrt{\hat{p}\hat{q}/n}$
$0.94 \pm 2.575\sqrt{(0.94)(0.06)/4276}$
$0.94 \pm 0.01$
$0.93 < p < 0.95$
Using STAT TESTS 1-PropZINT with x = (0.94)(4276) = 4019,
x : 4019
n : 4276
C-level: 0.99
0.931 < p < 0.949

Yes. Since 0.35 is not within the confidence interval, and is below it, we can be 99% certain that the percentage of households with telephones is now greater than the 35% rate of 1920.

NOTE: For exercises #41-#44, $z_{\alpha/2}$ was found using [unrounded] invNorm(1−α/2).

41. $\alpha = 0.01$ and E = 0.02; $\hat{p}$ unknown, use $\hat{p} = 0.5$
    $n = [(z_{\alpha/2})^2 \hat{p}\hat{q}]/E^2$
    $= [(2.576)^2(0.5)(0.5)]/(0.02)^2 = 4146.81$, rounded up to 4147

43. $\alpha = 0.02$ and E = 0.03; $\hat{p}$ unknown, use $\hat{p} = 0.5$
    $n = [(z_{\alpha/2})^2 \hat{p}\hat{q}]/E^2$
    $= [(2.326)^2(0.5)(0.5)]/(0.03)^2 = 1503.30$, rounded up to 1504

45. Of the 100 M&M's in Data Set 13, 27 are blue.
    Use STAT TESTS 1-PropZINT.
    x: 27
    n: 100
    C-level: 0.95
    0.183 < p < 0.357
    Yes. Since 0.24 is within the confidence interval, this result is consistent with the reported value.

## 112   CHAPTER 7   Estimates and Sample Sizes

47. a. There was precipitation on 16 of the 53 Wednesdays.
    Use STAT TESTS 1-PropZINT.
    x: 16
    n: 53
    C-level: 0.95
    $0.178 < p < 0.425$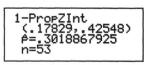

   b. There was precipitation on 15 of the 52 Sundays.
    Use STAT TESTS 1-PropZINT.
    x: 15
    n: 52
    C-level: 0.95
    $0.165 < p < 0.412$

   No. The confidence intervals are similar. Precipitation does not appear to occur more often on either day.

49. $\alpha = 0.05$ and $E = 0.04$; $\hat{p}$ unknown, use $\hat{p} = 0.5$

$$n = \frac{N\hat{p}\hat{q}[z_{\alpha/2}]^2}{\hat{p}\hat{q}[z_{\alpha/2}]^2 + (N-1)E^2}$$

$$= \frac{(1250)(0.5)(0.5)[1.96]^2}{(0.5)(0.5)[1.96]^2 + (1249)(0.04)^2} = \frac{1200.5}{2.9588} = 405.74, \text{ rounded up to } 406$$

51. $\alpha = 0.05$ and $\hat{p} = x/n = 3/8 = 0.3750$

   $\hat{p} \pm z_{\alpha/2}\sqrt{\hat{p}\hat{q}/n}$
   $0.3750 \pm 1.96\sqrt{(0.3750)(0.6250)/8}$
   $0.3750 \pm 0.3355$
   $0.0395 < p < 0.710$

   Yes. The results are "reasonably close" – being shifted down 4.5% from the correct interval $0.085 < p < 0.755$. But depending on the context, such an error could be serious.
   NOTE: STAT TESTS 1-PropZInt, as shown in the box above, gives the same results. Sometimes calculators and software automatically shift to an exact procedure for small samples for which the normal approximation is not appropriate, but the TI-83/84 Plus does not.

53. a. If $\hat{p} = x/n = 0/n = 0$, then
      (1) $np \approx 0 < 5$, and the normal approximation to the binomial does not apply.
      (2) $E = z_{\alpha/2}\sqrt{\hat{p}\hat{q}/n} = 0$, and there is no meaningful interval.
   b. Since $\hat{p} = x/n = 0/20 = 0$, use the upper limit $3/n = 3/20 = 0.15$ to produce the interval
      $0 \le p < 0.15$. NOTE: Do not use $0 < p < 0.15$, because the failure to observe any successes in the sample does not rule out $p=0$ as the true population proportion.

## 7-3 Estimating a Population Mean: σ Known

1. In repeated sampling, we expect the procedure we used to produce an interval that includes the value of the population mean 95% of the time – in particular, we have 95% confidence that the interval from 2.5 to 6.0 includes the value of the population mean.

3. No. The students at the science fair are a convenience sample that is not necessarily representative of the entire population. From a practical point of view, however, such a sample may be a reasonable staring point for such an investigation for the following reasons:
   (1) The students are present for a science fair, and taking such measurements would be a natural part of the day's agenda. Trying to take measurements from a representative random sample of general riders would be awkward and inconvenient.
   (2) Since there is concern about pressure on leg restraints, the ride is likely to be an active one not appropriate for young children or older patrons – i.e., the teenagers might be reasonably representative of the typical patrons of that particular ride.
   (3) Since the objective is probably to determine the maximum likely pressure on the leg restraints, it would make sense to take measurements from that part of the general population of riders that is strong and healthy and could reasonable be expected to exert larger amounts of pressure on the restraints – and teenagers meet that description.
   But while taking measurements from teenagers makes a certain amount of sense, the students attending a science fair may not be representative of the size, strength and general athleticism of the teenage population that patronizes the ride.

5. For 95% confidence, $\alpha = 1-0.95 = 0.05$ and $\alpha/2 = 0.05/2 = 0.025$.
   For the upper 0.025, A = 0.0250 and z = 1.96.
   $z_{\alpha/2} = 1.96$          invNorm(0.975) = 1.960

7. For 92% confidence, $\alpha = 1-0.92 = 0.08$ and $\alpha/2 = 0.08/2 = 0.04$.
   For the upper 0.04, A = 0.9600 [0.9599] and z = 1.75.
   $z_{\alpha/2} = 1.75$          invNorm(0.96) = 1.751

NOTE: For exercises #9-#12, $z_{\alpha/2}$ was found using [unrounded] invNorm(1–$\alpha$/2).

9. $\alpha = 0.05$
   $E = z_{\alpha/2} \sigma/\sqrt{n} = 1.960(15)/\sqrt{100} = 2.940$

11. $\alpha = 0.01$
    $E = z_{\alpha/2} \sigma/\sqrt{n} = 2.576(15)/\sqrt{9} = 12.879$

NOTE: The exercises in this section are worked using Table A-2 with the usual notation as given in the text AND using only the TI-83/84 Plus, employing the STAT TESTS ZInterval options as presented in the text.

13. $\sigma$ is known and n>30. Assuming that the data come from a random sample,
    $\alpha = 0.05$
    $\bar{x} \pm z_{\alpha/2} \sigma/\sqrt{n}$
    $67200 \pm 1.96(18277)/\sqrt{41}$
    $67200 \pm 5595$
    $\$61,605 < \mu < \$72,795$

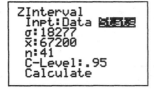
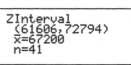

$\$61,606 < \mu < \$72,794$

15. $\sigma$ is known and n>30. Assuming that the data come from a random sample,
    $\alpha = 0.01$
    $\bar{x} \pm z_{\alpha/2} \sigma/\sqrt{n}$
    $688 \pm 2.575(68)/\sqrt{70}$
    $688 \pm 21$
    $667 < \mu < 709$

$667 < \mu < 709$

114   CHAPTER 7   Estimates and Sample Sizes

NOTE: For exercises #17-#20, $z_{\alpha/2}$ was found using [unrounded] invNorm($1-\alpha/2$).

17. $\alpha = 0.05$
   $n = [z_{\alpha/2} \cdot \sigma/E]^2 = [(1.960)(2.5)/(0.5)]^2 = 96.04$, rounded up to 97

19. $\alpha = 0.10$
   $n = [z_{\alpha/2} \cdot \sigma/E]^2 = [(1.645)(12)/(1)]^2 = 389.60$, rounded up to 390

21. $\bar{x} = 67.3849$ mph

23. First find the value of E as follows:   width = 68.1144 − 66.6554 = 2·E
   1.4590 = 2·E
   0.7295 = E

   Then use $\bar{x}$ and E to give the interval as: 67.3849 ± 0.7295 (mph)

25. Use STAT TESTS ZInterval with Input: Stats
   σ: 2.2
   $\bar{x}$: 4.8
   n: 80
   C-Level: 0.95

   ```
   ZInterval
   (4.3179,5.2821)
   x̄=4.8
   n=80
   ```

   4.3 yrs < μ < 5.3 yrs
   No. The proportion of the population falling below a certain value is determined by the shape of the distribution. The confidence interval makes a statement about the mean of the distribution, not about its shape. The most we can say is that since 4 falls below the confidence interval for μ, we expect less than half of the values to fall below 4 – and the 39% figure is consistent with that.

27. Use STAT TESTS ZInterval with Input: Stats
   σ: 9.5
   $\bar{x}$: 58.3
   n: 40
   C-Level: 0.95

   ```
   ZInterval
   (55.356,61.244)
   x̄=58.3
   n=40
   ```

   55.4 sec < μ < 61.2 sec
   Yes. Since the confidence interval includes 60, and since it does not extend very far below and above 60, it is likely that the mean value for the population of all statistics students is reasonably close to 60 seconds.

29. Use STAT TESTS ZInterval with Input: Stats
   σ: 10
   $\bar{x}$: (138+130+...+150)/14
   n: 14
   C-Level: 0.95

   ```
   ZInterval
   (128.69,139.17)
   x̄=133.9285714
   n=14
   ```

   128.7 mm < μ < 139.2 mm
   Ideally, there is a sense in which all the measurements should be the same – and in that case there would be no need for a confidence interval. It is unclear what the given σ = 10 represents in this situation. Is it the true standard deviation in the values of all people in the population (in which case it would not be appropriate in this context where only a single person is involved)? Is it the true standard deviation in momentary readings on a single person (due to constant biological fluctuations)? Is it the true standard deviation in readings from evaluator to evaluator (when they are supposedly evaluating the same thing)? Using the methods of this section and assuming σ = 10, the confidence interval would be 128.7 < μ < 139.2 as given above even if all the readings were the same.

31. Use STAT CALC 1-Var Stats CQPST to find $\bar{x} = 5.6392975$
Use STAT TESTS ZInterval with Input: Stats
σ: 0.068
$\bar{x}$: 5.6392975
n: 40
C-Level: 0.99

```
ZInterval
(5.6116,5.667)
x=5.6392975
n=40
```

$5.61161 \text{ g} < \mu < 5.66698 \text{ g}$
The proportion of the population falling between certain values is determined by the shape of the distribution. The confidence interval makes a statement about the mean of the distribution, not about its shape. Even though the mean weight is apparently well within the prescribed limits, there is no guarantee that this is true for individual quarters.
NOTE: Obtaining the usual confidence interval accuracy [one more decimal place than the original data] as given above requires working the problem $\bar{x} \pm z_{\alpha/2} \cdot \sigma/\sqrt{n}$ by hand.

NOTE: For exercises #33-#36, $z_{\alpha/2}$ was found using [unrounded] invNorm($1-\alpha/2$).

33. $\alpha = 0.05$
$n = [z_{\alpha/2} \cdot \sigma/E]^2 = [(1.960)(15)/(2)]^2 = 216.08$, rounded up to 217

35. $\alpha = 0.05$
$n = [z_{\alpha/2} \cdot \sigma/E]^2 = [(1.960)(10.6)/(0.25)]^2 = 6906.02$, rounded up to 6907

37. Using the range rule of thumb, $\sigma \approx$ (range)/4 = (70,000–12,000)/4 = 58,000/4 = 14,500.
    $\alpha = 0.05$
    $n = [z_{\alpha/2} \cdot \sigma/E]^2 = [(1.96)(14500)/(100)]^2 = 80,769.64$, rounded up to 80,770
No; this sample size is not practical. There are only 3 terms in the formula for n: $z_{\alpha/2}$ and E are set by the expectations of the researcher, while σ is determined by the variable under consideration. Lower the confidence level (to make $z_{\alpha/2}$ less than 1.96, thus decreasing n) and/or increase the margin of error (to make E larger than 100, thus decreasing n).

39. Since n/N = 35/110 = 0.318 > 0.05, use the finite population correction factor.
    $\alpha = 0.05$
    $\bar{x} \pm [z_{\alpha/2}\sigma/\sqrt{n}] \cdot \sqrt{(N-n)/(N-1)}$
    $110 \pm [1.96(15)/\sqrt{35}] \cdot \sqrt{(250-35)/(250-1)}$
    $110 \pm [4.9695] \cdot [0.9292]$
    $110 \pm 4.617$
    $105 < \mu < 115$

## 7-4 Estimating a Population Mean: σ Not Known

1. According to the point estimate ("average"), the parameter of interest is a population mean. But according to the margin of error ("percentage points"), the parameter of interest is a population proportion. It is possible that the margin of error the paper intended to communicate was 1% of $483 (or $4.83, which in a 95% confidence interval would correspond to a sample standard deviation of $226.57) – but the proper units for the margin of error in a situation like this are "dollars" and not "percentage points."

3. We are 99% confident that the interval from 114.4 mmHg to 123.4 mmHg includes the true mean systolic blood pressure for the population of males.

IMPORTANT NOTE: This manual uses the following conventions.
(1) The designation "df" stands for "degrees of freedom."
(2) Since the t value depends on the degrees of freedom, a subscript may be used to clarify which t distribution is being used. For df =15 and α/2 =0.025, for example, one may indicate $t_{15,\alpha/2} = 2.131$. As with the z distribution, it is also acceptable to use the actual numerical value within the subscript and indicate $t_{15,.025} = 2.131$.
(3) When using Table A-3 to find t values, use the closest entry – if the desired df is exactly halfway between the two nearest tabled values, be conservative and choose the one with the lower df. On the TI-84, one may use the invT function – e.g., invT(0.975,15) gives 2.131.
(4) As the degrees of freedom increase, the t distribution approaches the standard normal distribution – and the "large" row of the t table actually gives z values. Consequently the z score for certain "popular" α and α/2 values may be found by reading Table A-3 "frontwards" instead of Table A-2 "backwards." This is not only easier but also more accurate – since Table A-3 includes one more decimal place. Note the following examples.
   For "large" df and α/2 = 0.05, $t_{\alpha/2} = 1.645 = z_{\alpha/2}$ (as found in the z table).
   For "large" df and α/2 = 0.01, $t_{\alpha/2} = 2.326 = z_{\alpha/2}$ (more accurate than the 2.33 in the z table).
[For df = "large" and α/2 = 0.005, $t_{\alpha/2} = 2.576 \neq 2.575 = z_{\alpha/2}$ (as found in the z table). This is a discrepancy caused by using different mathematical approximation techniques to construct the tables, and not a true difference. The correct value, as given by the invNorm function on the TI-83/84 Plus calculator, is 2.576.] In general, this manual will continue to use the invNorm function to find z scores and the invT function to find t scores.

5. σ unknown and population approximately normal
   If the sample is a simple random sample, use t.
   α = 0.05 and df = 11
   $t_{\alpha/2} = 2.201$

7. σ known and population not approximately normal
   Since n<30, neither the z distribution nor the t distribution applies.

9. σ unknown and population approximately normal
   If the sample is a simple random sample, use t.
   α = 0.10 and df = 199 [200]
   $t_{\alpha/2} = 1.653$          TI: invT(0.95,199) = 1.653

11. σ known and population approximately normal
    If the sample is a simple random sample, use z.
    α = 0.02
    $z_{\alpha/2} = 2.326$

13. α = 0.05 and df = 39 [38]          TI: invT(0.975,39) = 2.023
    a. $E = t_{\alpha/2} \cdot s/\sqrt{n}$          b.  $\bar{x} \pm E$
       $= 2.024(4.9)/\sqrt{40}$                3.0 ± 1.6
       = 1.6 kg                                1.4 kg < μ < 4.6 kg

15. 113.583 < μ < 122.417
    We have 95% confidence that the interval from 113.583 to 122.417 contains the true mean IQ of the population of statistics students.

## Estimating a Population Mean: σ Not Known  SECTION 7-4

17. σ unknown and n>30, use STAT TESTS TInterval
    $\bar{x}$: 3106
    $s_x$: 696
    n: 186
    C-Level: 0.95
    $3005 < \mu < 3207$ (grams)

    ```
    TInterval
    (3005.3,3206.7)
    x=3106
    Sx=696
    n=186
    ```

    The confidence interval for the mean birth weight of babies born to mothers who used cocaine is $2608 < \mu < 2792$ (grams) – which is significantly lower than the interval above. Yes; cocaine use appears to be associated with lower birth weights – although the relationship might not be cause and effect due to the cocaine alone, as those who use cocaine may also have other dietary and healthcare issues.

19. a. σ unknown and n>30, use STAT TESTS TInterval
    $\bar{x}$: -1.3
    $s_x$: 4.7
    n: 35
    C-Level: 0.99
    $-3.5 < \mu < 0.9$ (°F)

    ```
    TInterval
    (-3.468,.86756)
    x=-1.3
    Sx=4.7
    n=35
    ```

    b. Yes, the confidence interval includes 0°F. Since the confidence interval includes zero, it is possible that the forecasts are correct on the average. While the meteorologist may want to do further investigation to see if the pattern of slightly high predictions prevails, the above interval does not provide evidence to support his claim.

21. σ unknown (and assuming the distribution is approximately normal), use STAT TESTS TInterval with the 6 values stored in list L1
    List: L1
    Freq: 1
    C-Level: 0.95
    $-0.471 < \mu < 3.547$ [which should be adjusted, since negative values are not possible]
    $0 < \mu < 3.547$ (micrograms/cubic meter)

    ```
    TInterval
    (-.4705,3.5472)
    x=1.538333333
    Sx=1.914203925
    n=6
    ```

    Yes. The fact that 5 of the 6 sample values are below $\bar{x}$ raises a question about whether the data meet the requirement that the underlying distribution is normal.

23. σ unknown (and assuming the distribution is approximately normal), use STAT TESTS TInterval with the 16 values stored in list L1
    List: L1
    Freq: 1
    C-Level: 0.99
    $589.7 < \mu < 731.0$ (FICO units)

    ```
    TInterval
    (589.67,730.96)
    x=660.3125
    Sx=95.89766681
    n=16
    ```

    No. The proportion of the population falling above a certain value is determined by the shape of the distribution. The confidence interval makes a statement about the mean of the distribution, not about its shape. If the distribution is approximately normal with $\mu = 660.3$ and $\sigma = 95.9$, then $P(x>620) \approx P(z>-0.42) = 1 - 0.3372 = 0.6628$, which is not "almost all."

25. σ unknown (and assuming the distribution is approximately normal), use STAT TESTS TInterval with the 7 values stored in list L1
    List: L1
    Freq: 1
    C-Level: 0.98
    $0.075 < \mu < 0.168$ (grams/mile)

    ```
    TInterval
    (.07521,.16765)
    x=.1214285714
    Sx=.0389138242
    n=7
    ```

    No. Since the confidence interval includes values greater than 0.165, there is a reasonable

118   CHAPTER 7   Estimates and Sample Sizes

possibility that the true mean emission amount is greater than that.

27. a. σ unknown and n>30, use STAT TESTS TInterval with list MPULS
    List: MPULS
    Freq: 1
    C-Level: 0.95
    65.8 < μ < 73.0  (beats/min)

    ```
    TInterval
    (65.787,73.013)
    x̄=69.4
    Sx=11.29737887
    n=40
    ```

  b. σ unknown and n>30, use STAT TESTS TInterval with list FPULS
    List: FPULS
    Freq: 1
    C-Level: 0.95
    72.3 < μ < 80.3  (beats/min)

    ```
    TInterval
    (72.303,80.297)
    x̄=76.3
    Sx=12.49861531
    n=40
    ```

  c. Since the two confidence intervals overlap, we cannot conclude that the two population means are different.

29. σ unknown (and assuming the distribution is approximately normal),
    use STAT TESTS TInterval with the 29 values stored in list L1
    List: L1
    Freq: 1
    C-Level: 0.95
    26.2 < μ < 96.3  (years)

    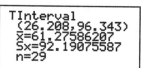

    This is a considerable change from the 42.6 < μ < 46.4 (years) of exercise #22. Confidence intervals can be quite sensitive to extreme outliers. If an outlier is known to be an error and the correct value cannot be determined, it should be discarded. If an outlier may be correct, one could construct confidence intervals with and without the outlier and then make an assessment.

31. assuming a large population                      using the finite population N = 465
    α = 0.05 and df = 99                             α = 0.05 and df = 99
    $E = t_{\alpha/2} \cdot s / \sqrt{n}$            $E = [t_{\alpha/2} \cdot s / \sqrt{n}\,] \times \sqrt{(N-n)/(N-1)}$
    $= 1.984(0.0518)/\sqrt{100}$                     $= [1.984(0.0518)/\sqrt{100}\,] \times \sqrt{365/464}$
    = 0.0103 g                                       = 0.0091 g
    $\bar{x} \pm E$                                  $\bar{x} \pm E$
    0.8565 ± 0.0103                                  0.8565 ± 0.0091
    0.8462 < μ < 0.8668 (grams)                      0.8474 < μ < 0.8656 (grams)

    The second confidence interval is narrower, reflecting the fact that there are more restrictions and less variability (and more certainty) in the finite population situation when n > .05N.

33. a. In general, one sample value gives no information about the variation of the variable. It is possible, however, that one value plus other considerations can give some insight. If one knows that 0 is a possible value, for example, then one large sample value would indicate a large variance. [For example: If you take a sample of n=1 of the daily snowfall in a US city and find that 10.0 feet of snow fell that day, you would assume that there are days with no snow and that there must be a large variability in the amounts of daily snowfall.]

  b. The formula for E requires a value for s and a t score with n-1 degrees of freedom. When n=1, the formula for s fails to produce a value [because there is an (n-1) in the denominator] and there is no df=0 row for the t statistic. No confidence interval can be constructed.

c.  $\bar{x} \pm 9.68|\bar{x}|$
   $12.0 \pm 9.68|12.0|$
   $12.0 \pm 116.2$
   $-104.2 < \mu < 128.2$ [which should be adjusted, since negative heights are not possible]
   $0 < \mu < 128.2$ (feet)

Is it likely that some other randomly selected Martian may be 50 feet tall? This question may be interpreted in two different ways.

No, if "likely" is understood to be "highly probable." The range for individual heights would be even larger than the 0 – 128 given for the mean. With so many possibilities over such a wide range, 50 (or any other individual value) is not highly probable.

Yes, if "likely" is understood to be "reasonable." Since the confidence interval includes the value 50, it is a reasonable possibility for the mean height of all Martians – and any possible mean height would be a possible individual height.

## 7-5 Estimating a Population Variance

1. We have 95% confidence that the interval from 2.25 to 3.52 (inches) contains the true value for the standard deviation of the heights of all women.

3. No. The researcher is dealing with the standard deviation of the $\bar{x}$'s, not with the standard deviation of the individual x's. The Central Limit Theorem states that $\sigma_{\bar{x}} = \sigma/\sqrt{n}$, and the standard deviation on which the researcher is placing a confidence interval would be too small by a factor of $\sqrt{n}$.

NOTE: Since the $\chi^2$ value depends on the degrees of freedom, a subscript may be used to clarify which $\chi^2$ distribution is being used. For df =26 and α/2 =0.025, for example, one may indicate $\chi^2_{26,R} = 41.923$. Note that $\chi^2_R = \chi^2_{\alpha/2}$ and $\chi^2_L = \chi^2_{1-\alpha/2}$. As with the z and t distributions, one may also use clarifying numerical values in the subscripts. For df =26 and α =0.05, for example, one may indicate $\chi^2_R = \chi^2_{26,.025} = 41.923$ and $\chi^2_L = \chi^2_{26,.975} = 13.844$. In general, the intervals in section will be calculated by hand, with the PRGM S2INT display results shown at the right.

5. α = 0.05 and df = 26     $\chi^2_L = 13.844$      $\chi^2_R = 41.923$

7. α = 0.01 and df = 40     $\chi^2_L = 20.707$      $\chi^2_R = 66.766$

9. α = 0.05 and df = 40
   $(n-1)s^2/\chi^2_R < \sigma^2 < (n-1)s^2/\chi^2_L$
   $(40)(18277)^2/59.342 < \sigma^2 < (40)(18277)^2/24.433$
   $225168501 < \sigma^2 < 546881233$
   $15006 < \sigma < 23385$ ($)

11. α = 0.01 and df = 69
    $(n-1)s^2/\chi^2_R < \sigma^2 < (n-1)s^2/\chi^2_L$
    $(69)(68)^2/104.215 < \sigma^2 < (69)(68)^2/43.275$
    $3061.52 < \sigma^2 < 7372.76$
    $55 < \sigma < 86$ (FICO units)  TI: $56 < \sigma < 87$

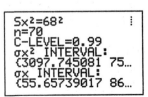

13. From the upper right section of Table 7-2, n = 767.

15. From the lower left section of Table 7-2, n = 1401.
    Whether or not this is a practical sample size depends on the nature of the problem – i.e., on how hard it is to select 1401 individuals from the population, and on how difficult and/or expensive it is to get the desired information/measurement from each individual.

17. $\alpha = 0.05$ and df = 189
    $$(n-1)s^2/\chi_R^2 < \sigma^2 < (n-1)s^2/\chi_L^2$$
    $$(189)(645)^2/228.9638 < \sigma^2 < (189)(645)^2/152.8222$$
    $$343411 < \sigma^2 < 514511$$
    $$586 < \sigma < 717 \text{ (grams)}$$
    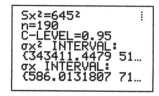
    No. Since the confidence interval includes 696, it is a reasonable possibility for $\sigma$.

19. $\alpha = 0.01$ and df = 105
    $$(n-1)s^2/\chi_R^2 < \sigma^2 < (n-1)s^2/\chi_L^2$$
    $$(105)(0.62)^2/140.169 < \sigma^2 < (105)(0.62)^2/67.328$$
    $$0.2880 < \sigma^2 < 0.5995$$
    $$0.54 < \sigma < 0.77 \text{ (°F)} \quad \text{TI: } 0.53 < \sigma < 0.75$$
    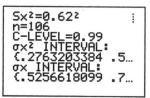
    Yes. Since the confidence interval does not include 2.10, and is so far below 2.10, we can safely conclude that the population standard deviation is less than 2.10°F.

NOTE: The remaining exercises in this section involving actual data (as opposed to summary statistics) are worked only using the TI-83/84 Plus. Two separate steps are required: one to obtain the sample value of s, another to obtain the confidence interval. In addition, the screen giving the confidence interval must be scrolled to the right to see all the digits of the upper limit.

21. Enter the data in list L1.
    Use STAT CALC 1-Var Stats.
    Use PRGM S2INT.
    $$(n-1)s^2/\chi_R^2 < \sigma^2 < (n-1)s^2/\chi_L^2$$
    $$4205.45 < \sigma^2 < 29982.17$$
    $$64.8 < \sigma < 173.2 \text{ (FICO units)}$$

23. Enter the data in list L1.
    Use STAT CALC 1-Var Stats.
    Use PRGM S2INT.
    $$(n-1)s^2/\chi_R^2 < \sigma^2 < (n-1)s^2/\chi_L^2$$
    $$1.4277 < \sigma^2 < 22.0412$$
    $$1.195 < \sigma < 4.695 \text{ (micrograms per cubic meter)}$$
    Yes. One of the requirements to use the methods of this section is that the original distribution be approximately normal, and the fact that 5 of the 6 sample values are less than the mean suggests that the original distribution is not normal.

25. a. Use the data in list MBMI.
   Use STAT CALC 1-Var Stats.
   Use PRGM S2INT.
   $(n-1)s^2/\chi_R^2 < \sigma^2 < (n-1)s^2/\chi_L^2$
   $7.0107 < \sigma^2 < 22.9562$
   $2.65 < \sigma < 4.79$

   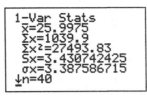 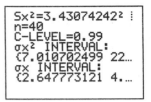

   b. Use the data in list FBMI.
   Use STAT CALC 1-Var Stats.
   Use PRGM S2INT.
   $(n-1)s^2/\chi_R^2 < \sigma^2 < (n-1)s^2/\chi_L^2$
   $2.6429 < \sigma^2 < 74.1431$
   $4.76 < \sigma < 8.61$

   c. The intervals appear not to overlap. It appears that the female BMI values are more variable than the male BMI values.
   NOTE: In some statistical packages these intervals overlap slightly at the 99% level, indicating the standard deviations could be the same and are not significantly different.

27. Applying the given formula yields the following $\chi_L^2$ and $\chi_R^2$ values.

   $\chi^2 = (1/2)[\pm z_{\alpha/2} + \sqrt{2(df) - 1}\,]^2$
   $= (1/2)[\pm 1.96 + \sqrt{2(189) - 1}\,]^2$
   $= (1/2)[\pm 1.96 + 19.416]^2$
   $= (1/2)[17.456]^2$ and $(1/2)[21.376]^2$
   $= 152.3645$ and $228.4771$

   These are close to the 152.8222 and 228.9638 given in exercise #17.

## Statistical Literacy and Critical Thinking

1. The critical value $z_{\alpha/2}=1.96$ is the z score with $\alpha/2 = 0.025$ of the distribution above it. Since the standard normal distribution of z scores is symmetric around 0, there is also 0.025 of the distribution below a z score of -1.96. And so the portion of the distribution between z scores of -1.96 and +1.96 is $1 - (0.025+0.025) = 1 - 0.05 = 0.95 = 95\%$.

2. The confidence interval gives an indication of the accuracy of the estimate. A wide confidence interval indicates that there is considerable variability in the problem and the point estimate obtained from the sample may not be close to the true value of the parameter in the population. A narrow confidence interval indicates that the point estimate obtained from the sample is likely close to the true value of the parameter in the population.

3. $0.65 \pm 0.03$ or $0.62 < p < 0.68$

4. We are 95% confident that the interval from 0.62 to 0.68 includes the true population value for the parameter p.

## 122  CHAPTER 7  Estimates and Sample Sizes

**Review Exercises**

1. NOTE: Since the actual value of x is not given, this problem is limited to the two decimal accuracy reported in 93%. Any $929 \leq x \leq 938$ rounds to $\hat{p} = x/n = x/1004 = 0.93$. All that can be said is that $0.92350 \leq \hat{p} \leq 0.93426$, and 3 decimal confidence interval accuracy is not possible.

   Let x = the number of adults who think that service should be refused as indicated.
   $\alpha = 0.05$ and $\hat{p} = x/n = x/1004 = 0.93$

   $\hat{p} \pm z_{\alpha/2} \sqrt{\hat{p}\hat{q}/n}$
   $0.93 \pm 1.96\sqrt{(0.93)(0.07)/1004}$
   $0.93 \pm .02$
   $0.91 < p < 0.95$
   Using STAT TESTS 1-PropZINT with $x = (0.93)(1004) = 933$,
   x : 933
   n : 1004
   C-level: 0.95
   $0.913 < p < 0.945$

   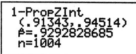

   We are 95% confident that the interval from 0.91 to 0.95 includes the true proportion of all adults who think that service should be refused as indicated.

2. $\alpha = 0.05$ and $E = 0.04$; $\hat{p}$ estimated to be 0.93
   $n = [(z_{\alpha/2})^2 \hat{p}\hat{q}]/E^2$
   $= [(1.960)^2(0.93)(0.07)]/(0.04)^2$
   $= 156.299$, rounded up to 157

NOTE: Recall that the manual continues to find $z_{\alpha/2}$ using invNorm($1-\alpha/2$) and using that unrounded value. That applies in exercise #2 above and in exercise #5 below.

3. $\sigma$ unknown (assuming the distribution is approximately normal), use STAT TESTS TInterval
   $\bar{x}$: 10.89
   $s_x$: 1.56
   n: 16
   C-Level: 0.95
   $10.06 < \mu < 11.72$  (oz.)

   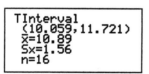

   No. Since the confidence interval does not include 12, we can be confident that the population mean is not as required and that the machines are not working properly. In addition, there is far too much variation in the individual amounts. It appears that a customer could easily (since these values appeared in a small sample of size n=16) get as little as 6.8 oz. or as much as 13.2 oz. (which is almost twice as much as 6.8!) – and that amount of variation is unacceptable.

4. summary statistics as given in exercise #3: n = 16, $\bar{x}$ = 10.89, s = 1.56
   [working the problem by hand (as a review)]
   $\alpha = 0.05$ and df = 15
   $(n-1)s^2/\chi_R^2 < \sigma^2 < (n-1)s^2/\chi_L^2$
   $(15)(1.56)^2/27.488 < \sigma^2 < (15)(1.56)^2/6.262$
   $1.3280 < \sigma^2 < 5.8294$
   $1.15 < \sigma < 2.41$  (oz.)

[working the problem using PRGM S2INT]

```
Sx²=1.56²                          Sx²=1.56²
n=16                               n=16
C-LEVEL=0.95                       C-LEVEL=0.95
σx² INTERVAL:                      σx² INTERVAL:
{1.327978692 5.…                   …92 5.829319187}
σx INTERVAL:                       σx INTERVAL:
{1.152379578 2.…                   …78 2.414398307}
```

and after scrolling to the right

$$1.3280 < \sigma^2 < 5.8293$$
$$1.15 < \sigma < 2.41 \text{ (oz.)}$$

No. Since the confidence interval does not include 0.25, it is not a reasonable possibility for the true population standard deviation. Modifications are required.

5. a. $\alpha = 0.05$ and $E = 0.02$; $\hat{p}$ unknown, use $\hat{p} = 0.5$

   $n = [(z_{\alpha/2})^2 \hat{p}\hat{q}]/E^2$
   $= [(1.960)^2(0.5)(0.5)]/(0.02)^2$
   $= 2400.912$, rounded up to 2401

   b. $\alpha = 0.05$ and $E = 0.5$; assume $\sigma = 8.7$
   $n = [z_{\alpha/2} \cdot \sigma/E]^2$
   $= [(1.960)(8.7)/(0.5)]^2$
   $= 1163.04$, rounded up to 1164

   c. To meet the requirements in both (a) and (b) simultaneously, choose the larger of the two required sample sizes – i.e., use n = 2401.

6. a. σ unknown (assuming the distribution is approximately normal),
   use STAT TESTS TInterval
   $\bar{x}$: 7.01
   $s_x$: 3.74
   n: 25
   C-Level: 0.95
   $5.47 < \mu < 8.55$ (years)

   ```
   TInterval
   (5.4662,8.5538)
   x̄=7.01
   Sx=3.74
   n=25
   ```

   b. $\alpha = 0.05$ and df = 24
   $(n-1)s^2/\chi_R^2 < \sigma^2 < (n-1)s^2/\chi_L^2$
   $(24)(3.74)^2/39.364 < \sigma^2 < (24)(3.74)^2/12.401$
   $8.5282 < \sigma^2 < 27.0706$
   $2.92 < \sigma < 5.20$ (years)

   ```
   Sx²=3.74²
   n=25
   C-LEVEL=0.95
   σx² INTERVAL:
   {8.52814102 27.…
   σx INTERVAL:
   {2.920298105 5.…
   ```

   c. $\alpha = 0.01$
   $n = [z_{\alpha/2} \cdot \sigma/E]^2$
   $= [(2.576)(3.74)/(0.25)]^2$
   $= 1484.90$, rounded up to 1485

   d. No. Since persons who purchase GM cars do not necessarily have the same expectations and experience as those who purchase other cars, the sample would not be representative of all car owners.

7. a. Use STAT TESTS 1-PropZInt with x = (0.183)(785) = 144
   x: 144
   n: 785
   C-Level: 0.98
   $0.151 < p < 0.216$

   ```
   1-PropZInt
   (.1513,.21557)
   p̂=.1834394904
   n=785
   ```

   b. Yes. Since the confidence interval does not include 0.27, the smoking rate for those with four years of college appears to be different than the rate for the general population.

124   CHAPTER 7   Estimates and Sample Sizes

8. a. summary statistics: n = 20, $\bar{x}$ = 9004, s = 5629
   σ unknown and the distribution is approximately normal,
   use STAT TESTS TInterval
   $\bar{x}$: 9004
   $s_x$: 5629
   n: 20
   5403 < μ < 12605  ($)

   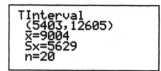

   b. A conservative (i.e., worst case scenario) estimate for the mean hospital costs for an accident victim wearing seat belts is $12,605 – but the worst case scenario estimate for an individual would be considerably higher.

**Cumulative Review Exercises**

1. For a thorough review of the techniques involved, begin by making a stem-and-leaf plot and calculating summary statistics.

   10|5
   11|
   11|599       n = 9
   12|3         Σx = 1089
   12|5788      Σx² = 132223

   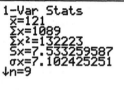

   Most of the answers may also be obtained by placing the weights in list L1 and using STAT CALC 1-Var Stats to obtain the two display screens given above.

   a. $\bar{x}$ = (Σx)/n = 1089/9 = 121.0 lbs

   b. $\tilde{x}$ = 123 lbs

   c. M = 119 lbs, 128 lbs  (bi-modal)

   d. m.r. = (105 + 128)/2 = 116.5 lbs

   e. R = 128 – 105 = 23 lbs

   f. $s^2$ = [n(Σx²) – (Σx)²]/[n(n-1)]
      = [9(132223) – (1089)²]/[9(8)]
      = 4086/72 = 56.75, rounded to 56.8 lbs²

   g. s = 7.5 lbs

   h. for $Q_1 = P_{25}$, L = (25/100)(9) = 2.25, round up to 3
      $Q_1 = x_3$ = 119 lbs
      TI: 117 lbs

   i. for $Q_2 = P_{50}$, L = (50/100)(9) = 4.50, round up to 5
      $Q_2 = x_5$ = 123 lbs

   j. for $Q_3 = P_{75}$, L = (75/100)(9) = 6.75, round up to 7
      $Q_3 = x_7$ = 127 lbs
      TI: 127.5 lbs

   k. Ratio, since differences are consistent and there is a meaningful zero.

1. The boxplot is given below.

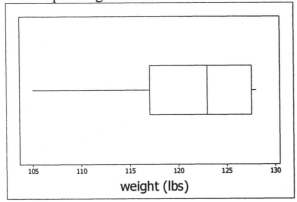

NOTE: Both the stem-and-leaf plot and the boxplot indicate that the weights do not appear to come from a normal distribution – but from a distribution that is truncated at some upper limit and skewed to the left. This means that the given sample of size n=9 should not be used with the methods of this chapter to construct confidence intervals for the mean and the standard deviation. The intervals constructed in parts (m) and (n) are given as a review of the techniques, and not because those techniques apply.

m. σ unknown (and assuming the distribution is approximately normal), use t
  α = 0.01 and df = 8          or use STAT TESTS TInterval
  $\bar{x} \pm t_{\alpha/2} \cdot s/\sqrt{n}$       List: L1
  $121.0 \pm 3.355(7.5)/\sqrt{9}$    Freq: 1
  $121.0 \pm 8.4$           C-Level: 0.99
  $112.6 < \mu < 129.4$ (lbs)   $112.6 < \mu < 129.4$ (lbs)

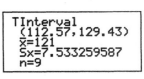

n. α = 0.01 and df = 8
  $(n-1)s^2/\chi_R^2 < \sigma^2 < (n-1)s^2/\chi_L^2$
  $(8)(56.75)/21.995 < \sigma^2 < (8)(56.75)/1.344$
  $20.68 < \sigma^2 < 337.80$
  $4.5 < \sigma < 18.4$ (lbs)

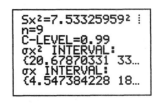

o. α = 0.01
  $n = [z_{\alpha/2} \cdot \sigma/E]^2$
  $= [(invNorm(0.995))(7.533259587)/2]^2$
  = 94.13, rounded up to 95

p. For the general female population, an usually low weight would be one below 85 lbs (i.e., more than 2σ below μ. Individually, none of the supermodels has an unusually low weight. As a group, however, their weights are well below the general population in both mean and standard deviation. Yes; supermodels tend to weigh less than the general female population – and they are a more homogeneous group than the general female population.

2. a. binomial: n=200 and p=0.25
   $P(x \geq 65) = 1 - P(x \leq 64)$
   $= 1 - \text{binomcdf}(200, 0.25, 64)$
   $= 0.0103$

   b. Use STAT TESTS 1-PropZInt
   x: 65
   n: 200
   C-Level: 0.95
   $0.260 < p < 0.390$

   c. No. The expert's value of 0.25 does not seem correct for two reasons.
   From part (a): if the expert is correct, then the probability of getting the sample obtained is very small – approximately 1%.
   From part (b): there is 95% confidence that the interval from 0.26 to 0.39 includes the true value – i.e., there is 95% confidence that the true value is not 0.25.

3. NOTE: Detailed information about the sampling procedure is not given. It is unclear whether the methods of this text and the given information can answer the specific questions asked in parts (a) and (b). We proceed under the following reasonable scenario. *On one particular day, 1233 persons taking a trip were randomly selected from all persons taking a trip that day. Of those selected, 111 indicated that the particular trip which they were taking involved a visit to a theme park.* A more thorough discussion is given in the EXTENDED NOTE appearing at the conclusion of the exercise.

   a. $\hat{p} = x/n = 111/1233 = 0.090$ or 9.0%

   b. Use STAT TESTS 1-PropZInt
   x: 111
   n: 1233
   C-Level: 0.95
   $0.074 < p < 0.106$ or $7.4\% < p < 10.6\%$

   c. $\alpha = 0.01$ and $E = 0.025$; $\hat{p}$ unknown, use $\hat{p} = 0.5$
   $n = [(z_{\alpha/2})^2 \hat{p}\hat{q}]/E^2$
   $= [(\text{invNorm}(.995))^2 (0.5)(0.5)]/(0.025)^2 = 2653.96$, rounded up to 2654

EXTENDED NOTE: The goal appears to be to estimate "the percentage of all people who visit a theme park when they take a trip." Even the reasonable scenario suggested above is limited in that the estimate would be valid only for the time of year the people were sampled – and the proportion of trips resulting in visits to theme parks varies according to the season of the year, whether schools are in session, etc.

The wording that "there were 1122 other *respondents*" implies that the data were gathered from people who responded. Since people on trips can't be required to answer surveys, the data may have come only from those who chose to respond – e.g., by people getting pre-stamped cards while they are on a trip and filling them out when they get home. If the sample was not a simple random sample (e.g., a self-selected sample or a convenience sample), then it is not representative of the population and the estimate is not valid.

To see what other difficulties are raised by various interpretations, consider the following population of 5 people and their trip data for the past year.

| person | total number of trips | number of trips involving theme parks |
|---|---|---|
| A | 4 | 1 |
| B | 8 | 0 |
| C | 3 | 1 |
| D | 5 | 3 |
| E | 5 | 3 |

One approach is to say that the 5 people took 25 trips, 8 of which involved visits to theme parks – and so p = 8/25 = 0.32. But what if D and E are a couple who always travel together? In that case there were really only 20 different trips, 5 of which involved theme parks – and so p = 5/20 = 0.25.

In the wording of the exercise for the above population, one could say that: *A survey includes 4 people (A,C,D,E) who took trips that included visits to theme parks, and there was 1 other respondent (B) who took trips without visits to a theme park.* That yields p = 4/5 = 0.80.

Were the people surveyed at a particular time of the year and/or with regard to one particular trip, or were they asked about all the trips they had taken last year? This exercise illustrates the importance of clearly defining the goal of the problem and method of sampling.

# Chapter 8

# Hypothesis Testing

## 8-2 Basics of Hypothesis Testing

1. In general, a survey that does not examine the entire population is always subject to the possibility of error – i.e., to the possibility that the sample does not provide an exact picture of the population. The only way to definitively "prove" that a value is correct is to examine the entire population. Furthermore, hypothesis testing cannot be used with survey data to conclude that a specific value correct – the proper statement would be that there is not enough evidence to conclude that the 50% figure is wrong.

3. The test statistic is determined by the sample data. The critical value is determined by the set-up of the test of hypothesis – before the sample is even taken. The decision about the null hypothesis is made by comparing the test statistic to the critical value.

5. If the claim were not true, and $p \leq 0.5$, then getting 11 heads in 20 tosses would not be an unusual event. There is not sufficient evidence to support the claim.

7. If the claim were not true, and $\mu \leq 60$, then getting a mean pulse rate of 69.4 in a sample of 400 would be an unusual event. There is sufficient evidence to support the claim.

9. original claim: $p > 0.25$
   $H_o$: $p = 0.25$
   $H_1$: $p > 0.25$

11. original claim: $\mu = 121$ lbs
    $H_o$: $\mu = 121$ lbs
    $H_1$: $\mu \neq 121$ lbs

13. original claim: $\sigma < 15$
    $H_o$: $\sigma = 15$
    $H_1$: $\sigma < 15$

15. original claim: $\mu \geq 0.8535$ g
    $H_o$: $\mu = 0.8535$ g
    $H_1$: $\mu < 0.8535$ g

17. Two-tailed test; place $\alpha/2$ in each tail.
    Use $A = 1-\alpha/2 = 1-0.0250 = 0.9750$ and $z = 1.96$.
    critical values: $\pm z_{\alpha/2} = \pm 1.96$

19. Right-tailed test; place $\alpha$ in the upper tail.
    Use $A = 1-\alpha = 1-0.0100 = 0.9900$ [closest entry = 0.9901] and $z = 2.33$.
    critical value: $+z_\alpha = +2.33$ [or 2.326 from the "large" row of the t table]

21. Two-tailed test; place $\alpha/2$ in each tail.
    Use $A = 1-\alpha/2 = 1-0.0500 = 0.9500$ and $z = 1.645$.
    critical values: $\pm z_{\alpha/2} = \pm 1.645$

23. Left-tailed test; place α in the lower tail.
    Use A = α = 0.0200 [closest entry = 0.0202] and z = -2.05.
    critical value: $-z_\alpha = -2.05$

25. $\hat{p} = x/n = 224/1018 = 0.220$

    $z_{\hat{p}} = (\hat{p} - p)/\sqrt{pq/n}$
    $= (0.220 - 0.250)/\sqrt{(0.25)(0.75)/1018}$
    $= -0.030/0.0136 = -2.21$

27. $\hat{p} = x/n = 516/580 = 0.890$

    $z_{\hat{p}} = (\hat{p} - p)/\sqrt{pq/n}$
    $= (0.890 - 0.750)/\sqrt{(0.75)(0.25)/580}$
    $= 0.140/0.0180 = 7.77$

29. P-value
    $= P(z>1.00)$
    $= 1 - P(z<1.00)$
    $= 1 - 0.8413$
    $= 0.1587$

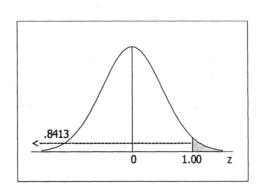

31. P-value
    $= 2 \cdot P(z>1.96)$
    $= 2 \cdot [1 - P(z<1.96)]$
    $= 2 \cdot [1 - 0.9750]$
    $= 2 \cdot [0.0250]$
    $= 0.0500$

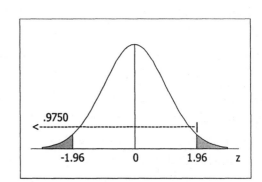

33. P-value
    $= P(z>1.50)$
    $= 1 - P(z<1.50)$
    $= 1 - 0.9332$
    $= 0.0668$

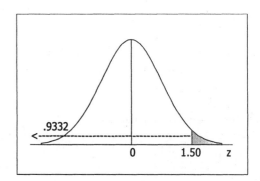

35. P-value
    = 2·P(z<-1.75)
    = 2·(0.0401)
    = 0.0802

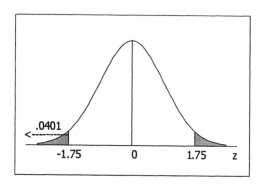

37. NOTE: It is unclear whether "male golfers" is intended to mean "males that are golfers" or "golfers that are male." The final conclusion is stated using the same unclear words given the statement of the original claim.
    original claim: $p < 0.5$
    competing idea: $p \geq 0.5$
        $H_o$: $p = 0.5$
        $H_1$: $p < 0.5$
    initial conclusion: reject $H_o$
    final conclusion: There is sufficient evidence to support the claim that the proportion of male golfers is less than 0.5.

39. original claim: $p \neq 0.13$
    competing idea: $p = 0.13$
        $H_o$: $p = 0.13$
        $H_1$: $p \neq 0.13$
    initial conclusion: fail to reject $H_o$
    final conclusion: There is not sufficient evidence to support the claim that the proportion of M&M's that is red is different from 0.13.

41. type I error: rejecting the hypothesized claim that the proportion of settled malpractice suits is 0.25, when 0.25 is the correct proportion.
    type II error: failing to reject the hypothesized claim that the proportion of settled malpractice suits is 0.25, when 0.25 is not the correct proportion.

43. type I error: rejecting the hypothesized claim that the proportion of murders cleared by arrests is 0.62, when 0.62 is the correct proportion.
    type II error: failing to reject the hypothesized claim that the proportion of murders cleared by arrests is 0.62, when 0.62 is not the correct proportion.

45. original claim: $p > 0.5$
    competing idea: $p \leq 0.5$
        $H_o$: $p = 0.5$
        $H_1$: $p > 0.5$
    $\hat{p} = x/n = x/491 = 0.27$
    $z_{\hat{p}} = (\hat{p} - p)/\sqrt{pq/n} = (0.27 - 0.50)/\sqrt{(0.50)(0.50)/491} = -0.23/0.0226 = -10.19$
    P-value = P(z>-10.19) = 1 − P(z<-10.19) = 1 - 0.0001 = 0.9999
    The claim is that p>.5. Consider the null and alternative hypotheses given above. Only $\hat{p}$ values so much larger than 0.5 that they are unlikely to occur by chance if p=0.5 is true give statistical support to the claim. A $\hat{p}$ smaller than 0.5 does not give any support to the claim.

47. The test of hypothesis is given below
and illustrated by the figure at the right.
$H_o$: $p = 0.50$
$H_1$: $p > 0.50$
$\alpha = 0.05$
C.V. $z = z_\alpha = 1.645$
The c corresponding to $z = 1.645$ is
found by solving $z_c = (c-p)/\sqrt{pq/n}$
for c as follows:

$c = p + z_c \cdot \sqrt{pq/n}$
$= 0.50 + (1.645) \cdot \sqrt{(0.50)(0.50)/64}$
$= 0.50 + (1.645) \cdot (0.0625)$
$= 0.6028$

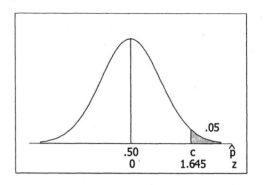

a. The power calculations are given below
and illustrated by the figure at the right.
power = P(rejecting $H_o$|$H_o$ is false)
$= P(\hat{p} > .6028|p=0.65)$
$= P(z > -0.79)$
$= 1 - P(z < -0.79)$
$= 1 - 0.2148 = 0.7852$

The z corresponding to $c = 0.6028$ is
found as follows:

$z_c = (c-p)/\sqrt{pq/n}$
$= (0.6028-0.65)/\sqrt{(0.65)(0.35)/64}$
$= -0.0472/0.0596$
$= -0.79$

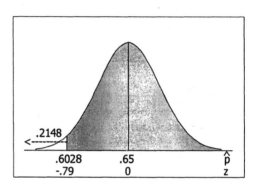

b. $\beta$ = P(type II error) = P( not rejecting $H_o$|$H_o$ is false)
$= 1 - $ P(rejecting $H_o$|$H_o$ is false)
$= 1 - 0.7852 = 0.2148$

## 8-3 Testing a Claim About a Proportion

NOTE: To reinforce the concept that all z scores are standardized rescalings obtained by subtracting the mean and dividing by the standard deviation, visualize the calculated test statistic as the "usual" z formula written to apply to $\hat{p}$'s

$z_{\hat{p}} = (\hat{p} - \mu_{\hat{p}})/\sigma_{\hat{p}}$.

When the normal approximation to the binomial applies, the $\hat{p}$'s are normally distributed with

$\mu_{\hat{p}} = p$ and $\sigma_{\hat{p}} = \sqrt{pq/n}$.

And so the formula for the z statistic may also be written as $z_{\hat{p}} = (\hat{p} - p)/\sqrt{pq/n}$.

1. The requirements of this section, that $np \geq 5$ and $nq \geq 5$, are precisely the requirements for using a normal distribution of x's to approximate a binomial distribution of x's. Since dividing each score by a constant does not change the shape of the distribution, the $\hat{p} = x/n$ values will also follow a normal distribution whenever those requirements are satisfied.

132    CHAPTER 8    Hypothesis Testing

3. No, for at least two reasons: Internet users are not necessarily representative of the general population, and those who answered constitute a voluntary-response sample.

5. At the right are the set-up and result screens for STAT TESTS 1-PropZTest. The exercise is worked below by hand, but the values may be read directly from the TI-83/84 Plus screen displays. [x = (.2492)(8023) = 2001]

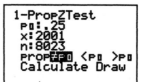 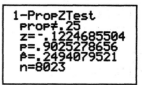

   a. $z\hat{p} = (\hat{p} - \mu\hat{p})/\sigma\hat{p}$
   
   $= (\hat{p} - p)/\sqrt{pq/n}$
   
   $= (0.2494 - 0.25)/\sqrt{(0.25)(0.75)/8023} = -0.0006/0.0048 = -0.12$  [TI: -0.1224]

   b. $z = \pm z_{\alpha/2} = \pm 1.96$
   
   c. P-value = $2 \cdot P(z<-0.12) = 2 \cdot (0.4522) = 0.9044$  [TI: 0.9025]
   
   d. Do not reject $H_o$; there is not sufficient evidence to conclude that $p \neq 0.25$.
   
   e. No. A hypothesis test will either "reject" or "fail to reject" a claim that a population parameter is equal to a specified value.

7. original claim: p > 0.29

   $\hat{p} = x/n = 31{,}969/109{,}857 = 0.2910$
   
   $H_o$: p = 0.29
   $H_1$: p > 0.29
   $\alpha = 0.05$ [assumed]
   C.V. $z = z_\alpha = 1.645$
   calculations:
   $z\hat{p} = (\hat{p} - \mu\hat{p})/\sigma\hat{p} = 0.73$
   P-value = 0.2313
   conclusion:
   Do not reject $H_o$; there is not sufficient evidence to conclude that p > 0.29.

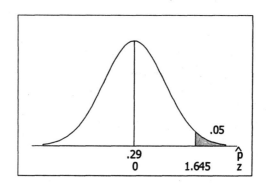

9. original claim: p > 0.50

   $\hat{p} = x/n = 295/325 = 0.9077$
   
   $H_o$: p = 0.50
   $H_1$: p > 0.50
   $\alpha = 0.01$
   C.V. $z = z_\alpha = 2.326$
   calculations:
   $z\hat{p} = (\hat{p} - \mu\hat{p})/\sigma\hat{p}$
   $= 14.700$
   P-value = P(z>14.700) = 3.40E-49 ≈ 0
   conclusion:
   Reject $H_o$; there is sufficient evidence to conclude that p > 0.50.
   Yes. The method appears to work.

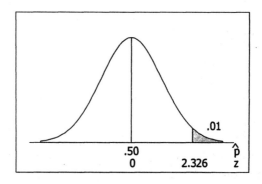

NOTE: The test of hypothesis is the key concept for the remainder of the course. In general, once the basic principles of a given test have been established, the manual will give the solutions to the exercises in a format similar to the one used for the remainder of this section. In particular, there will be the following:
– a statement of the null and alternative hypotheses, the level of significance, and the critical value(s)
– the final screen display from the TI-83/84 Plus calculator, and the visual graph of the test
– a statement of the calculated test statistic, the P-value, and the conclusion

Hints on entering the information into the TI-83/84 Plus calculator and/or the actual screen display from the set-up are given when appropriate. Critical values are taken from the tables in the text.

11. original claim: $p > 0.50$

$\hat{p} = x/n = 5720/11000 = 0.5200$

$H_o$: $p = 0.50$
$H_1$: $p > 0.50$
$\alpha = 0.01$
C.V. $z = z_\alpha = 2.326$
calculations:

$z_{\hat{p}} = (\hat{p} - \mu_{\hat{p}})/\sigma_{\hat{p}}$
$= 4.195$

P-value = $P(z > 4.195) = 1.36\text{E}{-5} = 0.00001$

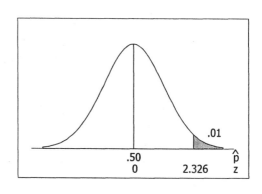

conclusion:
Reject $H_o$; there is sufficient evidence to conclude that $p > 0.50$.

Probably not. The insurance industry is closely monitored, and there would be little room for subjectivity or manipulation in the type of data being analyzed. When those sponsoring a survey have a vested financial interest in the results, however, there is always the possibility that the results might have been presented in a manner as favorable as possible to the sponsor.

13. original claim: $p > 0.15$

$\hat{p} = x/n = 149/880 = 0.1693$

$H_o$: $p = 0.15$
$H_1$: $p > 0.15$
$\alpha = 0.05$
C.V. $z = z_\alpha = 1.645$
calculations:

$z_{\hat{p}} = (\hat{p} - \mu_{\hat{p}})/\sigma_{\hat{p}}$
$= 1.605$

P-value = $P(z > 1.605) = 0.0543$

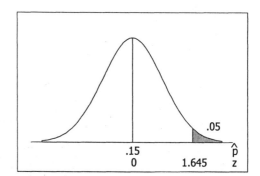

conclusion:
Do not reject $H_o$; there is not sufficient evidence to conclude that $p > 0.15$.

No. Technology is changing so rapidly that figures from 1997 are no longer valid today. By the time you read this, e-mail could be so common that essentially 100% of the households use it – or it could have been replaced by something newer so that no one still uses it!

134   CHAPTER 8   Hypothesis Testing

15. original claim: p = 0.000340
    $\hat{p} = x/n = 135/420{,}095 = 0.000321$
    $H_0$: p = 0.000340
    $H_1$: p ≠ 0.000340
    α = 0.005
    C.V. $z = \pm z_{\alpha/2} = \pm 2.81$
    calculations:
    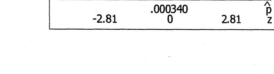
    $z_{\hat{p}} = (\hat{p} - \mu_{\hat{p}})/\sigma_{\hat{p}}$
    = -0.655
    P-value = 2·P(z<-0.655) = 0.5122
    conclusion:
    Do not reject $H_0$; there is not sufficient evidence to reject the claim that p = 0.000340.
    No. Based on these results, cell phone uses have no reason for such concern.

17. original claim: p = 0.01
    $\hat{p} = x/n = 20/1234 = 0.0162$
    $H_0$: p = 0.01
    $H_1$: p ≠ 0.01
    α = 0.05
    C.V. $z = \pm z_{\alpha/2} = \pm 1.96$
    calculations:
    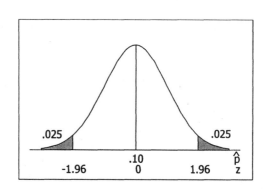
    $z_{\hat{p}} = (\hat{p} - \mu_{\hat{p}})/\sigma_{\hat{p}}$
    = 2.192
    P-value = 2·P(z>2.192) = 0.0284
    conclusion:
    Reject $H_0$; there is sufficient evidence to reject the claim that p = 0.01 and conclude that
    p ≠ 0.01 (in fact, that p > 0.01).
    No. Based on these results, consumers appear to be subjected to more overcharges than under
    the old pre-scanner system.

19. original claim: p < 1/3
    $\hat{p} = x/n = 312/976 = 0.3197$
    $H_0$: p = 1/3
    $H_1$: p < 1/3
    α = 0.05
    C.V. $z = -z_{\alpha} = -1.645$
    calculations:
    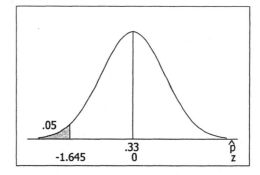
    $z_{\hat{p}} = (\hat{p} - \mu_{\hat{p}})/\sigma_{\hat{p}}$
    = -0.905
    P-value = P(z<-0.905) = 0.1826
    conclusion:
    Do not reject $H_0$; there is not sufficient evidence to conclude that p < 1/3.
    The question is worded appropriately, with nothing to encourage the respondents to answer
    untruthfully. It is difficult, however, to trust people to give honest responses – even in matters
    like height and weight, and especially regarding private behaviors. Here, a likely misstatement
    would be for people who drink "every day" to respond with "a few times a week."

21. NOTE: The value for x is not given. In truth, any $524 \le x \le 534$ rounds to the given $\hat{p} = x/1125 = 47\%$. The manual follows the advice in the text and uses
$x = (0.47)(1125) = 528.75$, rounded to 529.
original claim: $p = 0.50$
$H_o$: $p = 0.50$
$H_1$: $p \ne 0.50$
$\alpha = 0.05$
C.V. $z = \pm z_{\alpha/2} = \pm 1.96$
calculations:

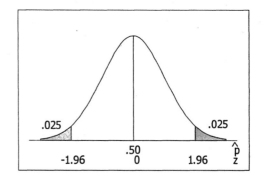

$z\hat{p} = (\hat{p} - \mu_{\hat{p}})/\sigma_{\hat{p}} = -1.998$
P-value = $2 \cdot P(z < -1.998) = 0.0458$
conclusion:
Reject $H_o$; there is sufficient evidence to reject the claim that $p = 0.50$ and conclude that $p \ne 0.50$ (in fact, that $p < 0.50$).
Yes. It is difficult to trust people to give accurate responses – even in matters like height and weight, and especially regarding private behaviors. In this situation, what it means to fly "never or rarely" is not precisely defined and is open to private interpretation.

23. NOTE: The value for x is not given. In truth, either x=34 or x=35 rounds to the given $\hat{p} = x/1655 = 2.1\%$. The manual follows the advice in the text and uses
$x = (0.021)(1655) = 34.755$, rounded to 35.
original claim: $p > 0.012$
$H_o$: $p = 0.012$
$H_1$: $p > 0.012$
$\alpha = 0.01$
C.V. $z = z_\alpha = 2.326$
calculations:

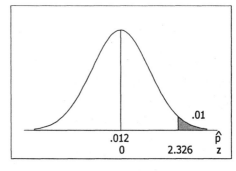

$z\hat{p} = (\hat{p} - \mu_{\hat{p}})/\sigma_{\hat{p}}$
$= 3.417$
P-value = $P(z > 3.417) = 3.16\text{E}{-4} = 0.0003$
conclusion:
Reject $H_o$; there is sufficient evidence to conclude that $p > 0.012$.
Yes. It appears that fatigue is an adverse reaction of Clarinex.

25. There are 100 M&M's listed in Data Set 13, and 27 of them are blue.
original claim: $p = 0.24$
$\hat{p} = x/n = 27/100 = 0.27$
$H_o$: $p = 0.24$
$H_1$: $p \ne 0.24$
$\alpha = 0.05$ [assumed]
C.V. $z = \pm z_{\alpha/2} = \pm 1.96$
calculations:

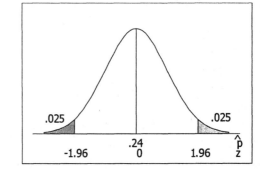

$z\hat{p} = (\hat{p} - \mu_{\hat{p}})/\sigma_{\hat{p}}$
$= 0.702$
P-value = $2 \cdot P(z > 0.702) = 0.4824$
conclusion:
Do not reject $H_o$; there is not sufficient evidence to reject the claim that $p = 0.24$.

27. There are data for 35 days in Data Set 8. For 9 of those days the difference between the actual high and one day forecasted high is more than 2°. This can be determined using
ACTHI − PHI1 → L1     MATH NUM abs(L1) → L1     STAT SortD(L1)     STAT Edit
original claim: p < 0.50
$\hat{p} = x/n = 9/35 = 0.2571$

$H_o$: p = 0.50
$H_1$: p < 0.50
$\alpha = 0.05$
C.V. $z = -z_\alpha = -1.645$
calculations:
$z\hat{p} = (\hat{p} - \mu\hat{p})/\sigma\hat{p}$
    = −2.874
P-value = P(z<−2.874) = 0.0020

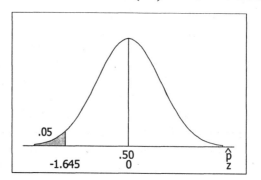

conclusion:
Reject $H_o$; there is sufficient evidence to conclude that p < 0.50.
The result suggests that the one-day high forecasts are fairly accurate in that they are off by more than 2° less than 50% of the time.

29. original claim: p = 0.10
$\hat{p} = x/n = 119/1000 = 0.119$

$H_o$: p = 0.10
$H_1$: p ≠ 0.10
$\alpha = 0.05$
C.V. $z = \pm z_{\alpha/2} = \pm 1.96$
calculations:
$z\hat{p} = (\hat{p} - \mu\hat{p})/\sigma\hat{p}$
    = 2.003
P-value = 2·P(z>2.003) = 0.0452
conclusion:
Reject $H_o$; there is sufficient evidence to reject the claim that p = 0.10 and conclude that p ≠ 0.10 (in fact, that p > 0.10).

a. As seen above, the traditional method leads to rejection of the claim that p=0.10 because the calculated z=2.003 is greater than the critical value of 1.96.

b. As seen above, the P-value method leads to rejection of the claim that p=0.10 because the calculated P-value=0.0452 is less than the level of significance of 0.05.

c. Let x = the number of the randomly selected last digits that are zero.
Use STAT TESTS 1-PropZINT.
x : 119
n : 1000
C-level: 0.95
0.0989 < p < 0.1391
Since 0.10 is inside the confidence interval, p=0.10 is a reasonable claim that should not be rejected.

d. The traditional method and the P-value method are mathematically equivalent and will always agree. As seen by this example, the confidence interval method does not always lead to the same conclusion as the other two methods.

31. a. The claim p>0.0025 results in the null hypothesis $H_o$: p = 0.0025.
Since np = (80)(0.0025) = 0.2 < 5, the normal approximation to the binomial is not appropriate and the methods of this section cannot be used.

b. binomial: n = 80 and p = 0.0025
$P(x) = [n!/(n-x)!x!]p^x q^{n-x}$
$P(x=0) = [80!/80!0!](0.0025)^0(0.9975)^{80} = 0.818525754280$
$P(x=1) = [80!/79!1!](0.0025)^1(0.9975)^{79} = 0.164115439455$
$P(x=2) = [80!/78!2!](0.0025)^2(0.9975)^{78} = 0.016247017189$
$P(x=3) = [80!/77!3!](0.0025)^3(0.9975)^{77} = 0.001058702874$
$P(x=4) = [80!/76!4!](0.0025)^4(0.9975)^{76} = 0.000051077770$
$P(x=5) = [80!/75!5!](0.0025)^5(0.9975)^{75} = 0.000001945820$
$P(x=6) = [80!/74!6!](0.0025)^6(0.9975)^{74} = \underline{0.000000060959}$
$\phantom{P(x=6) = [80!/74!6!](0.0025)^6(0.9975)^{74} = }0.999999998347$
$P(x \geq 7) = 1 - P(x \leq 6) = 1 - 0.999999998347 = 0.000000001653$

c. Based on the result from part (b), getting 7 or more males with this color blindness if p = 0.0025 would be an extremely rare event. This leads one to reject the claim that p = 0.0025. This is essentially the P-value approach to hypothesis testing for
$H_o$: p = 0.0025
$H_1$: p > 0.0025
α = 0.01
P-value = P(getting a result as extreme as or more extreme than x=7)
$\phantom{P-value }= P(x \geq 7)$
$\phantom{P-value }= 0.000000001653$

NOTE: This problem may also be worked using the classical approach to hypothesis testing. Using the binomial probabilities in part (b),
P(x > 0) = 1 − [P(x=0)]         = 1 − [0.818526]                             = 0.181474 > 0.01
P(x > 1) = 1 − [P(x=0)+P(x=1)]     = 1 − [0.818526+0.164115]        = 0.017359 > 0.01
P(x > 2) = 1 − [P(x=0)+P(x=1)+P(x=2)] = 1 − [0.81526+0.164115+0.016247] = 0.001112 ≤ 0.01

To place 0.01 (or as close to it as possible, without going over) in the upper tail, the C.V. must be x > 2, and the classical test [accompanied by a non-appropriate normal-looking figure for illustrative purposes only] is as follows.

$H_o$: p = 0.0025
$H_1$: p > 0.0025
α = 0.01
C.V. x = 2
calculations:
   x = 7
conclusion:
   Reject $H_o$; there is sufficient evidence to conclude p > 0.0025.

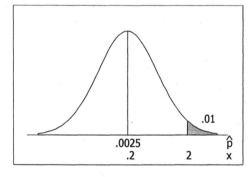

138  CHAPTER 8  Hypothesis Testing

33. The test of hypothesis is given below
and illustrated by the figure at the right.
$H_o$: p = 0.40
$H_1$: p < 0.40
α = 0.05 [assumed]
C.V. z = $-z_\alpha$ = -1.645
The c corresponding to z = -1.645 is
found by solving $z_c$ = (c-p)/$\sqrt{pq/n}$
for c as follows:

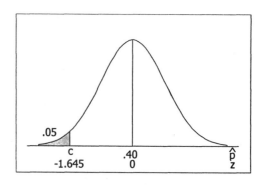

$$c = p + z_c \cdot \sqrt{pq/n}$$
$$= 0.40 + (-1.645)\sqrt{(.40)(.60)/50}$$
$$= 0.40 - (1.645)(0.0693)$$
$$= 0.2860$$

a. The power calculations are given below
and illustrated by the figure at the right.
power = P(rejecting $H_o$|$H_o$ is false)
  = P($\hat{p}$ < 0.2860|p=0.25)
  = P(z<0.59) = 0.7224

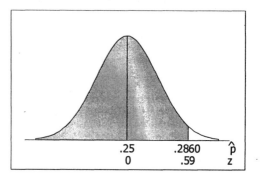

The z corresponding to c= 0.2860 is
found as follows:
$z_c$ = (c-p)/$\sqrt{pq/n}$ = (0.2860-0.25)/$\sqrt{(.25)(.75)/50}$ = 0.0360/0.0612 = 0.59

b. β = P(type II error) = P( not rejecting $H_o$|$H_o$ is false)
  = 1 – P(rejecting $H_o$|$H_o$ is false)
  = 1 – 0.7224 = 0.2776

c. No test of hypothesis can lead the researcher to the right conclusion 100% of the time. If $H_o$ is true, there is a 95% chance $H_o$ will (correctly) not be rejected; if $H_o$ is false and p=0.25 is true, there is a 72% chance that $H_o$ will (correctly) be rejected. These are reasonable probabilities of success, and the test is reasonably effective.

## 8-4 Testing a Claim About a Mean: σ Known

1. No, a sample of size n>30 is not necessary to use the methods of this section. If the x's are normally distributed, the $\bar{x}$'s are normally distributed with $\sigma_{\bar{x}}$ = σ/$\sqrt{n}$ for any value of n. If the x's are not normally distributed, then it must be true that n>30. In either case, the underlying assumption is that the data represent a simple random sample.

3. A usual two-sided confidence interval at the 1-α level places α/2 in each tail of the distribution. To correspond with a one-tailed test at the 0.01 level, a two-sided confidence interval should be constructed using α = 0.02 – i.e., using the 98% level of confidence.

5. Yes. The methods of this section apply to this situation.

7. No. Since σ is unknown, the methods of this section do not apply.

9. At the right are the set-up and result screens for STAT TESTS Z-Test. The exercises is worked below by hand, but the values may be read directly from the TI-83/84 Plus screen displays.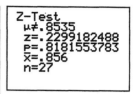

   a. $z_{\bar{x}} = (\bar{x} - \mu_{\bar{x}})/\sigma_{\bar{x}}$
   
   $= (\bar{x} - \mu)/(\sigma/\sqrt{n})$
   
   $= (0.8560 - 0.8535)/(0.0565/\sqrt{27}) = 0.0025/0.0109 = 0.23$ [TI:0.230]
   
   b. $z = \pm z_{\alpha/2} = \pm 1.96$
   
   c. P-value $= 2 \cdot P(z>0.23)$
   
   $= 2[1 - P(z<0.23)] = 2[1 - 0.5910] = 2[0.4090] = 0.8180$ [TI: 0.8182]
   
   d. Fail to reject $H_o$.
   
   e. There is not sufficient evidence to reject the claim that the true mean weight of blue M&M's is 0.8535 grams.

11. original claim: $\mu > 30.0$ °C
    
    $H_o$: $\mu = 30.0$ °C
    
    $H_1$: $\mu > 30.0$ °C
    
    $\alpha = 0.05$
    
    C.V. $z > z_{.05} = 1.645$
    
    calculations:
    
    $z_{\bar{x}} = (\bar{x} - \mu_{\bar{x}})/\sigma_{\bar{x}} = 1.732$
    
    P-value $= 0.0416$
    
    conclusion:
    
    Reject $H_o$; there is sufficient evidence to conclude that $\mu > 30.0$.

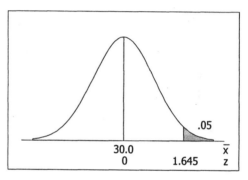

NOTE: The format for presenting the solutions for the remainder of the tests of hypotheses in this section, and for the entire course, is the one established in the previous section. Refer to the NOTE on page 133.

13. original claim: $\mu = 60$ sec
    
    $H_o$: $\mu = 60$ sec
    
    $H_1$: $\mu \neq 60$ sec
    
    $\alpha = 0.05$
    
    C.V. $z = \pm z_{\alpha/2} = \pm 1.96$
    
    calculations:
    
    $z_{\bar{x}} = (\bar{x} - \mu_{\bar{x}})/\sigma_{\bar{x}}$
    
    $= -1.132$
    
    P-value $= 2 \cdot P(z<-1.132) = 0.2577$

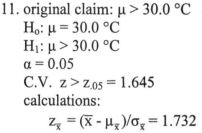

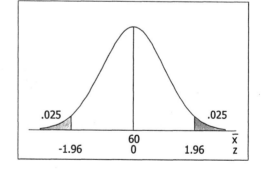

conclusion:

Do not reject $H_o$; there is not sufficient evidence to reject the claim that $\mu = 60$.

Yes. Based on this result, the overall perception of one minute appears to be reasonably accurate – although individual results might vary considerably.

140   CHAPTER 8   Hypothesis Testing

15. original claim: $\mu < 0$ lbs
    $H_0$: $\mu = 0$ lbs
    $H_1$: $\mu < 0$ lbs
    $\alpha = 0.05$
    C.V. $z = -z_\alpha = -1.645$
    calculations:
    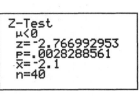
    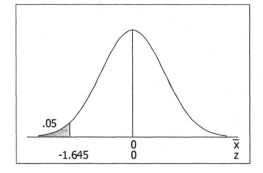
    $z_{\bar{x}} = (\bar{x} - \mu_{\bar{x}})/\sigma_{\bar{x}}$
    $= -2.767$
    P-value = $P(z<-2.767) = 0.0028$
    conclusion:
    Reject $H_0$; there is sufficient evidence to conclude that $\mu < 0$.
    Yes. Based on this result, the diet is effective. If the goal of the diet is to lose weight, the mean weight loss of only 2.1 pounds after one full year is not worth the effort. If the goal of the diet is to not add weight and to be a generally healthier person, then it is worth the effort.

17. Place the 14 values in list L1.
    original claim: $\mu < 140$ mmHg
    $H_0$: $\mu = 140$ mmHg
    $H_1$: $\mu < 140$ mmHg
    $\alpha = 0.05$
    C.V. $z = -z_\alpha = -1.645$
    calculations:
    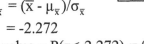

    $z_{\bar{x}} = (\bar{x} - \mu_{\bar{x}})/\sigma_{\bar{x}}$
    $= -2.272$
    P-value = $P(z<-2.272) = 0.0116$
    conclusion:
    Reject $H_0$; there is sufficient evidence to conclude that $\mu < 140$.
    Yes. It can be concluded with 95% confidence that the person does not have hypertension.

19. Use list CQPST.
    original claim: $\mu = 5.670$
    $H_0$: $\mu = 5.670$
    $H_1$: $\mu \neq 5.670$
    $\alpha = 0.01$
    C.V. $z = \pm z_{\alpha/2} = \pm 2.575$
    calculations:
    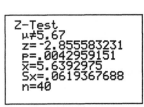
    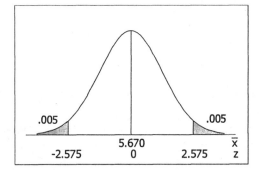
    $z_{\bar{x}} = (\bar{x} - \mu_{\bar{x}})/\sigma_{\bar{x}}$
    $= -2.856$
    P-value = $2 \cdot P(z<-2.856) = 0.0043$
    conclusion:
    Reject $H_0$; there is sufficient evidence to reject the claim that $\mu = 5.670$ and conclude that $\mu \neq 5.670$ (in fact, that $\mu < 5.670$). If the quarters were in uncirculated condition, they appear not to have been manufactured according to specifications – but if they were worn from use, all that can be said is that their mean original weight was at least 5.6393 grams.

21. The test of hypothesis is given below
and illustrated by the figure at the right.
$H_o$: $\mu = 100$
$H_1$: $\mu > 100$
$\alpha = 0.05$ [assumed]
C.V. $z = z_\alpha = 1.645$
The c corresponding to $z = 1.645$ is
found by solving $z_c = (c-\mu)/(\sigma/\sqrt{n})$
for c as follows.

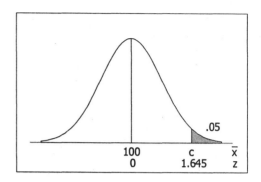

$c = \mu + z_c \cdot (\sigma/\sqrt{n})$
$= 100 + 1.645(15/\sqrt{40})$
$= 100 + 3.90$
$= 103.90$

a. The power calculations are given below
and illustrated by the figure at the right.
power = P(rejecting $H_o|H_o$ is false)
$= P(\bar{x}>103.90|\mu=108)$
$= P(z>-1.73)$
$= 1 - P(z<-1.73)$
$= 1 - 0.0418$
$= 0.9582$

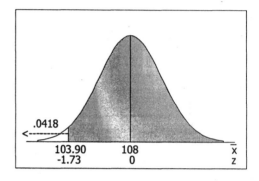

The z corresponding to c= 103.90 is
$z_c = (c-\mu)/(\sigma/\sqrt{n})$
$= (103.90-108)/(15/\sqrt{40})$
$= -4.099/2.372 = -1.73$

The test is very effective in recognizing that the mean is greater than 100 when the mean is actually equal to 108, for it will do so 95.8% of the time.

b. $\beta$ = P(type II error) = P( not rejecting $H_o|H_o$ is false)
$= 1 - $ P(rejecting $H_o|H_o$ is false)
$= 1 - 0.9582$
$= 0.0418$

## 8-5 Testing a Claim About a Mean: σ Not Known

1. When using $\bar{x}$ and s to make inferences about $\mu$, the degrees of freedom is df = n-1. If a sample consists of n = 5 values, then df = n-1 = 4. Given $\bar{x} = (\Sigma x)/n$, one can make up n-1 of the x values – but the final x value has to be whatever must be added to those n-1 values to produce $\Sigma x = n \cdot \bar{x}$. And so in this instance, 4 of the 5 values can be made up before the last one is determined by the restriction that the mean is 20.0.

3. The claim that $\mu$ is less than 12 is supported by the sample data only when $\bar{x}$ is significantly less than 12. The sample data definitely do not support the claim that $\mu$ is less than 12 when $\bar{x}$ is greater than or equal to 12.

5. Use t. When σ is unknown and the x's are normally distributed, use t.

7. Neither the z nor the t applies. When σ is unknown and the x's are not normally distributed, sample sizes n≤30 cannot be used with the techniques in this chapter.

142  CHAPTER 8  Hypothesis Testing

NOTE: Exercises #9–#12 may be worked as follows.
  table: find the correct df row in Table A-3, and see what values surround the given t.
  TI: use tcdf(lower bound, upper bound, df) to get the probability between 2 values

9. P-value = $P(t_6 > 3.500)$
   table for area in one tail: [3.707 > 3.500 > 3.143]  0.005 < P-value < 0.01
   TI: tcdf(3.500,99,6) = 0.00641

11. P-value = $2 \cdot P(t_{20} > 9.883)$
    table for area in two tails: [9.833 > 2.845]  P-value < 0.01
    TI: 2·tcdf(9.883,99,20) = 3.85E-10 = 0.00000000385

NOTE: In the following tests of hypotheses, the C.V. for t is taken from Table A-3 whenever possible – otherwise the invT function is used.

13. At the right are the set-up and result screens for STAT TESTS T-Test. The exercise is worked below by hand, but the values may be read directly from the TI-83/84 Plus screen displays.

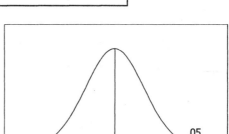

   original claim: $\mu > 120$
   $H_o$: $\mu = 120$
   $H_1$: $\mu > 120$
   $\alpha = 0.05$ and df = 11
   C.V.  $t = t_\alpha = 1.796$
   calculations:
       $t_{\bar{x}} = (\bar{x} - \mu)/s_{\bar{x}}$
          $= (132 - 120)/(12/\sqrt{12})$
          $= 12/3.464 = 3.464$
       P-value = $P(t_{11} > 3.464)$
          = tcdf(3.464,99,11) = 0.0026
   conclusion:
       Reject $H_o$; there is sufficient evidence to conclude that $\mu > 120$.

15. original claim: $\mu > 110$
    $H_o$: $\mu = 110$
    $H_1$: $\mu > 110$
    $\alpha = 0.05$ and df = 24
    C.V.  $t = t_\alpha = 1.711$
    calculations:
        $t_{\bar{x}} = (\bar{x} - \mu)/s_{\bar{x}}$
           = 3.74
        P-value = $P(t_{24} > 3.74) = 0.001$
    conclusion:
        Reject $H_o$; there is sufficient evidence to conclude that $\mu > 110$.

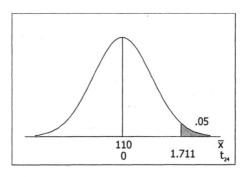

17. original claim: $\mu < 98.6$ °F
 $H_o$: $\mu = 98.6$ °F
 $H_1$: $\mu < 98.6$ °F
 $\alpha = 0.05$ and df = 105
 C.V.  t = $-t_\alpha$ = $-1.659$
 calculations:
 $t_{\bar{x}} = (\bar{x} - \mu)/s_{\bar{x}}$
   = $-6.642$
 P-value = $P(t_{105} < -6.642) = 7.02\text{E-}10 \approx 0$

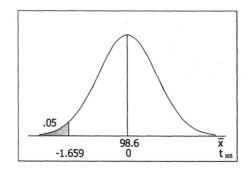

conclusion:
 Reject $H_o$; there is sufficient evidence to conclude that $\mu < 98.6$.
Yes. It appears that the commonly-used mean of 98.6°F is not correct.

19. original claim: $\mu < 3103$ g
 $H_o$: $\mu = 3103$ g
 $H_1$: $\mu < 3103$ g
 $\alpha = 0.01$ and df = 189
 C.V.  t = $-t_\alpha$ = $-2.346$
 calculations:
 $t_{\bar{x}} = (\bar{x} - \mu)/s_{\bar{x}}$
   = $-8.612$
 P-value = $P(t_{189} < -8.612) \approx 0$

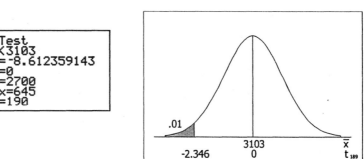

conclusion:
 Reject $H_o$; there is sufficient evidence to conclude that $\mu < 3103$.
Not necessarily. There is sufficient evidence to conclude that weights of babies born to cocaine uses are less than weights of babies born to non-users, but the association is not necessarily one of cause and effect. The birth weights might not be affected by the cocaine *per se*, but by other factors associated with cocaine users – e.g., poor diet, poor pre-natal health care, etc.

21. original claim: $\mu > 0$
 $H_o$: $\mu = 0$
 $H_1$: $\mu > 0$
 $\alpha = 0.01$ and df = 20
 C.V.  t = $t_\alpha$ = 2.528
 calculations:
 $t_{\bar{x}} = (\bar{x} - \mu)/s_{\bar{x}}$
   = 8.447
 P-value = $P(t_{20} > 8.447) = 2.49\text{E-}8 \approx 0$

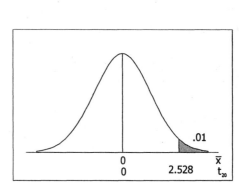

conclusion:
 Reject $H_o$; there is sufficient evidence to conclude that $\mu > 0$.
Yes. The treatment appears to be effective.

23. original claim: μ > 63.6 in
H₀: μ = 63.6 in
H₁: μ > 63.6 in
α = 0.01 and df = 8
C.V. t = t_α = 2.896
calculations:
$t_{\bar{x}} = (\bar{x} - \mu)/s_{\bar{x}}$
= 13.200
P-value = P(t₈ > 13.200) = 5.17E-7 ≈ 0
conclusion:
Reject H₀; there is sufficient evidence to conclude that μ > 63.6.
Yes. Even though the sample was only n = 9, the conclusion is valid. The sample size is built into the critical region – the smaller the sample size, the larger t must be in order to reject H₀.

25. Place the 6 values in list L1.
original claim: μ > 1.5 μg/m³
H₀: μ = 1.5 μg/m³
H₁: μ > 1.5 μg/m³
α = 0.05 and df = 5
C.V. t = t_α = 2.015
calculations:
$t_{\bar{x}} = (\bar{x} - \mu)/s_{\bar{x}}$
= 0.049
P-value = P(t₅ > 0.049) = 0.4814
conclusion:
Do not reject H₀; there is not sufficient evidence to conclude that μ > 1.5.
Yes. Since 5 of the 6 sample values are below the sample mean, there is reason to doubt that the population values are normally distributed.

27. Place the 15 values in list L1.
original claim: μ = 1.8 g
H₀: μ = 1.8 g
H₁: μ ≠ 1.8 g
α = 0.05 [assumed]
   and df = 14
C.V. t = ±t_{α/2} = ±2.145
calculations:
$t_{\bar{x}} = (\bar{x} - \mu)/s_{\bar{x}}$
= -1.297
P-value = 2·P(t₁₄ < -1.297) = 0.2155
conclusion:
Do not reject H₀; there is not sufficient evidence to reject the claim that μ = 1.8.

29. Use the list CQPST.
   original claim: $\mu = 5.670$
   $H_o$: $\mu = 5.670$
   $H_1$: $\mu \neq 5.670$
   $\alpha = 0.05$ [assumed]
      and df = 39
   C.V. $t = \pm t_{\alpha/2} = \pm 2.023$
   calculations:
      $t_{\bar{x}} = (\bar{x} - \mu)/s_{\bar{x}}$
         $= -3.135$
      P-value $= 2 \cdot P(t_{39} < -3.135) = 0.0033$
   conclusion:
      Reject $H_o$; there is sufficient evidence to reject the claim that $\mu = 5.670$ and conclude that $\mu \neq 5.670$ (in fact, that $\mu < 5.670$).
   NOTE: This does not mean that the quarters are not manufactured according to specifications, but only that they weigh less than the specified manufacturing weight when they are sampled.

31. Use the list MPULS.
   original claim: $\mu > 60$ beats/min
   $H_o$: $\mu = 60$ beats/min
   $H_1$: $\mu > 60$ beats/min
   $\alpha = 0.05$ [assumed]
      and df = 39
   C.V. $t = t_{\alpha} = 1.686$
   calculations:
      $t_{\bar{x}} = (\bar{x} - \mu)/s_{\bar{x}}$
         $= 5.262$
      P-value $= P(t_{39} > 5.262) = 2.73\text{E-}6 \approx 0$
   conclusion:
      Reject $H_o$; there is sufficient evidence to conclude that $\mu > 60$.

33. The two methods may be compared as follows.

   | method of this section | alternative method |
   |---|---|
   | $H_o$: $\mu = 100$ | $H_o$: $\mu = 100$ |
   | $H_1$: $\mu \neq 100$ | $H_1$: $\mu \neq 100$ |
   | $\alpha = 0.05$ and df = 31 | $\alpha = 0.05$ |
   | C.V. $t = \pm t_{\alpha/2} = \pm 2.040$ | C.V. $z = \pm z_{\alpha/2} = \pm 1.960$ |
   | calculations: | calculations: |
   | $t_{\bar{x}} = (\bar{x} - \mu)/s_{\bar{x}}$ | $z_{\bar{x}} = (\bar{x} - \mu)/\sigma_{\bar{x}}$ |
   | $= (105.3 - 100)/(15.0/\sqrt{32})$ | $= (105.3 - 100)/(15.0/\sqrt{32})$ |
   | $= 5.3/2.6517 = 1.999$ | $= 5.3/2.6517 = 1.999$ |
   | conclusion: | conclusion: |
   | Do not reject $H_o$; there is not enough evidence to reject the claim that $\mu = 100$. | Reject $H_o$; there is enough evidence to reject the claim that $\mu = 100$ and conclude that $\mu \neq 100$ (in fact, that $\mu > 100$). |

   The two methods lead to different conclusions – because of the unwarranted artificial precision created in the alternative method by assuming that the s value could be used for $\sigma$.

146  CHAPTER 8  Hypothesis Testing

35. The new test of hypothesis is given below. In this instance the presence of a drastic outlier drastically changed the summary statistics but did not have much of an impact on the overall test of hypothesis. In general, however, the effects of an outlier can be substantial. One could argue that the effect is not so great in this exercise because according to the $\bar{x}\pm 2s$ standard the original test already involved an outlier – so much larger than the others that it was the only value above $\bar{x}$.

Place the 6 values in L1.
original claim: $\mu > 1.5$
$H_o$: $\mu = 1.5$
$H_1$: $\mu > 1.5$
$\alpha = 0.05$ and df = 5
C.V. $t = t_\alpha = 2.015$
calculations:
$t_{\bar{x}} = (\bar{x} - \mu)/s_{\bar{x}}$
$= 0.992$
P-value = $P(t_5 > 0.992) = 0.1834$

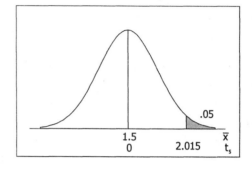

conclusion:
Do not reject $H_o$; there is not sufficient evidence to conclude that $\mu > 1.5$.

37. The set up for the basic test of hypothesis is as follows.
summary statistics: n = 15, $\Sigma x = 25.7$, $\Sigma x^2 = 44.97$, $\bar{x} = 1.713$, s = 0.2588
original claim: $\mu < 1.8$
$H_o$: $\mu = 1.8$
$H_1$: $\mu < 1.8$
$\alpha = 0.05$ [assumed] and df = 14
C.V. $t = -t_\alpha = -1.761$

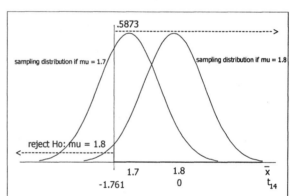

The set up for the test indicates that
$\alpha = P(\text{type I error})$
$= P(\text{rejecting } H_o | H_o \text{ is true}) = 0.05$.
In general,
$\beta = P(\text{type II error})$
$= P(\text{not rejecting } H_o | H_o \text{ is false})$
$= $ depends on how false $H_o$ really is.
In particular, Minitab has determined that
$\beta = P(\text{type II error})$
$= P(\text{not rejecting } H_o | \mu = 1.7) = 0.5873$.

a. power of the test = $1 - \beta = 1 - 0.5873 = 0.4127$

b. P(type II error) = $\beta = 0.5873$

c. No. The power of the test when $\mu = 1.7$ is not very high. The probability of properly concluding that $\mu$ is less than 1.8 when it is really 1.7 is only 0.4127. This is less than 50%, and a test should have a power of at least 80% to be considered effective.

## 8-6 Testing a Claim About a Standard Deviation or Variance

1. If a test is *not robust* against departures from normality, then the original population must be very close to a normal distribution in order for the statistical theory and methods to apply. Since the chi-square test of this section is not robust, a stricter set of guidelines should be applied than in previous sections when checking whether the original population is normal.

3. No. The chi-square test of this section is not robust against departures from normality, and the original population of x values from a fair die follows a uniform distribution.

NOTE: Following the pattern used with the z and t distributions, this manual find the critical values by using the closest entry from Table A-4 for the $\chi^2$ as if it were the precise value necessary and does not use interpolation. This sacrifices little accuracy – and even interpolation does not yield precise values. When more accuracy is needed, refer either to more complete tables or to computer-produced values. The P-value portion of exercises #5-#8 may be worked as follows.
 table: find the correct df row in Table A-4, and see what values surround the calculated $\chi^2$
 TI-83/84 Plus: use $\chi^2$cdf(lower bound, upper bound, df) to get the probability between 2 values
For the remainder of this manual, exact P-values for the $\chi^2$ distribution USING THE UNROUNDED VALUES of $\chi^2$ will be given in TI-83/84 Plus calculator format as described above.

5. a. test statistic: $\chi^2 = (n-1)s^2/\sigma^2 = (9)(3.00)^2/(2.00)^2 = 20.250$
 b. critical values for $\alpha = 0.05$ and df = 9: $\chi^2 = \chi^2_{1-\alpha/2} = 2.700$
 $\chi^2 = \chi^2_{\alpha/2} = 19.023$
 c. P-value limits: $19.023 < 20.250 < 21.666$
 $0.02 < \text{P-value} < 0.05$
 P-value exact: $2 \cdot \chi^2\text{cdf}(20.250, 999, 9) = 0.0329$
 d. conclusion: Reject $H_o$; there is sufficient to conclude that $\sigma \neq 2.00$ (in fact, that $\sigma > 2.00$).

NOTE: As with the t distribution a subscript (df) may be used to identify which $\chi^2$ distribution to use in the tables. As before, it is also acceptable to use the actual numerical values in place of the generic $\alpha$, $\alpha/2$ or $1-\alpha/2$. In exercise #5, for example, one could indicate
 $\chi^2_{9, 0.975} = 2.700$ [the $\chi^2$ value with df = 9 and $1-\alpha/2 = 0.975$ in the upper tail]
 $\chi^2_{9, 0.025} = 19.023$ [the $\chi^2$ value with df – 9 and $\alpha/2 = 0.025$ in the upper tail]

7. a. test statistic: $\chi^2 = (n-1)s^2/\sigma^2 = (20)(10)^2/(15)^2 = 8.889$
 b. critical value for $\alpha = 0.01$ and df = 20: $\chi^2 = \chi^2_{1-\alpha} = 8.260$
 c. P-value limits: $8.260 < 8.889 < 9.591$
 $0.01 < \text{P-value} < 0.025$
 P-value exact: $\chi^2\text{cdf}(0, 8.889, 20) = 0.0158$
 d. conclusion: Do not reject $H_o$; there is not sufficient evidence to conclude that $\sigma < 15$.

NOTE: In the remaining exercises, the manual takes the critical value(s) from Table A-4. When the exact df needed in the problem does not appear in the table, the closest $\chi^2$ value is used to determine the critical region. Since the formula $\chi^2 = (n-1)s^2/\sigma^2$ used in the calculations contains df = n-1, some instructors recommend using the same df in the calculations that were used to determine the C.V. This manual typically uses the closest entry to determine the C.V. and the n from the problem in the calculations – even though this introduces a slight discrepancy.
 This manual takes the calculated $\chi^2$ and the P-value from the S2TEST program and shows the appropriate display screens.

9. original claim: σ ≠ 696 g
   $H_0$: σ = 696 g
   $H_1$: σ ≠ 696 g
   α = 0.05 and df = 189 $\chi^2_{39}$
   C.V. $\chi^2 = \chi^2_{1-\alpha/2} = 152.8222$
   $\chi^2 = \chi^2_{\alpha/2} = 228.9638$
   calculations:
   $\chi^2 = (n-1)s^2/\sigma^2$
   = 162.317
   P-value = $2 \cdot P(\chi^2_{189} > 162.317) = 0.1591$

   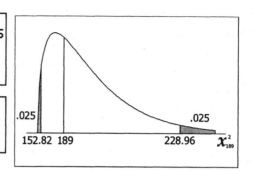

   conclusion:
   Do not reject $H_0$; there is not sufficient evidence to conclude that σ ≠ 696.
   Based on this result, the variation in the weights of babies born to mothers who use cocaine is not significantly different from the variation in the weights for mothers who don't use cocaine.

GRAPHICS NOTE: While the $\chi^2$ distribution is neither symmetric nor centered at zero, there are some important guidelines that help to keep $\chi^2$ values in perspective. Loosely speaking, it "bunches up" around its df (which is actually its expected value) – and so the lower critical value will be less than df and the upper critical will be greater than df. Similarly, a calculated $\chi^2$ value less than df falls in the lower tail and a calculated $\chi^2$ value greater than df falls in the upper tail.

To illustrate $\chi^2$ tests of hypotheses, this manual uses a "generic" figure resembling a $\chi^2$ distribution with df=4. Actually $\chi^2$ distributions with df=1 and df=2 have no upper limit and approach the y axis asymptotically, while $\chi^2$ distributions with df>30 are essentially symmetric and normal-looking. Because the $\chi^2$ distribution is positively skewed, the expected value df is noted slightly to the right of the figure's peak.

11. original claim: σ > 0.056 g
    $H_0$: σ = 0.056 g
    $H_1$: σ > 0.056 g
    α = 0.01 and df = 40
    C.V. $\chi^2 = \chi^2_\alpha = 63.691$
    calculations:
    $\chi^2 = (n-1)s^2/\sigma^2$
    = 1225.765
    P-value = $P(\chi^2_{40} > 1225.765) \approx 0$

    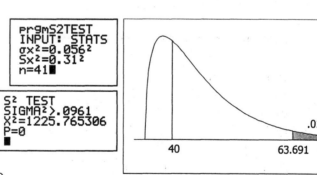

    conclusion:
    Reject $H_0$; there is sufficient evidence to conclude that σ > 0.056.
    The peanut candy weights should have a larger standard deviation because adding a peanut introduces more variability – peanuts do not all have the same weight.

13. Place the 16 values in list L1.
    original claim: $\sigma \neq 83$
    $H_0$: $\sigma = 83$
    $H_1$: $\sigma \neq 83$
    $\alpha = 0.05$ and df = 15
    C.V. $\chi^2 = \chi^2_{1-\alpha/2} = 6.262$
    $\chi^2 = \chi^2_{\alpha/2} = 27.488$
    calculations:
    $\chi^2 = (n-1)s^2/\sigma^2$
    $= 20.024$
    P-value $= 2 \cdot P(\chi^2_{15} > 20.024) = 0.3420$
    conclusion:
    Do not reject $H_0$; there is not sufficient evidence to conclude that $\sigma \neq 83$.
    No. Based on these results, applicants from the new branch do not appear to have credit ratings that vary more than those at the main bank. .

    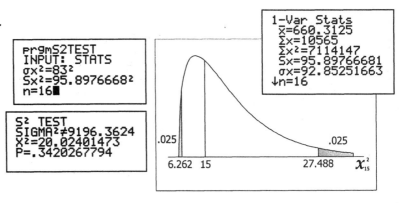

15. Place the 9 values in list L1.
    original claim: $\sigma < 29$ lbs
    $H_0$: $\sigma = 29$ lbs
    $H_1$: $\sigma < 29$ lbs
    $\alpha = 0.01$ and df = 8
    C.V. $\chi^2 = \chi^2_{1-\alpha} = 1.646$
    calculations:
    $\chi^2 = (n-1)s^2/\sigma^2$
    $= 0.540$
    P-value $= P(\chi^2_8 < 0.540) = 0.0002$
    conclusion:
    Reject $H_0$; there is sufficient evidence to conclude that $\sigma < 29$.

    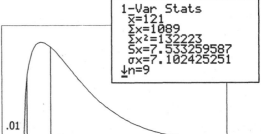

17. Place the 6 values in List L1.
    original claim: $\sigma > 0.4$ µg/m³
    $H_0$: $\sigma = 0.4$ µg/m³
    $H_1$: $\sigma > 0.4$ µg/m³
    $\alpha = 0.05$ and df = 5
    C.V. $\chi^2 = \chi^2_{\alpha} = 11.071$
    calculations:
    $\chi^2 = (n-1)s^2/\sigma^2$
    $= 114.506$
    P-value $= P(\chi^2_5 > 114.506) \approx 0$
    conclusion:
    Reject $H_0$; there is sufficient evidence to conclude that $\sigma > 0.4$.
    Yes. Since 5 of the n=6 sample values are below the sample mean, the data do not appear to come from a population with a normal distribution. Considering that this test is non-robust, the methods of this section should not be used on these data.

19. Use list CQPST.
    original claim: σ = 0.068 g
    $H_0$: σ = 0.068 g
    $H_1$: σ ≠ 0.068 g
    α = 0.05 and df = 39
    C.V. $\chi^2 = \chi^2_{1-\alpha/2} = 24.433$
    $\chi^2 = \chi^2_{\alpha/2} = 59.342$
    calculations:
    $\chi^2 = (n-1)s^2/\sigma^2$
        = 32.355
    P-value = $2 \cdot P(\chi^2_{39} < 32.359) = 0.4695$

    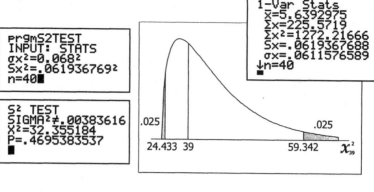

    conclusion:
        Do not reject $H_0$; there is not sufficient evidence to reject the claim that σ = 0.068.
        Yes. The variation of the weights appears to be as desired.

21. lower $\chi^2 = \frac{1}{2}(-z_{\alpha/2} + \sqrt{2 \cdot df - 1})^2$
        = $\frac{1}{2}(-1.96 + \sqrt{2 \cdot (189) - 1})^2$
        = $\frac{1}{2}(-1.96 + \sqrt{377})^2$
        = $\frac{1}{2}(304.7290)$
        = 152.3645, which is close to the STATDISK value of 152.8222
    upper $\chi^2 = \frac{1}{2}(z_{\alpha/2} + \sqrt{2 \cdot df - 1})^2$
        = $\frac{1}{2}(1.96 + \sqrt{2 \cdot (189) - 1})^2$
        = $\frac{1}{2}(1.96 + \sqrt{377})^2$
        = $\frac{1}{2}(456.9542)$
        = 228.4771, which is close to the STATDISK value of 228.9638

23. Yes, although the effect may not be "dramatic" if the sample is extremely large. Because this section tests the "spread" of the scores, and by definition an outlier is one that is "spread" away from the others, an outlier directly affects the property being measured and tested.

**Statistical Literacy and Critical Thinking**

1. Yes; the mean weight loss is statistically significant. No; the mean weight loss does not suggest that this diet is practical. No; the diet should not be recommended for weight loss. NOTE: This illustrates the difference between statistical significance and practical significance. In one sense, every null hypothesis is false. Consider, for example, the claim that μ = 50. No matter what the story problem, it is unlikely that the true population mean is <u>exactly</u> 50 (i.e., to 100 decimal places) – and the true mean in the population might be, for example, 50.000013. With a large enough sample that difference can be discovered and declared statistically significant – even though such a difference is likely of no practical consequence.

    Mathematically in a test about a population mean, $t = (\bar{x} - \mu)/(s/\sqrt{n}) = \sqrt{n} \cdot [(\bar{x} - \mu)/s]$ and increasing n by a factor of 100, for example, multiplies the t by 10. The same $\bar{x}$ and s that give an insignificant t = 1.00 for a particular value of n give a significant t = 10.00 for a sample 100 times as large. In general, taking a large enough sample can make any result statistically significant. This is the phenomenon behind the given weight loss scenario.

2. The data gathered applies only to those who had access to the survey and cared enough to respond. The population is precisely those who responded. The "sample" is not representative of any larger group. Since the data comes from a voluntary-response sample it cannot be used to make inferences using the techniques presented in the text.

3. The smaller the P-value, the less likely it is that the observed results occurred by chance alone. Prefer the smallest given P-value of 0.001.

4. The null and alternative hypotheses are as follows.
$H_o: \mu = 12$
$H_1: \mu > 12$
$\alpha = 0.05$ [assumed]
Since P-value = 0.250 > 0.05, the correct conclusion is as follows.
Do not reject $H_o$; there is not sufficient evidence to conclude that $\mu > 12$.
The P-value indicates that if the null hypothesis is true and the population mean is really 12, one expects 25% of the time to get a sample mean as large as or larger than the one obtained.

**Review Exercises**

1. a. $H_1: \mu < 10{,}000$
sampling distribution: t, with df = 749
   b. $H_1: p > 0.50$
sampling distribution: z
   c. $H_1: \mu \neq 100$
sampling distribution: z
   d. $H_1: \sigma > 1.8$
sampling distribution: none of the methods in this chapter apply
   e. $H_1: \sigma < 15$
sampling distribution: $\chi^2$, with df = 23

2. NOTE: The value for x is not given, but x = 66 is the only value that rounds to $\hat{p} = 44\%$.
original claim: $p < 0.50$
$\hat{p} = 66/150 = 0.44$

$H_o: p = 0.50$
$H_1: p < 0.50$
$\alpha = 0.05$
C.V. $z = -z_\alpha = -1.645$
calculations:
$z_{\hat{p}} = (\hat{p} - \mu_{\hat{p}})/\sigma_{\hat{p}}$
$= -1.470$
P-value = P(z<-1.470) = 0.0708

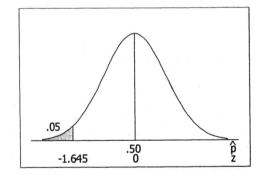

conclusion:
Do not reject $H_o$; there is not sufficient evidence to conclude that $p < 0.50$.

152 CHAPTER 8 Hypothesis Testing

3. NOTE: The value for x is not given. In truth, any $1488 \le x \le 1512$ rounds to the given $\hat{p} = x/2500 = 60\%$. The manual follows the advice in the text and uses
$x = (0.60)(2500) = 1500$.
original claim: $p < 0.62$
$H_o$: $p = 0.62$
$H_1$: $p < 0.62$
$\alpha = 0.01$
C.V. $z = -z_\alpha = -2.326$
calculations:
$z_{\hat{p}} = (\hat{p} - \mu_{\hat{p}})/\sigma_{\hat{p}}$
$= -2.060$
P-value $= P(z < -2.060) = 0.0197$

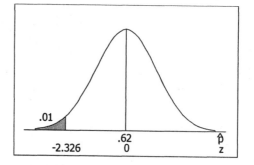

conclusion:
Do not reject $H_o$; there is not sufficient evidence to conclude that $p < 0.62$.
If the results were obtained from a voluntary-response survey, then they cannot be used with the methods of this chapter to make inferences.

4. original claim: $\mu = 3.5$ g
$H_o$: $\mu = 3.5$ g
$H_1$: $\mu \ne 3.5$ g
$\alpha = 0.05$ [assumed]
and df = 69
C.V. $t = \pm t_{\alpha/2} = \pm 1.994$
calculations:
$t_{\bar{x}} = (\bar{x} - \mu)/s_{\bar{x}}$
$= 9.723$
P-value $= 2 \cdot P(t_{69} > 9.723) = 1.46\text{E-}14 \approx 0$

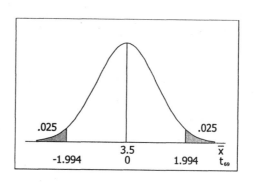

conclusion:
Reject $H_o$; there is sufficient evidence to reject the claim that $\mu = 3.5$ and conclude that $\mu \ne 3.5$ (in fact, that $\mu > 3.5$).
No. The extra 0.086 grams amounts to $0.086/3.5 = 2.5\%$ more than the stated amount – which is not large enough to create a health hazard for unsuspecting consumers.

5. original claim: $\mu < 12$ oz
$H_o$: $\mu = 12$ oz
$H_1$: $\mu < 12$ oz
$\alpha = 0.05$ [assumed]
and df = 23
C.V. $t = -t_\alpha = -1.714$
calculations:
$t_{\bar{x}} = (\bar{x} - \mu)/s_{\bar{x}}$
$= -4.741$
P-value $= P(t_{23} < -4.741) = 4.44\text{E-}5 = 0.00004$
conclusion:
Reject $H_o$; there is sufficient evidence to conclude that $\mu < 12$.
No. The claim that the sample is too small has no validity. The test incorporates the sample size into the decision-making by requiring a larger test statistic t for the smaller degrees of freedom associated with the smaller values of n.

6. original claim: $p > 0.019$
   $\hat{p} = x/n = 19/863 = 0.022$
   $H_0$: $p = 0.019$
   $H_1$: $p > 0.019$
   $\alpha = 0.01$
   C.V. $z = z_\alpha = 2.326$
   calculations:
   $z\hat{p} = (\hat{p} - \mu\hat{p})/\sigma\hat{p}$
   $= 0.649$
   P-value = $P(z > 0.649) = 0.2582$

   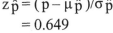

   conclusion:
   Do not reject $H_0$; there is not sufficient evidence to conclude that $p > 0.019$.
   No. There is not sufficient evidence to say flu symptoms are an adverse treatment reaction.

7. original claim: $\sigma = 15$
   $H_0$: $\sigma = 15$
   $H_1$: $\sigma \neq 15$
   $\alpha = 0.05$ and df = 12
   C.V. $\chi^2 = \chi^2_{1-\alpha/2} = 4.404$
   $\chi^2 = \chi^2_{\alpha/2} = 23.337$
   calculations:
   $\chi^2 = (n-1)s^2/\sigma^2$
   $= 2.7648$
   P-value = $2 \cdot P(\chi^2_{12} < 2.7648) = 0.0060$

   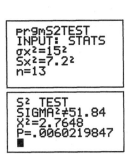
   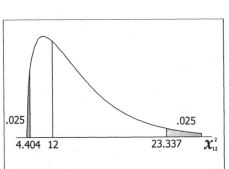

   conclusion:
   Reject $H_0$; there is sufficient evidence to reject the claim that $\sigma = 15$ and conclude that $\sigma \neq 15$ (in fact, that $\sigma > 15$).
   Based on this result, we conclude that the standard deviation of IQ scores for statistics professors is less than that of the general population.

8. original claim: $\mu = 100$
   $H_0$: $\mu = 100$
   $H_1$: $\mu \neq 100$
   $\alpha = 0.10$
   C.V. $z = \pm z_{\alpha/2} = \pm 1.645$
   calculations:
   $z_{\bar{x}} = (\bar{x} - \mu_{\bar{x}})/\sigma_{\bar{x}}$
   $= -0.754$
   P-value = $2 \cdot P(z < -0.754) = 0.4507$

   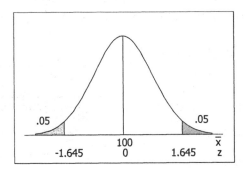

   conclusion:
   Do not reject $H_0$; there is not sufficient evidence to reject the claim that $\mu = 100$.
   Based on these results, it appears that the random number generator is working correctly.

154   CHAPTER 8   Hypothesis Testing

9. Place the 12 values in list L1.
   original claim: $\mu < 98.6$ °F
   $H_0$: $\mu = 98.6$ °F
   $H_1$: $\mu < 98.6$ °F
   $\alpha = 0.05$ and df = 11
   C.V. $t = -t_\alpha = -1.796$
   calculations:
   $t_{\bar{x}} = (\bar{x} - \mu)/s_{\bar{x}}$
   $= -1.349$
   P-value = $P(t_{11} < -1.349) = 0.1023$
   conclusion:
   Do not reject $H_0$; there is not sufficient evidence to conclude that $\mu < 98.6$.

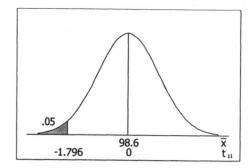

10. original claim: $\sigma = 2.52$ in
    $H_0$: $\sigma = 2.52$ in
    $H_1$: $\sigma \neq 2.52$ in
    $\alpha = 0.05$ and df = 39
    C.V. $\chi^2 = \chi^2_{1-\alpha/2} = 24.433$
    $\chi^2 = \chi^2_{\alpha/2} = 59.342$
    calculations:
    $\chi^2 = (n-1)s^2/\sigma^2$
    $= 46.107$
    P-value = $2 \cdot P(\chi^2_{39} > 46.107) = 0.4038$
    conclusion:
    Do not reject $H_0$; there is not sufficient evidence to reject the claim that $\sigma = 2.52$. Equipment is generally designed to accommodate all but a certain extreme percentage (which depends on the particular equipment) of the population. If the population has more variability than is planned for, then the portion of the population not accommodated by the equipment will be larger than proposed.

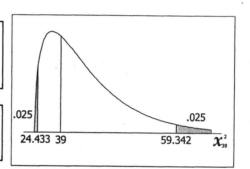

11. original claim: $p > 0.610$
    $\hat{p} = x/n = 2231/3581 = 0.623$
    $H_0$: $p = 0.610$
    $H_1$: $p > 0.610$
    $\alpha = 0.05$
    C.V. $z = z_\alpha = 1.645$
    calculations:
    $z_{\hat{p}} = (\hat{p} - \mu_{\hat{p}})/\sigma_{\hat{p}}$
    $= 1.596$
    P-value = $P(z > 1.596) = 0.0552$
    conclusion:
    Do not reject $H_0$; there is not sufficient evidence to conclude that $p > 0.610$.

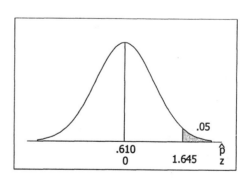

12. Place the 10 values in list L1.
 original claim: μ > 40 mg
 $H_0$: μ = 40 mg
 $H_1$: μ > 40 mg
 α = 0.01 and df = 9
 C.V. t = $t_α$ = 2.821
 calculations:
 $t_{\bar{x}} = (\bar{x} - μ)/s_{\bar{x}}$
 = 2.746
 P-value = P($t_9$ > 2.746) = 0.0113
 conclusion:
  Do not reject $H_0$; there is not sufficient evidence to conclude that μ > 40.

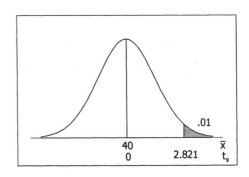

## Cumulative Review Exercises

In the spirit of a true review, these exercises are worked by hand. To use the TI-83/84 Plus, enter the 17 values into list L1 and use the appropriate commands. The display from the 1-Var Stats is given below, and other display screens and TI answers are given as relevant.

1. scores in order: 67 69 69.5 70 70 70.5 71 71 71.5 72 72 72 73 74 74 74.5 75
 summary statistics: n = 17, Σx = 1216.0, Σx² = 87053.00

 a. $\bar{x}$ = (Σx)/n
   = (1216.0)/17 = 71.53 in

 b. $\tilde{x}$ = 71.5 in

 c. $s^2 = [n(Σx^2) - (Σx)^2]/[n(n-1)]$
   $= [17(87053.00) - (1216.0)^2]/[17(16)] = 4.577$
  s = 2.139, rounded to 2.1 in

 d. $s^2$ = 4.577, rounded to 4.6 in²

 e. R = 75 – 67 = 8 in

 f. σ unknown and n ≤ 30: assuming the data come from a normal population, use t
   α = 0.05 and df = 16
   $\bar{x} ± t_{α/2} \cdot s/\sqrt{n}$
   71.53 ± 2.120·(2.139)/√17
   71.53 ± 1.10
   70.43 < μ < 72.63 (inches)

g. original claim: $\mu > 69.0$ in
   $H_0$: $\mu = 69$ in
   $H_1$: $\mu > 69$ in
   $\alpha = 0.05$ and df = 16
   C.V. $t = t_\alpha = 1.746$
   calculations:

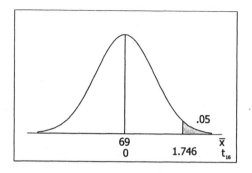

$$t_{\bar{x}} = (\bar{x} - \mu)/s_{\bar{x}}$$
$$= (71.53 - 69)/(2.139/\sqrt{17})$$
$$= 2.53/0.5189 = 4.875$$

P-value = $P(t_{16} > 4.875)$ = tcdf(4.875,99,16) = 0.0001 [TI: 8.43E-5 = .00008]

conclusion:
   Reject $H_0$; there is sufficient evidence to conclude that $\mu > 69$.
   Yes. Presidents appear to be taller than the typical man.

2. The normal quantile plot is given below, to the right of the columns used in the calculations.
   As in section 6-7, x = the original scores arranged in order
   cp = the cumulative probability values 1/2n, 3/2n, 5/2n,…,(2n-1)/2n
   z = the normal z scores that have cp to their left

| x | cp | z |
|---|---|---|
| 67 | 0.029 | -1.89 |
| 69 | 0.088 | -1.35 |
| 69.5 | 0.147 | -1.05 |
| 70 | 0.206 | -0.82 |
| 70 | 0.265 | -0.63 |
| 70.5 | 0.324 | -0.46 |
| 71 | 0.382 | -0.30 |
| 71 | 0.441 | -0.15 |
| 71.5 | 0.500 | 0.00 |
| 72 | 0.559 | 0.15 |
| 72 | 0.618 | 0.30 |
| 72 | 0.676 | 0.46 |
| 73 | 0.735 | 0.63 |
| 74 | 0.794 | 0.82 |
| 74 | 0.853 | 1.05 |
| 74.5 | 0.912 | 1.35 |
| 75 | 0.971 | 1.89 |

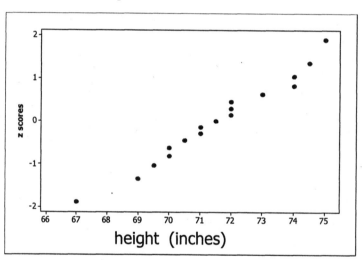

Yes. Based on the normal quantile plot, which approximates a straight line, the heights appear to come from a normal distribution. Using the TI-83/4 Plus, and STATPLOT

followed by ZOOM 9,

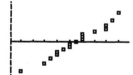

3. a. normal distribution
   $\mu = 496$
   $\sigma = 108$
   $P(x>500)$
   $= 1 - P(x<500)$
   $= 1 - P(z<0.04)$
   $= 1 - 0.5160$
   $= 0.4840$
   TI: normalcdf(500,9999,496,108)
   $= 0.4852$

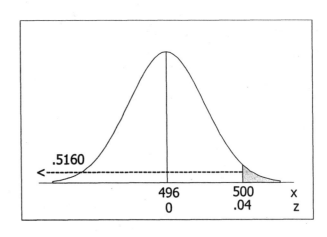

b. let A = a selected score is above 500
   $P(A) = 0.4852$, for each selection
   $P(A_1$ and $A_2$ and $A_3$ and $A_4$ and $A_5)$
   $= P(A_1) \cdot P(A_2) \cdot P(A_3) \cdot (A_4) \cdot P(A_5)$
   $= (0.4852)^5$
   $= 0.0269$

c. normal distribution, since the original distribution is so
   $\mu_{\bar{x}} = \mu = 496$
   $\sigma_{\bar{x}} = \sigma/\sqrt{n} = 108/\sqrt{5} = 48.39$
   $P(\bar{x}>500)$
   $= 1 - P(\bar{x}<500)$
   $= 1 - P(z<0.08)$
   $= 1 - 0.5319$
   $= 0.4681$
   TI: normalcdf(500,9999,496,108/√5)
   $= 0.4670$

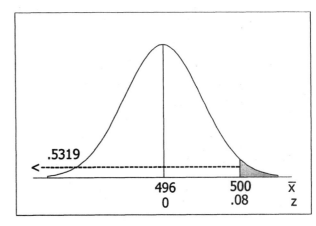

d. For $P_{90}$, A = 0.9000 [0.8997] and z = 1.28. [or 1.282 from the "large" df row of the t table]
   $x = \mu + z \cdot \sigma$
   $= 496 + (1.282)(108)$
   $= 496 + 138$
   $= 634$
   TI: 496 + (invNorm(0.90))×108 = 634

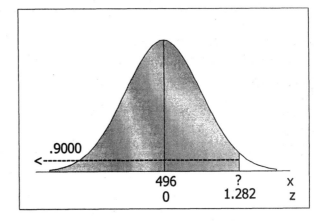

# Chapter 9

# Inferences from Two Samples

## 9-2 Inferences About Two Proportions

1. The pooled proportion for two samples is given by the symbol $\bar{p}$. It is the total number of successes from both samples divided by the total number of trials from both samples. It is literally the result of conceptually pooling both samples into one large sample.

3. We are 95% confident that the interval from 0.200 to 0.300 contains the true difference $p_1 - p_2$, where $p_1$ and $p_2$ are the true proportions in two populations.

5. $x = (0.23)(3250) = 747.50 \approx 748$     [NOTE: Any $712 \leq x \leq 763$ rounds to $x/3250 = 23\%$.]

7. $x = (0.07)(976) = 68.32 \approx 68$     [NOTE: Any $64 \leq x \leq 73$ rounds to $x/976 = 7\%$.]

9. At the right are the set-up and result screens for STAT TESTS 2-PropZTest. The exercises is worked below by hand, but the values may be read directly from the TI-83/84 Plus screen displays.

   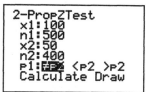

   $\hat{p}_1 = x_1/n_1 = 100/500 = 0.200$     $\hat{p}_1 - \hat{p}_2 = 0.200 - 0.125 = 0.075$
   $\hat{p}_2 = x_1/n_1 = 50/400 = 0.125$

   a. $\bar{p} = (x_1+x_2)/(n_1+n_2) = (100+50)/(500+400) = 150/900 = 0.167$
   b. $z_{\hat{p}_1-\hat{p}_2} = (\hat{p}_1-\hat{p}_2 - \mu_{\hat{p}_1-\hat{p}_2})/\sigma_{\hat{p}_1-\hat{p}_2}$
      $= (0.075 - 0)/\sqrt{(0.167)(0.833)/500 + (0.167)(0.833)/400} = 0.075/0.025 = 3.00$
   c. for $\alpha = 0.05$, the critical values are $z = \pm z_{\alpha/2} = \pm 1.96$
   d. P-value $= 2 \cdot P(z>3.00) = 2(0.0013) = 0.0026$     [TI: 0.0027]

11. Let the basketball games be group 1.
    original claim: $p_1-p_2 = 0$
    $H_o$: $p_1-p_2 = 0$
    $H_1$: $p_1-p_2 \neq 0$
    $\alpha = 0.05$ [assumed]
    C.V. $z = \pm z_{\alpha/2} = \pm 1.96$
    calculations:
    $z_{\hat{p}_1-\hat{p}_2} = (\hat{p}_1-\hat{p}_2 - \mu_{\hat{p}_1-\hat{p}_2})/\sigma_{\hat{p}_1-\hat{p}_2}$
    $= 1.099$
    P-value = 0.2719

    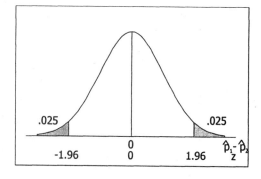

    conclusion:
       Do not reject $H_o$; there is not sufficient evidence to reject the claim that $p_1-p_2 = 0$.
    No. There does not appear to be a significant difference between the proportions of home wins for basketball and football. Since $\bar{p} = 0.6195$, the home field advantage seems to be real.

13. Let the XSORT babies be group 1.
   a. $0.0224 < p_1 - p_2 < 0.264$, as shown by the following set-up and result screens

   ```
   2-PropZInt       2-PropZInt
    x1:295           (.02239,.26358)
    n1:325           p̂1=.9076923077
    x2:39            p̂2=.7647058824
    n2:51            n1=325
    C-Level:.95      n2=51
    Calculate
   ```

   b. Yes. Since the confidence interval does not include 0, there does appear to be a difference. While both the XSORT and YSORT methods appear to be effective, the XSORT method appears to be more effective than the YSORT method.

15. Let those contacted for the standard survey be group 1.

    $x_1$: 720   $n_1$: 1720   $x_2$: 429   $n_2$: 1640

    original claim: $p_1 - p_2 > 0$
    $H_o$: $p_1 - p_2 = 0$
    $H_1$: $p_1 - p_2 > 0$
    $\alpha = 0.01$
    C.V. $z = z_\alpha = 2.326$
    calculations:

    $z_{\hat{p}_1 - \hat{p}_2} = (\hat{p}_1 - \hat{p}_2 - \mu_{\hat{p}_1 - \hat{p}_2})/\sigma_{\hat{p}_1 - \hat{p}_2}$
    $= 9.591$
    P-value = $P(z > 9.591) = 4.45\text{E-}22 \approx 0$
    conclusion:
       Reject $H_o$; there is sufficient evidence to conclude that $p_1 - p_2 > 0$.

    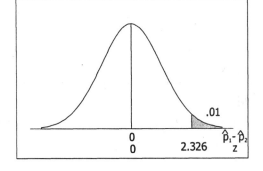

    Yes. Since the rigorous survey has a lower refusal weight, it is more likely to produce the better results.

17. Let the workers be group 1.

    $x_1$: 192   $n_1$: 436   $x_2$: 40   $n_2$: 121

    original claim: $p_1 - p_2 > 0$
    $H_o$: $p_1 - p_2 = 0$
    $H_1$: $p_1 - p_2 > 0$
    $\alpha = 0.05$
    C.V. $z = z_\alpha = 1.645$
    calculations:

    $z_{\hat{p}_1 - \hat{p}_2} = (\hat{p}_1 - \hat{p}_2 - \mu_{\hat{p}_1 - \hat{p}_2})/\sigma_{\hat{p}_1 - \hat{p}_2}$
    $= 2.167$
    P-value = $P(z > 2.167) = 0.0151$
    conclusion:
       Reject $H_o$; there is sufficient evidence to conclude that $p_1 - p_2 > 0$.

    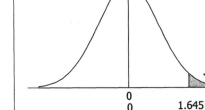

160   CHAPTER 9   Inferences from Two Samples

19. Let the employees of hospitals with the smoking ban be group 1.
   $x_1$: 56   $n_1$: 843   $x_2$: 27   $n_2$: 703
   original claim: $p_1-p_2 \neq 0$
   $H_0$: $p_1-p_2 = 0$
   $H_1$: $p_1-p_2 \neq 0$
   $\alpha = 0.05$
   C.V. $z = \pm z_{\alpha/2} = \pm 1.96$
   calculations:
   
   $z_{\hat{p}_1-\hat{p}_2} = (\hat{p}_1-\hat{p}_2 - \mu_{\hat{p}_1-\hat{p}_2})/\sigma_{\hat{p}_1-\hat{p}_2}$
   $= 2.434$
   P-value = $2 \cdot P(z>2.434) = 0.0149$
   conclusion:

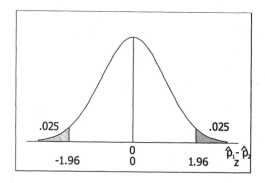

   Reject $H_0$; there is sufficient evidence to conclude that $p_1-p_2 \neq 0$ (in fact, that $p_1-p_2 > 0$). Since P-value = 0.0150 > 0.01, the difference is not significant at the 0.01 level.  Even though one cannot be 99% certain, it appears that the hospital employees with the smoking ban had a higher quit smoking rate – BUT, the difference in quit smoking rates could be a result of the working place (hospital vs. non-hospital) or the policy (ban vs. no-ban) or a combination.

NOTE: The information given for exercises #21, #22, #25, #26, #27, #28 and #35 is not sufficient to determine the exact values of $x_1$ and $x_2$. The manual follows the suggestion of the text and uses $x_i = p_i n_i$, rounded to the nearest integer. Be aware that this can seriously compromise the reported accuracy in the results.

21. Let the workers be group 1.
   $x_1$: (0.021)(1655) = 35   $n_1$: 1655   $x_2$: (0.012)(1652) = 20   $n_2$: 1652
   original claim: $p_1-p_2 > 0$
   $H_0$: $p_1-p_2 = 0$
   $H_1$: $p_1-p_2 > 0$
   $\alpha = 0.05$
   C.V. $z = z_\alpha = 1.645$
   calculations:
   
   $z_{\hat{p}_1-\hat{p}_2} = (\hat{p}_1-\hat{p}_2 - \mu_{\hat{p}_1-\hat{p}_2})/\sigma_{\hat{p}_1-\hat{p}_2}$
   $= 2.032$
   P-value = $P(z>2.032) = 0.0210$
   conclusion:

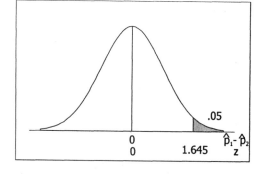

   Reject $H_0$; there is sufficient evidence to conclude that $p_1-p_2 > 0$.
   While those who use Clarinex are more likely to experience fatigue, a difference of 0.009 = 0.9% does not pose a "major concern."

23. Let the homeowners be group 1.
   $0.106 < p_1-p_2 < 0.179$, as shown from the following set-up and result screens
   ```
   2-PropZInt          2-PropZInt
   x1:880              (.10645,.17871)
   n1:1068             p1=.8239700375
   x2:725              p2=.6813909774
   n2:1064             n1=1068
   C-Level:.95         n2=1064
   Calculate
   ```
   Yes. Since the confidence interval does not include 0, there seems to be a significant difference between the proportions. One major factor might be income levels: homeowners are more likely to have the higher income necessary to maintain a vehicle for commuting –

Inferences About Two Proportions  SECTION 9-2  161

which often necessitates at least two vehicles in the household. Another major factor might be the availability of public transportation: renters are more likely to live in the more densely populated areas typically served by public transportation – homeowners in suburban/rural areas might not have convenient access to public transportation even if they wanted to use it.

25. Let the males be group 1.
   $x_1$: $(0.72)(2200) = 1584$   $n_1$: 2200   $x_2$: $(0.84)(2380) = 1999$   $n_2$: 2380
   original claim: $p_1-p_2 = 0$
   $H_o$: $p_1-p_2 = 0$
   $H_1$: $p_1-p_2 \neq 0$
   $\alpha = 0.05$
   C.V. $z = \pm z_{\alpha/2} = \pm 1.96$
   calculations:

   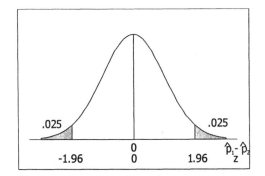

   $z_{\hat{p}_1-\hat{p}_2} = (\hat{p}_1-\hat{p}_2 - \mu_{\hat{p}_1-\hat{p}_2})/\sigma_{\hat{p}_1-\hat{p}_2}$
   $= -9.825$
   P-value = $2 \cdot P(z<-9.825) = 8.94\text{E}-23 \approx 0$
   conclusion:
   Reject $H_o$; there is sufficient evidence to reject the claim that $p_1-p_2 = 0$ and conclude that $p_1-p_2 \neq 0$ (in fact, that $p_1-p_2 < 0$).
   Yes. There does appear to be a gender gap in that males have a lower seat belt usage rate.

27. Let the single women be group 1.
   $x_1$: $(0.24)(205) = 49$   $n_1$: 205   $x_2$: $(0.27)(260) = 70$   $n_2$: 260
   $-0.135 < p_1-p_2 < 0.0742$, as shown from the following set-up and result screens

   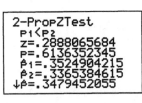

   No. Since the confidence interval contains 0, there does not appear to be a gender gap on this issue.

29. Combine the 261 Mon-Fri values into one list and the 104 Sat-Sun values into another list. Then sort the two resulting columns in descending order to quickly count the numbers of days with precipitation. For the TI-83/84 Plus, use
   LIST  OPS  augment(RNMON,RNTUE,RNWED,RNTHU,RNFRI)→L1  STAT  SortD(L1)
   LIST  OPS  augment(RNSAT,RNSUN)→L2  STAT  SortD(L2)
   STAT  EDIT
   Let the weekdays be group 1.
   $x_1$: 92   $n_1$: 261   $x_2$: 35   $n_2$: 104
   original claim: $p_1-p_2 < 0$
   $H_o$: $p_1-p_2 = 0$
   $H_1$: $p_1-p_2 < 0$
   $\alpha = 0.05$ [assumed]
   C.V. $z = -z_\alpha = -1.645$
   calculations:

   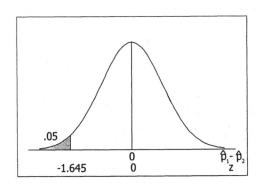

   $z_{\hat{p}_1-\hat{p}_2} = (\hat{p}_1-\hat{p}_2 - \mu_{\hat{p}_1-\hat{p}_2})/\sigma_{\hat{p}_1-\hat{p}_2}$
   $= 0.289$
   P-value = $P(z<0.289) = 0.6136$

162    CHAPTER 9    Inferences from Two Samples

conclusion:
  Do not reject $H_o$; there is not sufficient evidence to conclude that $p_1-p_2 < 0$.
  No. The Boston data do not support the claim that is rains more often on weekends.

31. Let the movies tested for alcohol be group 1.
    $x_1$: 25   $n_1$: 50   $x_2$: 28   $n_2$: 50
    original claim: $p_1-p_2 < 0$
    $H_o$: $p_1-p_2 = 0$
    $H_1$: $p_1-p_2 < 0$
    $\alpha = 0.05$ [assumed]
    C.V. $z = -z_\alpha = -1.645$
    calculations:

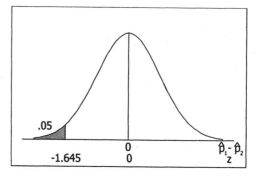

    $z_{\hat{p}_1-\hat{p}_2} = (\hat{p}_1-\hat{p}_2 - \mu_{\hat{p}_1-\hat{p}_2})/\sigma_{\hat{p}_1-\hat{p}_2}$
        $= -0.601$
    P-value = $P(z<-0.601) = 0.2739$
conclusion:
  Do not reject $H_o$; there is not sufficient evidence to conclude that $p_1-p_2 < 0$.
  No. One requirement for this section is that the two samples be independent. The data in Data Set 5 are not from independent samples – but from the same sample. This is analogous to testing whether the proportion of people who declare bankruptcy is different from the proportion of people who have more than $10,000 credit card debt – and using the same 50 people.

33. For all parts of this exercise,
    $x_1$: 112   $n_1$: 200   $x_2$: 88   $n_2$: 200

    a. $0.0227 < p_1-p_2 < 0.217$, as shown from the following set-up and result screens
    ```
    2-PropZInt       2-PropZInt
    x1:112           (.02271,.21729)
    n1:200           p1=.56
    x2:88            p2=.44
    n2:200           n1=200
    C-Level:.95      n2=200
    Calculate
    ```
    Since the interval does not include 0, conclude that $p_1$ and $p_2$ are different.
    Since the interval lies entirely above 0, conclude that $p_1 - p_2 > 0$.

    b. for group 1                            for group 2
       $\hat{p} \pm z_{\alpha/2} \sqrt{\hat{p}\hat{q}/n}$                    $\hat{p} \pm z_{\alpha/2} \sqrt{\hat{p}\hat{q}/n}$
       $0.560 \pm 1.96 \sqrt{(0.56)(0.44)/200}$     $0.440 \pm 1.96 \sqrt{(0.44)(0.56)/200}$
       $0.560 \pm 0.069$                      $0.440 \pm 0.069$
       $0.491 < p < 0.629$                    $0.371 < p < 0.509$
    The corresponding TI-83/84 Plus display screens are as follows.
    ```
    1-PropZInt            1-PropZInt
    (.37121,.50879)       (.49121,.62879)
    p=.44                 p=.56
    n=200                 n=200
    ```

    Since the intervals overlap, the implication is that $p_1$ and $p_2$ could have the same value.

c. original claim: $p_1-p_2 = 0$
   $H_0$: $p_1-p_2 = 0$
   $H_1$: $p_1-p_2 \neq 0$
   $\alpha = 0.05$
   C.V. $z = \pm z_\alpha = \pm 1.96$
   calculations:

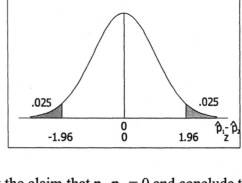

$$z_{\hat{p}_1-\hat{p}_2} = (\hat{p}_1-\hat{p}_2 - \mu_{\hat{p}_1-\hat{p}_2})/\sigma_{\hat{p}_1-\hat{p}_2}$$
$$= 2.400$$
P-value = $2 \cdot P(z>2.400) = 0.0164$
conclusion:
   Reject $H_0$; there is sufficient evidence to reject the claim that $p_1-p_2 = 0$ and conclude that $p_1-p_2 \neq 0$ (in fact, that $p_1-p_2 > 0$)

d. Based on the preceding results, conclude that $p_1$ and $p_2$ are unequal and that $p_1 > p_2$. The overlapping interval method of part (b) appears to be the least effective method for comparing two populations.

35. Let the vinyl gloves be group 1.
    $\hat{p}_1 = x_1/n_1 = x_1/240 = 0.63$  [as any $151 \leq x_1 \leq 152$ gives $\hat{p}_1 = 0.63$, use 0.63 in all calculations]
    $\hat{p}_2 = x_2/n_2 = x_2/240 = 0.07$  [as any $16 \leq x_2 \leq 17$ gives $\hat{p}_2 = 0.07$, use 0.07 in all calculations]
    $\hat{p}_1 - \hat{p}_2 = 0.63 - 0.07$
    $\quad\quad = 0.56$
    $\bar{p} = [(0.63)(240) + (0.07)(240)]/[240 + 240]$
    $\quad = 168/480$
    $\quad = 0.35$
    original claim: $p_1-p_2 = 0.50$
    $H_0$: $p_1-p_2 = 0.50$
    $H_1$: $p_1-p_2 \neq 0.50$
    $\alpha = 0.05$
    C.V. $z = \pm z_{\alpha/2} = \pm 1.96$
    calculations:

$$z_{\hat{p}_1-\hat{p}_2} = (\hat{p}_1-\hat{p}_2 - \mu_{\hat{p}_1-\hat{p}_2})/\sigma_{\hat{p}_1-\hat{p}_2}$$
$$= (0.56 - 0.50)/\sqrt{(0.63)(0.37)/240 + (0.07)(0.93)/240}$$
$$= 0.06/0.0352$$
$$= 1.702$$
   [See the accuracy comments at the beginning of the exercise.]
   P-value = $2 \cdot P(z>1.702) = 2 \times \text{normalcdf}(1.702, 999) = 0.0888$
conclusion:
   Do not reject $H_0$; there is not sufficient evidence to reject the claim that $p_1-p_2 = 0.50$.

37. a.

| # | sample | $\hat{p}$ | $\hat{p}-\mu_{\hat{p}}$ | $(\hat{p}-\mu_{\hat{p}})^2$ |
|---|---|---|---|---|
| 1 | HH | 1.0 | 0.5 | 0.25 |
| 2 | HT | 0.5 | 0.0 | 0.00 |
| 3 | TH | 0.5 | 0.0 | 0.00 |
| 4 | TT | 0.0 | -0.5 | 0.25 |
|   |    | 2.0 | 0.0 | 0.50 |

$\mu_{\hat{p}} = (\Sigma \hat{p})/N = 2.0/5 = 0.5$

$\sigma_{\hat{p}}^2 = \Sigma(\hat{p}-\mu_{\hat{p}})^2/N = 0.50/4 = 0.125$

b. The 16 possible $\hat{p}_D - \hat{p}_Q$ values are given at the right, identified by the sample #'s used to generate them.

$\mu_{D-Q} = \Sigma(\hat{p}_D - \hat{p}_Q)/N$
$= 0/16 = 0$

$\sigma_{D-Q}^2 = \Sigma[(\hat{p}_D - \hat{p}_Q) - \mu]^2/N$
$= \Sigma[(\hat{p}_D - \hat{p}_Q) - 0]^2/N$
$= \Sigma(\hat{p}_D - \hat{p}_Q)^2/N$
$= 4.00/16 = 0.25$

| sample | $\hat{p}_D - \hat{p}_Q$ | $(\hat{p}_D - \hat{p}_Q)^2$ |
|---|---|---|
| 1-1 | 0 | 0 |
| 1-2 | 0.5 | 0.25 |
| 1-3 | 0.5 | 0.25 |
| 1-4 | 1.0 | 1.00 |
| 2-1 | -0.5 | 0.25 |
| 2-2 | 0 | 0 |
| 2-3 | 0 | 0 |
| 2-4 | 0.5 | 0.25 |
| 3-1 | -0.5 | 0.25 |
| 3-2 | 0 | 0 |
| 3-3 | 0 | 0 |
| 3-4 | 0.5 | 0.25 |
| 4-1 | -1.0 | 1.00 |
| 4-2 | -0.5 | 0.25 |
| 4-3 | -0.5 | 0.25 |
| 4-4 | 0 | 0 |
|     | 0 | 4.00 |

c. $\sigma_{D-Q}^2 = \sigma_D^2 + \sigma_Q^2$
$= 0.125 + 0.125$
$= 0.25$

## 9-3 Inferences About Two Means: Independent Samples

1. The requirements of this section are not satisfied. The age of 80 would be an outlier for the sample data and suggest that the original population is not normal, but positively skewed.

3. The critical value t = 1.717 is "more conservative" than t = 1.682 because it requires a larger difference between the two means before rejecting the hypothesis that they are equal.
NOTE: Formula 9-1 yields a df value between $\min(df_1, df_2)$ and $(df_1 + df_2)$. The smaller the df, the more variability in the t distribution, and the larger (i.e., the farther from 0) the t score for any level of significance. A larger critical t score requires a greater deviation from the null hypothesis before rejecting the null hypothesis – making the test more conservative. Choosing $df = \min(df_1, df_2)$ selects the smallest reasonable df and uses the most conservative standard.

5. Independent samples, since there is no connection between the two groups – i.e., the two groups are selected and analyzed separately.

7. Matched pairs, since each pair of before and after weights is gathered from the same subject.

NOTE: To be consistent with the previous notation, reinforcing patterns and concepts presented in those sections, the manual uses the usual t formula written to apply to $\bar{x}_1 - \bar{x}_2$'s

$t_{\bar{x}_1 - \bar{x}_2} = (\bar{X}_1 - \bar{X}_2 - \mu_{\bar{x}_1 - \bar{x}_2})/s_{\bar{x}_1 - \bar{x}_2}$, with $\mu_{\bar{x}_1 - \bar{x}_2} = \mu_1 - \mu_2$ and $s_{\bar{x}_1 - \bar{x}_2} = \sqrt{s_1^2/n_1 + s_2^2/n_2}$.

## Inferences About Two Means: Independent Samples  SECTION 9-3

9. [by hand] Let the children given the echinacea treatment be group 1.
   $\bar{x}_1 - \bar{x}_2 = 0.81 - 0.64 = 0.17$
   original claim: $\mu_1 - \mu_2 \neq 0$ days
   $H_o$: $\mu_1 - \mu_2 = 0$ days
   $H_1$: $\mu_1 - \mu_2 \neq 0$ days
   $\alpha = 0.05$ and df = 336
   C.V. $t = \pm t_{\alpha/2} = \pm 1.968$
   calculations:
   $t_{\bar{x}_1-\bar{x}_2} = (\bar{x}_1-\bar{x}_2 - \mu_{\bar{x}_1-\bar{x}_2})/s_{\bar{x}_1-\bar{x}_2}$
   $= (0.17 - 0)/\sqrt{(1.50)^2/337 + (1.16)^2/370}$
   $= 0.17/0.1016 = 1.674$
   P-value = $2 \cdot \text{tcdf}(1.674,99,336) = 0.0951$
   conclusion:
   Do not reject $H_o$; there is not sufficient evidence to conclude that $\mu_1 - \mu_2 \neq 0$.
   No. Based on these results, echinacea does not appear to have an effect.

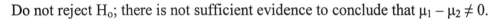

NOTE: Exercise #9 above was worked by hand. This NOTE comments on using the TI-83/84 Plus calculator and compares the "by hand" and "by calculator" results. Following the NOTE, exercise #9 is re-worked in the format used in this manual.
   Below are the two set-up screens and two result screens for using STAT TESTS 2-SampTTest.

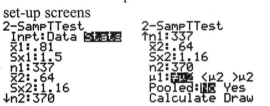

The calculated t = 1.674 agrees with the hand calculations, but the P-value = .0946 does not. This occurs because the hand calculations use df = min(df$_1$,df$_2$) = 336, while the calculator uses a more sophisticated df = function(df$_1$,df$_2$) = 631.2. In this section, this manual typically displays the first of the TI-83/Plus result screens and uses
(1) df = min(df$_1$,df$_2$), the conservative method given in the text, and the closest entry in Table A-3 to determine the critical value(s). [One could also determine the critical value(s) using the more sophisticated df determined by the calculator and/or the invT function.]
(2) the calculated t statistic and P-value from the calculator.

9. [by calculator] Let the children given the echinacea treatment be group 1.
   original claim: $\mu_1 - \mu_2 \neq 0$ days
   $H_o$: $\mu_1 - \mu_2 = 0$ days
   $H_1$: $\mu_1 - \mu_2 \neq 0$ days
   $\alpha = 0.05$ and df = 336
   C.V. $t = \pm t_{\alpha/2} = \pm 1.968$
   calculations:
   $t_{\bar{x}_1-\bar{x}_2} = (\bar{x}_1-\bar{x}_2 - \mu_{\bar{x}_1-\bar{x}_2})/s_{\bar{x}_1-\bar{x}_2}$
   $= 1.674$
   P-value = $2 \cdot P(t_{631.2} > 1.674) = 0.0946$

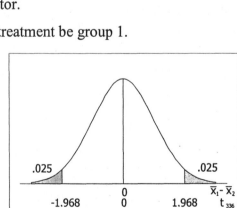

   conclusion:
   Do not reject $H_o$; there is not sufficient evidence to conclude that $\mu_1 - \mu_2 \neq 0$.
   No. Based on these results, echinacea does not appear to have an effect.

**166   CHAPTER 9  Inferences from Two Samples**

11. Let the children with low birth weights be group 1.
    $-12.1 < \mu_1 - \mu_2 < -6.7$, as shown from the following set-up and result screens

    Yes. Since the confidence interval is entirely below 0, it appears that children with extremely low birth weights do have lower IQ's at age 8.

13. Let the light users be group 1.
    original claim: $\mu_1 - \mu_2 > 0$ items
    $H_o$: $\mu_1 - \mu_2 = 0$ items
    $H_1$: $\mu_1 - \mu_2 > 0$ items
    $\alpha = 0.01$ and df = 63
    C.V. $t = t_\alpha = 2.385$
    calculations:
    $t_{\bar{x}_1-\bar{x}_2} = (\bar{x}_1 - \bar{x}_2 - \mu_{\bar{x}_1-\bar{x}_2})/s_{\bar{x}_1-\bar{x}_2}$
    $= 2.790$
    P-value = $P(t_{121.9} > 2.790) = 0.0031$
    conclusion:
    Reject $H_o$; there is sufficient evidence to conclude that $\mu_1 - \mu_2 > 0$.
    Yes. Based on these results, heavy marijuana use is associated with poor performance.

15. Let the placebo users be group 1.
    $-0.61 < \mu_1 - \mu_2 < 2.99$, as shown from the following set-up and result screens
    No. Since the confidence interval includes 0, we cannot be 95% certain that the two populations have different means. There is not enough evidence to make this the generally recommended treatment for bipolar depression.

17. Let those receiving the magnet treatment be group 1.
    original claim: $\mu_1 - \mu_2 > 0$
    $H_o$: $\mu_1 - \mu_2 = 0$
    $H_1$: $\mu_1 - \mu_2 > 0$
    $\alpha = 0.05$ and df = 19
    C.V. $t = t_\alpha = 1.729$
    calculations:
    $t_{\bar{x}_1-\bar{x}_2} = (\bar{x}_1 - \bar{x}_2 - \mu_{\bar{x}_1-\bar{x}_2})/s_{\bar{x}_1-\bar{x}_2}$
    $= 0.132$
    P-value = $P(t_{33.6} > 0.132) = 0.4480$
    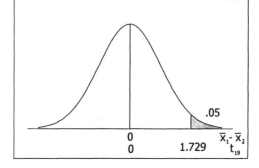
    conclusion:
    Do not reject $H_o$; there is not sufficient evidence to conclude that $\mu_1 - \mu_2 > 0$.
    No. It does not appear that magnets are effective in treating back pain. If a much larger sample size achieved these same results, the calculated t could fall in the critical region and

appear to provide evidence that the treatment is effective – but the observed difference would still be 0.05, and one would have to decide whether that statistically significant difference is of any practical significance.

19. Let the control subjects be group 1.
    $0.01 < \mu_1 - \mu_2 < 0.21$ (mL), as shown from the following set-up and result screens

    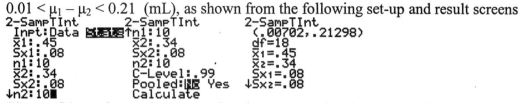

    The confidence interval suggests that the two samples do not come from populations with the same mean. Based on this result, one can be 99% certain that there is such a biological basis for the disorders.

21. Let the treatment subjects be group 1.
    $1.47 < \mu_1 - \mu_2 < 3.51$ (errors)

    Yes. Since high scores indicate the presence of more errors and the confidence interval falls entirely above 0, the results support the common belief that drinking alcohol is hazardous for the operators of passenger vehicles – and for their passengers.

23. Place the 5 oceanfront values in list L1 and the 5 oceanside values in list L2.
    Let the oceanfront homes be group 1.
    original claim: $\mu_1 - \mu_2 > 0$
    $H_o$: $\mu_1 - \mu_2 = 0$ ($1000's)
    $H_1$: $\mu_1 - \mu_2 > 0$ ($1000's)
    $\alpha = 0.05$ and df = 4
    C.V. $t = t_\alpha = 2.132$

    calculations:
    $$t_{\bar{x}_1-\bar{x}_2} = (\bar{x}_1-\bar{x}_2 - \mu_{\bar{x}_1-\bar{x}_2})/s_{\bar{x}_1-\bar{x}_2}$$
    $$= 3.992$$
    P-value = $P(t_{6.0} > 3.992) = 0.0071$

    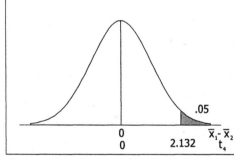

    conclusion:
    Reject $H_o$; there is sufficient evidence to conclude that $\mu_1 - \mu_2 > 0$.
    Yes. The fact that there are only five values in each sample is built into the critical region of the test. When the test indicates there is sufficient evidence to reject the null hypothesis, that conclusion is valid regardless of the sample size.

25. Place the 21 filtered values in list L1 and the 8 nonfiltered values in list L2.
Let the filtered cigarettes be group 1.
original claim: $\mu_1 - \mu_2 < 0$
$H_0$: $\mu_1 - \mu_2 = 0$ mg
$H_1$: $\mu_1 - \mu_2 < 0$ mg
$\alpha = 0.05$ and df = 7
C.V. $t = -t_\alpha = -1.895$
calculations:
$$t_{\bar{x}_1-\bar{x}_2} = (\bar{x}_1-\bar{x}_2 - \mu_{\bar{x}_1-\bar{x}_2})/s_{\bar{x}_1-\bar{x}_2}$$
$$= -10.585$$
P-value = $P(t_{25.9} < -10.585) = 3.31\text{E-}11 \approx 0$
conclusion:
Reject $H_0$; there is sufficient evidence to conclude that $\mu_1 - \mu_2 < 0$.

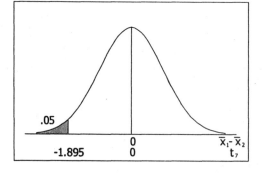

27. Use the CQPRE and CQPST lists.
Let the pre-1964 quarters be group 1.
original claim: $\mu_1 - \mu_2 = 0$
$H_0$: $\mu_1 - \mu_2 = 0$ g
$H_1$: $\mu_1 - \mu_2 \neq 0$ g
$\alpha = 0.05$ [assumed]
and df = 39
C.V. $t = \pm t_{\alpha/2} = \pm 2.024$
calculations:
$$t_{\bar{x}_1-\bar{x}_2} = (\bar{x}_1-\bar{x}_2 - \mu_{\bar{x}_1-\bar{x}_2})/s_{\bar{x}_1-\bar{x}_2}$$
$$= 32.773$$
P-value = $2 \cdot P(t_{70.5} > 32.773) = 2.19\text{E-}44 \approx 0$
conclusion:
Reject $H_0$; there is sufficient evidence to reject the claim that $\mu_1 - \mu_2 = 0$ and conclude that $\mu_1 - \mu_2 \neq 0$ (in fact, that $\mu_1 - \mu_2 > 0$).

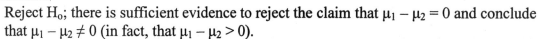

Yes. The conclusion is valid even though the population sizes are extremely large. If the requirements for using the test are satisfied, the size of the original population is irrelevant – except in cases where n>.05N and the finite population factor should be employed.

29. When assuming $\sigma_1 = \sigma_2$, df = $df_1 + df_2$.
Let the people using the Weight Watchers diet be group 1.
$-1.3 < \mu_1 - \mu_2 < 3.1$ (lbs), as shown from the following set-up and result screens

To one decimal accuracy, the results are the same as in exercise #12. In this case the results are not affected by the assumption of equal standard deviations.

31. When assuming $\sigma_1 = \sigma_2$, df = $df_1 + df_2$.
Let those receiving the magnet treatment be group 1.
original claim: $\mu_1 - \mu_2 > 0$
$H_o$: $\mu_1 - \mu_2 = 0$
$H_1$: $\mu_1 - \mu_2 > 0$
$\alpha = 0.05$ and df = 38
C.V. $t = t_\alpha = 1.686$
calculations:

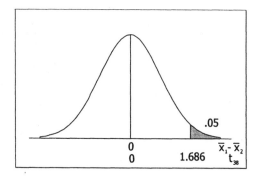

$t_{\overline{x}_1-\overline{x}_2} = (\overline{x}_1 - \overline{x}_2 - \mu_{\overline{x}_1-\overline{x}_2})/s_{\overline{x}_1-\overline{x}_2}$
= 0.132
P-value = $P(t_{38} > 0.132)$ = 0.4479
conclusion:
Do not reject $H_o$; there is not sufficient evidence to conclude that $\mu_1 - \mu_2 > 0$.
The conclusion is the same as in exercise #17, and the values in the problem changed very little. The calculated t statistic is exactly the same, as will be the case whenever $n_1 = n_2$.

33. Place the 28 car values in list L1 and the 20 taxi values in list L2.
Let the cars be group 1.
original claim: $\mu_1 - \mu_2 \neq 0$
$H_o$: $\mu_1 - \mu_2 = 0$ yrs
$H_1$: $\mu_1 - \mu_2 \neq 0$ yrs
$\alpha = 0.05$ and df = 19
C.V. $t = \pm t_{\alpha/2} = \pm 2.093$
calculations:

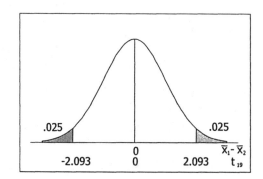

$t_{\overline{x}_1-\overline{x}_2} = (\overline{x}_1 - \overline{x}_2 - \mu_{\overline{x}_1-\overline{x}_2})/s_{\overline{x}_1-\overline{x}_2}$
= 0.697
P-value = $2 \cdot P(t_{33.8} > 0.697)$ = 0.4904
conclusion:
Do not reject $H_o$; there is not sufficient evidence to conclude that $\mu_1 - \mu_2 \neq 0$.
The conclusion is the same. The large change in $\overline{x}_1$ does not affect the problem because of the corresponding increase in the variability in the problem.

35. a. x = 5,10,15
$\mu = (\Sigma x)/N = 30/3 = 10$
$\sigma^2 = \Sigma(x-\mu)^2/N$
$= [(-5)^2 + (0)^2 + (5)^2]/3 = 50/3$

b. y = 1,2,3
$\mu = (\Sigma y)/N = 6/3 = 2$
$\sigma^2 = \Sigma(y-\mu)^2/N$
$= [(-1)^2 + (0)^2 + (1)^2]/3 = 2/3$

c. z = x-y = 4,3,2,9,8,7,14,13,12
$\mu = (\Sigma z)/N = 72/9 = 8$
$\sigma^2 = \Sigma(z-\mu)^2/N$
$= [(-4)^2 + (-5)^2 + (-6)^2 + (1)^2 + (0)^2 + (-1)^2 + (6)^2 + (5)^2 + (4)^2]/9 = 156/9 = 52/3$

d. $\sigma^2_{x-y} = \sigma^2_x + \sigma^2_y$
52/3 = 50/3 + 2/3
52/3 = 52/3
The variance of the differences is equal to the sum of the variances.

e. Let R stand for range.
$R_{x-y}$ = highest$_{x-y}$ − lowest$_{x-y}$
= (highest x − lowest y) − (lowest x − highest y)
= (highest x − lowest x) + (highest y − lowest y)
= $R_x + R_y$

Just as the variance of the differences is equal to the sum of the variances, so is the range of the differences equal to the sum of the ranges.

NOTE: This entire problem refers to all possible x-y difference (where $n_x$ and $n_y$ might even be unequal), and not to x-y differences for paired data.

37. $A = s_1^2/n_1 = 0.0064/10 = 0.00064$   $B = s_2^2/n_2 = 0.0064/10 = 0.00064$
$df = (A+B)^2/(A^2/df_1 + B^2/df_2)$
$= (0.00064 + 0.00064)^2/(0.00064^2/9 + 0.00064^2/9)$
$= 0.000001638/0.000000091 = 18$

When $s_1^2 = s_2^2$ and $n_1 = n_2$, Formula 9-1 yields $df = df_1 + df_2$.

In exercises #19 and #20, for which $\alpha = 0.01$, the tabled t value changes from $t_{9,\alpha/2} = 3.250$ to $t_{18,\alpha/2} = 2.878$. In general, the larger df signifies a "tighter" t distribution that is closer to the z distribution. In the test of hypothesis, the sampling distribution will be "tighter" and the critical t will be smaller – and since the calculated t is not affected, the P-value will be smaller. In the confidence interval, the tabled t will be smaller – and since the other values are not affected, the interval will be narrower. Using $df = \min(df_1, df_2)$ is more conservative in that such action will not reject $H_o$ as often and will lead to wider confidence intervals – i.e., it allows for a wider range of possible values for the parameter.

## 9-4 Inferences from Matched Pairs

1. No. The methods of this section apply to the differences between values that are matched in some way – but the difference between values with different units is not a meaningful concept.

3. When n d values represent the differences (in some specified order) between the two values in n matched pair of values, then $\bar{d}$ denotes the mean of those differences and $s_d$ denotes the standard deviation of those differences.

NOTE: To be consistent with the notation of the previous sections, thus reinforcing patterns and concepts presented in those sections, the manual uses the usual t formula written to apply to $\bar{d}$'s
$t_{\bar{d}} = (\bar{d} - \mu_{\bar{d}})/s_{\bar{d}}$, with $\mu_{\bar{d}} = \mu_d$ and $s_{\bar{d}} = s_d/\sqrt{n}$

As before, obtain critical values using Table A-3 whenever possible – otherwise use DISTR invT.

5. To review and reinforce concepts, this exercise is worked below by hand. One could also place the x values in list L1, place the y values in list L2, and use L1 − L2 → L3. Then use STAT TESTS T-Test on L3 to generate the display screen given at the right.

d = x−y: 0  2  −2  −3  1
summary: n = 5   Σd = −2   Σd² = 18
a. $\bar{d} = (\Sigma d)/n = -2/5 = -0.4$
b. $s_d^2 = [n(\Sigma d^2) - (\Sigma d)^2]/[n(n-1)]$
$= [5(18) - (-2)^2]/[5(4)] = 86/20 = 4.3$
$s_d = 2.07$

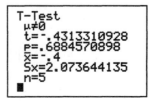

Inferences from Matched Pairs  SECTION 9-4   171

   c. $t_{\bar{d}} = (\bar{d} - \mu_{\bar{d}})/s_{\bar{d}}$
      $= (-0.4 - 0)/(2.07/\sqrt{5}) = -0.4/0.927 = -0.431$
   d. with df=4 and $\alpha$=0.05, the critical values are $t = \pm t_{\alpha/2} = \pm 2.776$

7. Place the x values in list L1, place the y values in list L2, and use L1 – L2 → L3. Then use STAT TESTS TInterval on L3.
   $-3.0 < \mu_d < 2.2$, as shown from the following set-up and result screens

   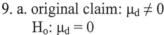

9. a. original claim: $\mu_d \neq 0$
      $H_0$: $\mu_d = 0$
      $H_1$: $\mu_d \neq 0$
      $\alpha = 0.05$ and df =9
      C.V.  $t = \pm t_{\alpha/2} = \pm 2.262$
      calculations:
         $t_{\bar{d}} = (\bar{d} - \mu_{\bar{d}})/s_{\bar{d}}$
         $= -0.41$
         P-value = 0.6910
      conclusion:
         Do not reject $H_0$; there is not sufficient evidence to conclude that $\mu_d \neq 0$.
         No. Based on this result, the medication does not appear to be effective.

   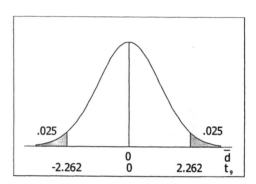

   b. If the medication is effective, the subject will be able to endure more head movement "after" than "before" – and so we expect d = ("before" – "after") < 0.
      original claim: $\mu_d < 0$
      $H_0$: $\mu_d = 0$
      $H_1$: $\mu_d < 0$
      $\alpha = 0.05$ and df = 9
      C.V.  $t = -t_\alpha = -1.833$
      calculations:
         $t_{\bar{d}} = (\bar{d} - \mu_{\bar{d}})/s_{\bar{d}}$
         $= -0.41$
         P-value = ½(0.6910) = 0.3455
      conclusion:
         Do not reject $H_0$; there is not sufficient evidence to conclude that $\mu_d < 0$.

   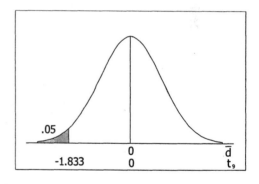

11. Place the actual values in list L1 and the forecast values in list L2. Use L1 – L2 → L3.
    a. original claim: $\mu_d = 0$
       $H_0$: $\mu_d = 0$ °F
       $H_1$: $\mu_d \neq 0$ °F
       $\alpha = 0.05$ and df = 4
       C.V.  $t = \pm t_{\alpha/2} = \pm 2.776$
       calculations
          $t_{\bar{d}} = (\bar{d} - \mu_{\bar{d}})/s_{\bar{d}}$
          $= 0.218$

   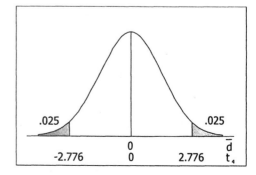

P-value = 2·P(t₄>0.218) = 0.8379
conclusion:
 Do not reject H₀; there is not sufficient evidence to reject the claim that $\mu_d = 0$.
 The results suggest that, on the average, the one-day forecasts are on target.

b. $-2.3 < \mu_d < 2.7$ (°F), as shown from the following set-up and result screens

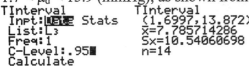

Since the confidence interval contains 0, there is no reason to reject the notion that the one-day predictions are correct on the average.

13. Place the day 1 values in list L1 and the day 2 values in list L2. Use L1 – L2 → L3.
 $1.7 < \mu_d < 13.9$ (mmHg), as shown from the following set-up and result screens

```
TInterval          TInterval
Inpt:Data Stats    (1.6997,13.872)
List:L3            x̄=7.785714286
Freq:1             Sx=10.54060698
C-Level:.95        n=14
Calculate
```

The results suggest either that blood pressure is not a stable and/or consistent quantity, or that it cannot (with present technology) be measured with accuracy and/or consistency.

15. Place the second values in list L1 and the first values in list L2. Use L1 – L2 → L3.
 original claim: $\mu_d = 0$
 $H_0$: $\mu_d = 0$
 $H_1$: $\mu_d \neq 0$
 $\alpha = 0.05$ [assumed]
  and df = 8
 C.V. $t = \pm t_{\alpha/2} = \pm 2.306$
 calculations
  $t_{\bar{d}} = (\bar{d} - \mu_{\bar{d}})/s_{\bar{d}}$
   = -0.132
  P-value = 2·P(t₈<-0.132) = 0.8981

```
T-Test
μ≠0
t=.1321637201
p=.8981184015
x̄=1.111111111
Sx=25.22124325
n=9
```

conclusion:
 Do not reject H₀; there is not sufficient evidence to reject the claim that $\mu_d = 0$.
 The results suggest that there is not much change between the first and second scores of those repeating the SAT with no intermediate preparatory course.

17. Place the before values in list L1 and the after values in list L2. Use L1 – L2 → L3.
 a. $0.69 < \mu_d < 5.56$ (cm), as shown from the following set-up and result screens

```
TInterval          TInterval
Inpt:Data Stats    (.69098,5.559)
List:L3            x̄=3.125
Freq:1             Sx=2.911430674
C-Level:.95        n=8
Calculate
```

b. original claim: $\mu_d > 0$
 $H_o$: $\mu_d = 0$ cm
 $H_1$: $\mu_d > 0$ cm
 $\alpha = 0.05$ and df = 7
 C.V. $t = t_\alpha = 1.895$
 calculations
 $t_{\bar{d}} = (\bar{d} - \mu_{\bar{d}})/s_{\bar{d}}$
  $= 3.036$
 P-value = $P(t_7 > 3.036) = 0.0095$
 conclusion:
  Reject $H_o$; there is sufficient evidence to conclude that $\mu_d > 0$.

c. Yes, hypnotism appears to be effective in reducing pain.

19. Place the regular values in list L1 and the kiln values in list L2. Use L1 − L2 → L3.
 a. original claim: $\mu_d = 0$
  $H_o$: $\mu_d = 0$ lbs/acre
  $H_1$: $\mu_d \neq 0$ lbs/acre
  $\alpha = 0.05$ and df = 10
  C.V. $t = \pm t_{\alpha/2} = \pm 2.228$
  calculations
  $t_{\bar{d}} = (\bar{d} - \mu_{\bar{d}})/s_{\bar{d}}$
   $= -1.690$
  P-value = $2 \cdot P(t_{10} < -1.690) = 0.1218$
  conclusion:
   Do not reject $H_o$; there is not sufficient evidence to reject the claim that $\mu_d = 0$.

 b. $-78.2 < \mu_d < 10.7$ (lbs/acre), as shown from the following set-up and result screens

 c. We cannot be 95% certain that either type of seed is better. If there are no differences in cost, or any other considerations, choose the kiln dried – even though we can't be sure that it's generally better, it did have the higher yield in this particular trial.

21. Use ACTLO − PLO5 → L1.
 a. original claim: $\mu_d = 0$
  $H_o$: $\mu_d = 0$ °F
  $H_1$: $\mu_d \neq 0$ °F
  $\alpha = 0.05$ and df = 34
  C.V. $t = \pm t_{\alpha/2} = \pm 2.032$
  calculations
  $t_{\bar{d}} = (\bar{d} - \mu_{\bar{d}})/s_{\bar{d}}$
   $= 0.155$
  P-value = $2 \cdot P(t_{34} > 0.155) = 0.8775$
  conclusion:
   Do not reject $H_o$; there is not sufficient evidence to reject the claim that $\mu_d = 0$.

174   CHAPTER 9   Inferences from Two Samples

b. $-1.7 < \mu_d < 2.0$ (°F), as shown from the following set-up and result screens
```
TInterval              TInterval
 Inpt:Data Stats        (-1.726,2.0117)
 List:L1                x̄=.1428571429
 Freq:1                 Sx=5.440279281
 C-Level:.95            n=35
 Calculate
```

c. The results are the same as in the example. It appears that, on the average, the forecast low temperatures are accurate.

23. Use HMSP – HMLST → L1.
   a. original claim: $\mu_d = 0$
   $H_o$: $\mu_d = \$0$
   $H_1$: $\mu_d \neq \$0$
   $\alpha = 0.05$ and df = 39
   C.V. $t = \pm t_{\alpha/2} = \pm 2.023$
   calculations
   ```
   T-Test
    μ≠0
    t=-5.353786153
    p=4.0796524E-6
    x̄=-7265
    Sx=8582.317839
    n=40
   ```
   $t_{\bar{d}} = (\bar{d} - \mu_{\bar{d}})/s_{\bar{d}}$
   $= -5.354$
   P-value = $2 \cdot P(t_{39} < -5.354) = 4.04\text{E-}6 \approx 0$
   conclusion:
   Reject $H_o$; there is sufficient evidence to reject the claim that $\mu_d = 0$ and conclude that $\mu_d \neq 0$ (in fact, that $\mu_d < 0$ – i.e., that the selling price is less than the list price).

   b. $-10{,}010 < \mu_d < -4{,}520$ (\$), as shown from the following set-up and result screens
   ```
   TInterval              TInterval
    Inpt:Data Stats        (-10010,-4520)
    List:L1                x̄=-7265
    Freq:1                 Sx=8582.317839
    C-Level:.95            n=40
    Calculate
   ```

25. The points on the graph are as given below.
         before: 460  470  490  490  510  510  600  610  620
         after:  480  510  500  610  590  630  630  690  660
   d = after – before:  20   40   10  120   80  120   30   80   40
   Place the before scores in list L1 and the after scores in list L2. Use L2 – L1 → L3
   original claim: $\mu_d > 0$
   $H_o$: $\mu_d = 0$
   $H_1$: $\mu_d > 0$
   $\alpha = 0.05$ [assumed]
      and df = 8
   C.V. $t = t_\alpha = 1.860$
   calculations
   ```
   T-Test
    μ>0
    t=4.333890711
    p=.0012494264
    x̄=60
    Sx=41.53311931
    n=9
   ```
   $t_{\bar{d}} = (\bar{d} - \mu_{\bar{d}})/s_{\bar{d}}$
   $= 4.334$
   P-value = $P(t_8 > 4.334) = 0.0012$
   conclusion:
   Reject $H_o$; there is sufficient evidence to conclude that $\mu_d > 0$.
   NOTE: If one defines the difference in terms of "before – after", then the original claim is "$\mu_d < 0$" and the C.V. and calculated t are the negative of the above values. The choice is arbitrary.

Inferences from Matched Pairs  SECTION 9-4  175

27. Place the x scores in list L1 and the y scors in list L2.
   a. Use L1 − L2 → L3 and STAT TESTS T-Test on L3.
      original claim: $\mu_d > 0$
      $H_o$: $\mu_d = 0$
      $H_1$: $\mu_d > 0$
      $\alpha = 0.05$ and df = 9
      C.V. $t = t_\alpha = 1.833$
      calculations
      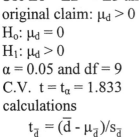
      $t_{\bar{d}} = (\bar{d} - \mu_{\bar{d}})/s_{\bar{d}}$
      $= 1.861$
      P-value $= P(t_9 > 1.861) = 0.0479$
      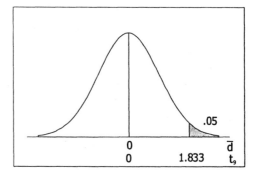
      conclusion:
         Reject $H_o$; there is sufficient evidence to conclude that $\mu_d > 0$.

   b. Use STAT TESTS 2-SampTTest on L1 and L2.
      original claim: $\mu_1 - \mu_2 > 0$
      $H_o$: $\mu_1 - \mu_2 = 0$
      $H_1$: $\mu_1 - \mu_2 > 0$
      $\alpha = 0.05$ and df = 9
      C.V. $t = t_\alpha = 1.833$
      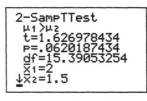
      calculations:
      $t_{\bar{x}_1 - \bar{x}_2} = (\bar{x}_1 - \bar{x}_2 - \mu_{\bar{x}_1 - \bar{x}_2})/s_{\bar{x}_1 - \bar{x}_2}$
      $= 1.627$
      P-value $= P(t_{15.4} > 1.627) = 0.06201$
      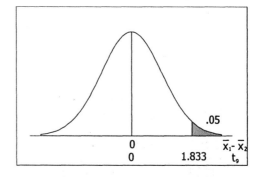
      conclusion:
         Do not reject $H_o$; there is not sufficient evidence to conclude that $\mu_1 - \mu_2 > 0$.

   c. Parts (a) and (b) do not have the same conclusion. Since different methods can give different results, it is important that the correct method be used.


## 9-5 Comparing Variation in Two Samples

1. The F test described in this section is not robust – i.e., it is very sensitive to departures from normality. Two alternatives that perform well even when the original populations are not normal are (1) the count five test and (2) the Levene-Brown-Forsythe test.

3. The F distribution is the sampling distribution formed by the ratio of the sample variances from two independent normal distributions with equal variances.

NOTE: The following conventions are used in this manual regarding the F test.
• The set of scores with the larger sample variance is designated group 1.
• Even though designating the scores with the larger sample variance as group 1 makes lower tail critical values unnecessary in two-tailed tests, the lower critical vales are calculated [see exercise #23] and included in the accompanying graphs for completeness and consistency with the other tests. The F distribution always "bunches up" around 1.0 regardless of the df values.
• Since the F value depends on two degrees of freedom, the df for group 1 (numerator) and group 2 (denominator) may be used with the F as a superscript and a subscript respectively [or as a first and second subscript to clarify which F distribution is being used. Accordingly, the value in the "5" column and "10" row on the first page of Table A-5 may be given either as

$F^5_{10,.025} = 4.2361$ or $F_{5,10,.025} = 4.2361$.

* If the desired df does not appear in Table A-5, the closest entry is used. For a desired entry exactly halfway between two tabled values, the conservative approach of using the smaller df is used – and 120 is used for all df larger than 120.
* F is statistically defined to be $F = (s_1^2/\sigma_1^2)/(s_2^2/\sigma_2^2)$. Since all hypotheses in the text question the equality of $\sigma_1^2$ and $\sigma_2^2$, the calculation of F may be shortened to $F = s_1^2/s_2^2$.
* Some problems are stated in terms of variance, and some are stated in terms of standard deviation – and some are stated simply in terms of "variation." Since $\sigma_1 = \sigma_2$ is equivalent to $\sigma_1^2 = \sigma_2^2$, this manual simply states all claims and hypotheses and conclusions using the variance.

5. [by hand] Let the treatment group be group 1.

original claim: $\sigma_1^2 \neq \sigma_2^2$

$H_o: \sigma_1^2 = \sigma_2^2$

$H_1: \sigma_1^2 \neq \sigma_2^2$

$\alpha = 0.05$ and $df_1 = 15$, $df_2 = 40$

C.V. $F = F_{\alpha/2} = 2.1819$

calculations:

$F = s_1^2/s_2^2$

$= (0.80)^2/(0.40)^2 = 4.000$

P-value = $2 \cdot Fcdf(4.000, 99, 15, 40) = 0.0005$

conclusion:

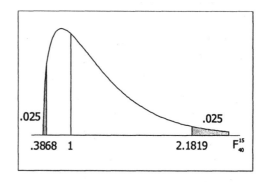

Reject $H_o$; there is sufficient evidence to conclude that $\sigma_1^2 \neq \sigma_2^2$ (in fact, that $\sigma_1^2 > \sigma_2^2$).

NOTE: Exercise #5 above was worked by hand. Below are the set-up screen and the two result screens for using STAT TESTS 2-SampFTest on the TI-83/84 Plus. Following the NOTE, exercise #5 is re-worked in the format used in this manual.

```
set-up screen              result screens
2-SampFTest                2-SampFTest        2-SampFTest
 Inpt:Data Stats            σ1≠σ2              σ1≠σ2
 Sx1:.8                    F=4                ↑p=4.5358316E-4
 n1:16                     p=4.5358316E-4     Sx1=.8
 Sx2:.4                    Sx1=.8             Sx2=.4
 n2:41                     Sx2=.4             n1=16
 σ1:≠σ2 <σ2 >σ2            ↓n1=16             n2=41
 Calculate Draw
```

5. [by calculator] Let the treatment group be group 1.

original claim: $\sigma_1^2 \neq \sigma_2^2$

$H_o: \sigma_1^2 = \sigma_2^2$

$H_1: \sigma_1^2 \neq \sigma_2^2$

$\alpha = 0.05$

and $df_1 = 15$, $df_2 = 40$

C.V. $F = F_{\alpha/2} = 2.1819$

calculations:

$F = s_1^2/s_2^2 = 4.0000$

P-value = $2 \cdot P(F_{15,40} > 4.0000) = 0.0005$

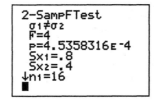

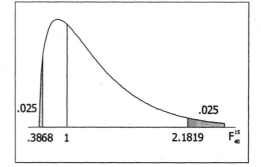

conclusion:

Reject $H_o$; there is sufficient evidence to conclude that $\sigma_1^2 \neq \sigma_2^2$ (in fact, that $\sigma_1^2 > \sigma_2^2$).

7. Let the regular Coke be group 1.

original claim: $\sigma_1^2 \neq \sigma_2^2$

$H_0: \sigma_1^2 = \sigma_2^2$

$H_1: \sigma_1^2 \neq \sigma_2^2$

$\alpha = 0.05$

and $df_1 = 35$, $df_2 = 35$

C.V.  $F = F_{\alpha/2} = 2.0739$

calculations:

$F = s_1^2/s_2^2$

$= (0.0075)^2/(0.0044)^2 = 2.9233$

P-value = 0.0021

conclusion:

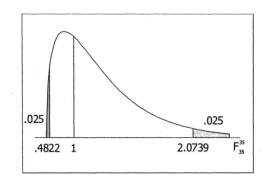

Reject $H_o$; there is sufficient evidence to conclude that $\sigma_1^2 \neq \sigma_2^2$ (in fact, that $\sigma_1^2 > \sigma_2^2$). The addition of sugar to the regular Coke may account for the larger variation. In general, each additional ingredient contributes a certain amount of its own variability – in particular, it would be difficult to ensure there is precisely the same amount of sugar in each can.

9. Let the diet Coke be group 1.

original claim: $\sigma_1^2 \neq \sigma_2^2$

$H_0: \sigma_1^2 = \sigma_2^2$

$H_1: \sigma_1^2 \neq \sigma_2^2$

$\alpha = 0.05$

and $df_1 = 35$, $df_2 = 35$

C.V.  $F = F_{\alpha/2} = 2.0739$

calculations:

$F = s_1^2/s_2^2 = 1.0133$

P-value = 2· P($F_{35,35}$>1.0133) = 0.9690

conclusion:

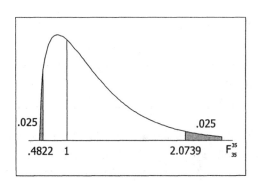

Do not reject $H_o$; there is not sufficient evidence to conclude that $\sigma_1^2 \neq \sigma_2^2$.

11. Let the sham group be group 1.

original claim: $\sigma_1^2 > \sigma_2^2$

$H_0: \sigma_1^2 = \sigma_2^2$

$H_1: \sigma_1^2 > \sigma_2^2$

$\alpha = 0.05$

and $df_1 = 19$, $df_2 = 19$

C.V.  $F = F_\alpha = 2.1555$

calculations:

$F = s_1^2/s_2^2 = 2.1267$

P-value = P($F_{19,19}$>2.1267) = 0.0543

conclusion:

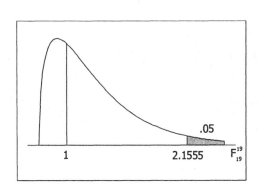

Do not reject $H_o$; there is not sufficient evidence to conclude that $\sigma_1^2 > \sigma_2^2$.

178   CHAPTER 9   Inferences from Two Samples

13. Let the treatment group be group 1.

original claim: $\sigma_1^2 > \sigma_2^2$

$H_0: \sigma_1^2 = \sigma_2^2$
$H_1: \sigma_1^2 > \sigma_2^2$
$\alpha = 0.05$
    and $df_1 = 21$, $df_2 = 21$
C.V.  $F = F_\alpha = 2.0960$
calculations:
    $F = s_1^2/s_2^2 = 9.3364$
    P-value = $P(F_{21,21} > 9.3364) = 1.68\text{E-}6 = 0.000002$
conclusion:
    Reject $H_0$; there is sufficient evidence to conclude that $\sigma_1^2 > \sigma_2^2$.

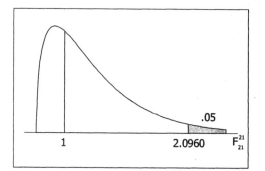

15. Let the pre-1964 quarters be group 1.

original claim: $\sigma_1^2 = \sigma_2^2$

$H_0: \sigma_1^2 = \sigma_2^2$
$H_1: \sigma_1^2 \neq \sigma_2^2$
$\alpha = 0.05$
    and $df_1 = 39$, $df_2 = 39$
C.V.  $F = F_{\alpha/2} = 1.8752$
calculations:
    $F = s_1^2/s_2^2 = 1.9729$
    P-value = $2 \cdot P(F_{39,39} > 1.9729) = 0.0368$
conclusion:
    Reject $H_0$; there is sufficient evidence to conclude that $\sigma_1^2 \neq \sigma_2^2$ (in fact, that $\sigma_1^2 > \sigma_2^2$).

No. It appears that the new quarters are <u>less</u> variable than the old ones, and so machines that accept quarters will not need adjustments to handle extra variability.
NOTE:  The most likely reason for the larger variability in the older quarters is larger variation in wear over time – and not necessarily a difference in the production process or materials.

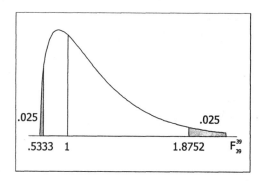

17. Let the easy to difficult group be group 1.  Input $\sqrt{47.020}$ at the "Sx1:" prompt, etc.

original claim: $\sigma_1^2 = \sigma_2^2$

$H_0: \sigma_1^2 = \sigma_2^2$
$H_1: \sigma_1^2 \neq \sigma_2^2$
$\alpha = 0.05$
    and $df_1 = 24$, $df_2 = 15$
C.V.  $F = F_{\alpha/2} = 2.7006$
calculations:
    $F = s_1^2/s_2^2 = 2.5906$
    P-value = $2 \cdot P(F_{24,15} > 2.5906) = 0.0599$
conclusion:
    Do not reject $H_0$; there is not sufficient evidence to reject the claim that $\sigma_1^2 = \sigma_2^2$.

19. Use lists RNWED and RNSUN.
    a. Let the Sunday amounts be group 1.
       original claim: $\sigma_1^2 = \sigma_2^2$
       $H_0: \sigma_1^2 = \sigma_2^2$
       $H_1: \sigma_1^2 \neq \sigma_2^2$
       $\alpha = 0.05$
          and $df_1 = 51$, $df_2 = 52$
       C.V. $F = F_{\alpha/2} = 1.6668$
       calculations:
          $F = s_1^2/s_2^2 = 2.2080$
          P-value $= 2 \cdot P(F_{51,52} > 2.2080) = 0.0052$
       conclusion:
          Reject $H_0$; there is sufficient evidence to reject the claim that $\sigma_1^2 = \sigma_2^2$ and conclude
          that $\sigma_1^2 \neq \sigma_2^2$ (in fact, that $\sigma_1^2 > \sigma_2^2$).

    b. For Sundays, 37 of the 52 observations are 0; for Wednesdays, 37 of the 53 observations are 0. The data for both days are very positively skewed. The rainfall amounts do not come from populations with normal distributions.

    c. Because the original populations are not normally distributed, the test in part (a) is not valid – which is good news, since there is no reason why rainfall amounts should be more variable on Sundays than on Wednesdays.

21. Use the following commands.
        MATH NUM abs(CRGWT − 0.81682) → L1   and then   STAT SortD(L1)
        MATH NUM abs(PRGWT − 0.82410) → L2   and then   STAT SortD(L2)
    The command STAT EDIT display indicates that the largest Coke MAD 0.02672 is the only value larger than the largest Pepsi MAD 0.01600 – and so $c_1 = 1$ and $c_2 = 0$.
    original claim: Coke and Pepsi have equal variation

    Parts (a)-(d) may be answered in the usual test of hypothesis format as follows.
    $H_0$: Coke and Pepsi have equal variation
    $H_1$: Coke and Pepsi do not have equal variation
    $\alpha = 0.05$
    C.V. $c = 5$.
    calculations:
       $c_1 = 1$
       $c_2 = 0$
    conclusion:
       Do not reject $H_0$; there is not sufficient evidence to reject the claim that Coke and Pepsi have equal variation.

23. a. $\alpha = 0.05$   $F_L = F^9_{9,1-\alpha/2} = 1/F^9_{9,\alpha/2} = 1/4.0260 = 0.2484$   $F_R = F^9_{9,\alpha/2} = 4.0260$
    b. $\alpha = 0.05$   $F_L = F^9_{6,1-\alpha/2} = 1/F^6_{9,\alpha/2} = 1/4.3197 = 0.2315$   $F_R = F^9_{6,\alpha/2} = 5.5234$
    c. $\alpha = 0.05$   $F_L = F^6_{9,1-\alpha/2} = 1/F^9_{6,\alpha/2} = 1/5.5234 = 0.1810$   $F_R = F^6_{9,\alpha/2} = 4.3197$

# 180   CHAPTER 9   Inferences from Two Samples

## Statistical Literacy and Critical Thinking

1. The statement "he is preferred by more male voters than female voters" is not clear.
   * If the statement intends to say "most of the voters who prefer him are male", then the claim is $p > 0.5$ and the techniques of section 8.3 are appropriate.
   * If the statement intends to say "the proportion of voters who prefer him is larger among males than among females", then the claim is $p_1 - p_2 > 0$ and the techniques of section 9.2 are appropriate.

2. Yes. Since every group of 200 creditors has the same chance of being selected, this satisfies the requirement for being a simple random sample of creditors.
   NOTE: There are two ambiguities in this problem that need to be resolved before proceeding.
   * Is the intent to compare the debt of men who live in the state to the debt of women who live in the state – no matter in what state the money is owed? Or is the intent to compare the debt owed within the state by men to the debt owed within the state by women – no matter in what state the debtors reside?
   * The interest appears to be in the debtors, but the simple random sample was taken from among the creditors. Since a single debtor can owe money to more than one creditor, debtors with multiple creditors are more likely to appear in the survey – and a simple random sample of creditors will not produce a simple random sample of debtors.

3. No, this procedure is not acceptable. The procedure described counts the mean income from each state equally, but the mean income from a large population state has a greater influence on the national mean than does the mean income from a small population state. The researcher should use weighted means.

4. If two samples are independent, then the inclusion of one observation in one of the samples in no way affects or determines the inclusion of any observation in the other sample.

## Review Exercises

1. Let the black drivers be group 1.
   a. original claim: $p_1 - p_2 > 0$
      $H_0$: $p_1 - p_2 = 0$
      $H_1$: $p_1 - p_2 > 0$
      $\alpha = 0.05$
      C.V. $z = z_\alpha = 1.645$
      calculations:
      $$z_{\hat{p}_1 - \hat{p}_2} = (\hat{p}_1 - \hat{p}_2 - \mu_{\hat{p}_1 - \hat{p}_2})/\sigma_{\hat{p}_1 - \hat{p}_2}$$
      $= 0.642$
      P-value = $P(z > 0.642) = 0.2603$
      conclusion:
      Do not reject $H_0$; there is not sufficient evidence to conclude that $p_1 - p_2 > 0$.

      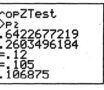
      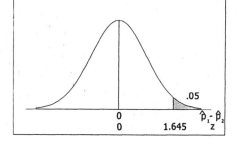

   b. $-0.025 < p_1 - p_2 < 0.055$, as shown by the following set-up and result screens

      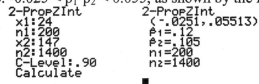

      Since the interval includes 0, there is not sufficient evidence to conclude that $p_1 - p_2 > 0$.

2. Place the reported values in list L1 and the measured values in list L2. Use L1 − L2 → L3.
   original claim: $\mu_d > 0$ in
   $H_o$: $\mu_d = 0$ in
   $H_1$: $\mu_d > 0$ in
   $\alpha = 0.05$ [assumed]
     and df = 10
   C.V. $t = t_\alpha = 1.812$
   calculations
   $t_{\bar{d}} = (\bar{d} - \mu_{\bar{d}})/s_{\bar{d}}$
       $= 2.701$
       P-value = $P(t_{10} > 2.701) = 0.0111$
   conclusion:
       Reject $H_o$; there is sufficient evidence to conclude that $\mu_d > 0$.

3. Place the Rowling values in list L1 and the Tolstoy values in list L2.
   Let the Rowling scores be group 1.
   original claim: $\mu_1 - \mu_2 > 0$
   $H_o$: $\mu_1 - \mu_2 = 0$
   $H_1$: $\mu_1 - \mu_2 > 0$
   $\alpha = 0.05$ and df = 11
   C.V. $t = t_\alpha = 1.796$
   calculations:
   $t_{\bar{x}_1 - \bar{x}_2} = (\bar{x}_1 - \bar{x}_2 - \mu_{\bar{x}_1 - \bar{x}_2})/s_{\bar{x}_1 - \bar{x}_2}$
       $= 5.529$
       P-value = $P(t_{17.9} > 5.529) = 1.51\text{E-}5 = 0.00002$
   conclusion:
       Reject $H_o$; there is sufficient evidence to conclude that $\mu_1 - \mu_2 > 0$.
       Yes. It was expected that the Harry Potter book would be easier to read than *War and Peace*.

4. Place the Rowling values in list L1 and the Tolstoy values in list L2.
   Let the Tolstoy scores be group 1.
   original claim: $\sigma_1^2 = \sigma_2^2$
   $H_o$: $\sigma_1^2 = \sigma_2^2$
   $H_1$: $\sigma_1^2 \neq \sigma_2^2$
   $\alpha = 0.05$
     and $df_1 = 11$, $df_2 = 11$
   C.V. $F = F_{\alpha/2} = 3.5257$
   calculations:
   $F = s_1^2/s_2^2 = 2.8176$
       P-value = $2 \cdot P(F_{11,11} > 2.8176) = 0.1000$
   conclusion:
       Do not reject $H_o$; there is not sufficient evidence to reject the claim that $\sigma_1^2 = \sigma_2^2$.

182  CHAPTER 9  Inferences from Two Samples

5. Let those that were warmed be group 1.
   a. original claim: $p_1 - p_2 < 0$
      $H_o$: $p_1 - p_2 = 0$
      $H_1$: $p_1 - p_2 < 0$
      $\alpha = 0.05$
      C.V. $z = -z_\alpha = -1.645$
      calculations:
      $$z_{\hat{p}_1 - \hat{p}_2} = (\hat{p}_1 - \hat{p}_2 - \mu_{\hat{p}_1 - \hat{p}_2})/\sigma_{\hat{p}_1 - \hat{p}_2}$$
      $$= -2.822$$
      P-value = $P(z < -2.822) = 0.0024$
      conclusion:
         Reject $H_o$; there is sufficient evidence to conclude that $p_1 - p_2 < 0$.
         Yes. If these results are verified, surgical patients should be routinely warmed.

   b. The test in part (a) places 0.05 in the lower tail. The appropriate two-sided interval for testing the claim in part (a) places 0.05 in each tail – i.e., is a 90% confidence interval.

   c. $-0.205 < p_1 - p_2 < -0.054$, as shown by the following set-up and result screens

   d. Since the test of hypothesis and the confidence interval use different estimates for $\sigma_{\hat{p}_1 - \hat{p}_2}$, they are not mathematically equivalent – and so it is possible that they may lead to different conclusions.

6. Let the obsessive-compulsive patients be group 1.
   a. $-17.32 < \mu_1 - \mu_2 < 260.56$, as shown by the following set-up and result screens

   b. original claim: $\mu_1 - \mu_2 = 0$
      $H_o$: $\mu_1 - \mu_2 = 0$
      $H_1$: $\mu_1 - \mu_2 \neq 0$
      $\alpha = 0.05$ and df = 9
      C.V. $t = \pm t_{\alpha/2} = \pm 2.262$
      calculations:
      $$t_{\bar{x}_1 - \bar{x}_2} = (\bar{x}_1 - \bar{x}_2 - \mu_{\bar{x}_1 - \bar{x}_2})/s_{\bar{x}_1 - \bar{x}_2}$$
      $$= 1.841$$
      P-value = $2 \cdot P(t_{17.7} > 1.841) = 0.0824$
      conclusion:
         Do not reject $H_o$; there is not sufficient evidence to reject the claim that $\mu_1 - \mu_2 = 0$.
   c. No. It does not appear that the total brain volume can be used as a reliable indicator.

7. Let the obsessive-compulsive patients be group 1.
   original claim: $\sigma_1^2 \neq \sigma_2^2$
   $H_0: \sigma_1^2 = \sigma_2^2$
   $H_1: \sigma_1^2 \neq \sigma_2^2$
   $\alpha = 0.05$
   and $df_1 = 9$, $df_2 = 9$
   C.V. $F = F_{\alpha/2} = 4.0260$
   calculations:
   $F = s_1^2/s_2^2 = 1.2922$
   P-value = $2 \cdot P(F_{9,9} > 1.2922) = 0.7087$
   conclusion:
   Do not reject $H_0$; there is not sufficient evidence to conclude that $\sigma_1^2 \neq \sigma_2^2$.

   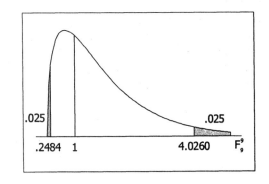

8. Place the regular values in list L1 and the kiln values in list L2. Use L1 – L2 → L3.
   a. original claim: $\mu_d = 0$
      $H_0: \mu_d = 0$ cwt/acre
      $H_1: \mu_d \neq 0$ cwt/acre
      $\alpha = 0.05$ and $df = 10$
      C.V. $t = \pm t_{\alpha/2} = \pm 2.228$
      calculations
      $t_{\bar{d}} = (\bar{d} - \mu_{\bar{d}})/s_{\bar{d}}$
      $= -1.532$
      P-value = $2 \cdot P(t_{10} < -1.532) = 0.1565$
      conclusion:
      Do not reject $H_0$; there is not sufficient evidence to reject the claim that $\mu_d = 0$.

      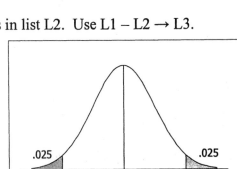

   b. $-2.68 < \mu_d < 0.50$, as shown by the following set-up and result screens

   c. We cannot be 95% certain that either type of seed is better. If there are no differences in cost, or any other considerations, choose the kiln dried – even though we can't be sure that it's generally better, it did have the higher yield in this particular trial.

184   CHAPTER 9   Inferences from Two Samples

**Cumulative Review Exercises**

1. Place the north values in list L1 and the south values in list L2.
   The following commands generate the additional necessary lists.
   L1 − L2 → L3
   LIST  OPS  augment(L1,L2) → L4

   a. Use STAT  CALC  1-Var Stats on L1 to obtain the screens below.
      $\bar{x}$ = 69.5 mph
      $\tilde{x}$ = 69.5 mph
      s = 3.4 mph
      $s^2$ = 11.63 $mi^2/hr^2$
      R = 74 − 65 = 9 mph

   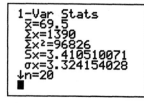 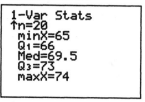

   b. Use STAT  TESTS  T-Test on L4.
      original claim: μ > 65 mph
      $H_o$: μ = 65 mph
      $H_1$: μ > 65 mph
      α = 0.05 [assumed]
         and df = 39
      C.V.  t = $t_α$ = 1.686
      calculations:
         $t_{\bar{x}}$ = ($\bar{x}$ − μ)/$s_{\bar{x}}$
              = 3.765
         P-value = P($t_{39}$>3.765) = 0.0003
      conclusion:
         Reject $H_o$; there is sufficient evidence to conclude that μ > 65.

   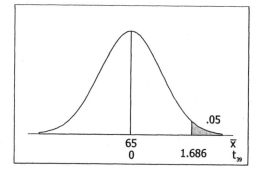

   c. Use STATPLOT on L1 to obtain the histogram at the right.
      No. The northbound speeds appear to be uniformly distributed.

   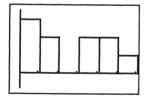

   d. Use STAT  TESTS  2-SampTTest on L1 and L2.
      original claim: $μ_1 − μ_2$ = 0
      $H_o$: $μ_1 − μ_2$ = 0
      $H_1$: $μ_1 − μ_2$ ≠ 0
      α = 0.05 [assumed]
         and df = 19
      C.V.  t = ±$t_{α/2}$ = ±2.093
      calculations:
         $t_{\bar{x}_1-\bar{x}_2}$ = ($\bar{x}_1-\bar{x}_2$ − $μ_{\bar{x}_1-\bar{x}_2}$)/$s_{\bar{x}_1-\bar{x}_2}$
              = 1.265
         P-value = 2·P($t_{27.1}$>1.265) = 0.2167
      conclusion:

   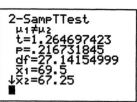 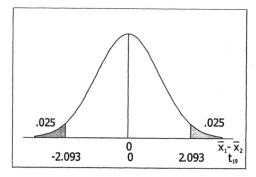

   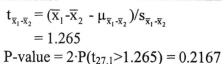

   Do not reject $H_o$; there is not sufficient evidence to reject the claim that $μ_1 − μ_2$ = 0.
   Yes, the hypothesis test is likely to be valid despite the departure from normality noted in
   part (c) – since the uniform distribution is symmetric with no outliers, and the sample size is
   reasonable.

2. a. P(all 9 heads) = P($H_1$ and $H_2$ and…$H_9$)
   = P($H_1$)·P($H_2$)·…·P($H_9$)
   = (½)·(½)·…(½)
   = (½)$^9$ = 1/512 = 0.00195

   The probability of obtaining all 9 heads by chance alone is so small that it is reasonable to conclude that factors other than chance are involved – whether psychic abilities or deceitful procedures.

   b. Independent. The outcomes are not matched or paired in any way, nor does the outcome for one coin affect the outcome for the other.

   c. original claim: p > 0.50
   $\hat{p}$ = x/n = 14/18 = 0.778
   $H_o$: p = 0.50
   $H_1$: p > 0.50
   α = 0.01
   C.V. z = $z_α$ = 2.326
   calculations:
   $z\hat{p}$ = ($\hat{p}$ − $μ\hat{p}$)/$σ\hat{p}$
   = 2.357
   P-value = P(z>2.357) = 0.0092

   ```
   1-PropZTest
   prop>.5
   z=2.357022604
   p=.0092110461
   p̂=.7777777778
   n=18
   ```

   conclusion:
   Reject $H_o$; there is sufficient evidence to conclude that p > 0.50.
   NOTE: The conclusion is that the probability the given coins come up heads is greater than ½ – not that the person has such psychic powers. Being the eternal skeptic, the writer wonders whether the quarter is an altered coin (or even a two-headed quarter) provided by the illusionist while the penny is an ordinary coin supplied from the audience.

3. There is a problem with the reported results, and no statistical analysis would be appropriate. Since there were 100 drivers in each group, and the number in each group owning a cell phone must be a whole number between 0 and 100 inclusive, the sample proportion for each group must be a whole percent. The reported values of 13.7% and 10.6% are not mathematical possibilities for the sample success rates of groups of 100.

# Chapter 10

# Correlation and Regression

## 10-2 Correlation

1. The first requirement listed in this section is that the (x,y) pairs be a random sample of independent quantitative data. Since social security numbers are not quantitative data, it would not be appropriate to use the methods of this section.

3. Correlation is the existence of a relationship between two variables – so that knowing the value of one of the variables allows a researcher to make a reasonable inference about the value of the other. A lurking variable is a third variable that has a direct influence on the two variables being studied – and that may even be responsible for the correlation between them. There may be a correlation between the value of a house and the number of telephones in a house, for example, but those variables really don't affect each other directly – instead they are both a function of a lurking variable, the income of the household.

5. a. From Table A-6 for n = 8, C.V. = ±0.707. Therefore r = 0.993 indicates a significant (positive) linear correlation.
   b. The proportion of the variation in weights that can be explained by the variation in chest sizes is $r^2 = (0.993)^2 = 0.986$, or 98.6%.

7. a. From Table A-6 for n = 21, C.V. = ±0.444. Therefore r = -0.133 does not indicate a significant linear correlation.
   b. The proportion of the variation in Super Bowl points that can be explained by the variation in DJIA values is $r^2 = (-0.133)^2 = 0.017$, or 1.7%.

NOTE: In addition to the value of n, calculation of r requires five sums: $\Sigma x$, $\Sigma y$, $\Sigma x^2$, $\Sigma y^2$ and $\Sigma xy$. The next problem gives the chart to find these sums. As the sums can usually be found conveniently using a calculator and without constructing the chart, subsequent problems typically give only the values of the sums and do not show a chart.
In addition, calculation of r involves three subcalculations.
  (1) $n(\Sigma xy) - (\Sigma x)(\Sigma y)$ determines the sign of r. If large values of x are associated with large values of y, it will be positive. If large values of x are associated with small values of y, it will be negative. If not, a mistake has been made.
  (2) $n(\Sigma x^2) - (\Sigma x)^2$ cannot be negative. If it is, a mistake had been made.
  (3) $n(\Sigma y^2) - (\Sigma y)^2$ cannot be negative. If it is, a mistake had been made.
Finally, r must be between -1 and 1 inclusive. If not, a mistake has been made. If this or any of the previous mistakes occurs, stop immediately and find the error – continuing is a waste of effort.
   Exercise #9 on the following page is worked by hand and using the TI-83/Plus calculator as described above and in the text. In general, the manual uses the TI-83/84 Plus and refers to hand calculations only when they are necessary or informative.

9. [by hand]

| x | y | xy | $x^2$ | $y^2$ |
|---|---|----|----|-----|
| 1 | 3 | 3 | 1 | 9 |
| 0 | 1 | 0 | 0 | 1 |
| 5 | 15 | 75 | 25 | 225 |
| 2 | 6 | 12 | 4 | 36 |
| 3 | 8 | 24 | 9 | 64 |
| 11 | 33 | 114 | 39 | 335 |

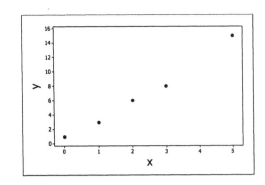

$n(\Sigma xy) - (\Sigma x)(\Sigma y) = 5(114) - (11)(33) = 207$
$n(\Sigma x^2) - (\Sigma x)^2 = 5(39) - (11)^2 = 74$
$n(\Sigma y^2) - (\Sigma y)^2 = 5(335) - (33)^2 = 586$

a. From the scatterplot, it appears that the points are very close to being on a straight line with positive slope. We expect a correlation very close to $r = +1.00$.

b. $r = [n(\Sigma xy) - (\Sigma x)(\Sigma y)]/[\sqrt{n(\Sigma x^2) - (\Sigma x)^2} \sqrt{n(\Sigma y^2) - (\Sigma y)^2}]$

$= 207/[\sqrt{74}\sqrt{586}] = 0.994$

From Table A-6 for n = 5, assuming $\alpha = 0.05$, C.V. = $\pm 0.878$. Therefore r = 0.994 indicates a significant (positive) linear correlation. This agrees with the interpretation of the scatterplot.

9. [using the TI-83/84 Plus] Place the x values in list L1 and the y values in list L2.
Use STATPLOT with L1 and L2 and ZOOM 9.
Use STAT TESTS LinRegTTest with L1 and L2.
The set-up and resulting display screens are given below.
In general, only the resulting screens will be given for subsequent exercises.

 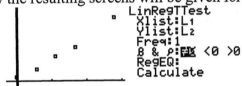 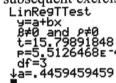

a. From the scatterplot, it appears that the points are very close to being on a straight line with positive slope. We expect a correlation very close to $r = +1.00$.

b. r = 0.994
P-value = 5.51E-4 = 0.0006 < 0.05 indicates a significant (positive) correlation. This agrees with the interpretation of the scatterplot.

11. The following x and y values may be read from the display.
    x: 1 1 1 2 2 2 3 3 3 10
    y: 1 2 3 1 2 3 1 2 3 10
Place the 10 x values in list L1 and the 10 y values in list L2.
Omitting the (10,10) outlier, place the remaining 9 x and y values in lists L3 and L4.

a. There appears to be a strong positive linear correlation, with r close to 1.
b. Use STAT TESTS LinRegTTest with L1 and L2.
  r = 0.906
  P-value = 3.08E-4 = 0.0003 < 0.05 indicates a significant (positive) linear correlation. This agrees with the interpretation of the scatterplot.

## 188   CHAPTER 10   Correlation and Regression

part (b) screen results [all 10 points]

```
LinRegTTest        LinRegTTest
 y=a+bx             y=a+bx
 β≠0 and ρ≠0        β≠0 and ρ≠0
 t=6.041441045      ↑b=.9056603774
 P=3.0883124E-4     s=1.19551047
 df=8               r²=.8202207191
 ↓a=.2641509434     r=.9056603774
```

part (c) screen results [point (10,10) removed]

```
LinRegTTest        LinRegTTest
 y=a+bx             y=a+bx
 β≠0 and ρ≠0        β≠0 and ρ≠0
 t=0                t=0
 P=1                P=1
 df=7               df=7
 ↓a=2               ↓a=2
```

c. There appears to be no linear correlation, with r close to 0.
   Use STAT TESTS LinRegTTest with L3 and L4.
   r = 0.000
   P-value = 1.0000 > 0.05 does not indicate a significant linear correlation. This agrees with the interpretation of the scatterplot.

d. The effect of a single pair of values can be dramatic, changing the conclusion entirely.

NOTE: In each of exercises #13-#28 the first variable listed is designated x, and the second variable listed is designated y. In correlation problems the designation of x and y is arbitrary – so long as a person remains consistent after making the designation. In each test of hypothesis, the C.V. and test statistic are given in terms of t using Method 1. The usual t formula written for r is

$$t_r = (r - \mu_r)/s_r, \text{ where } \mu_r = \rho = 0 \text{ and } s_r = \sqrt{(1-r^2)/(n-2)} \text{ and } df = n-2.$$

One advantage of Method 1 is that using the t statistic allows the calculation of exact P-values. <u>For those using Method 2, the C.V. in terms of r is indicated on the accompanying graph – and the test statistic is simply r.</u> The two methods are mathematically equivalent and always agree.

13. Place the heights in list L1 and the intervals in list L2.
    Use STATPLOT ZOOM 9 and STAT TESTS LinRegTTest.

```
LinRegTTest        LinRegTTest
 y=a+bx             y=a+bx
 β≠0 and ρ≠0        β≠0 and ρ≠0
 t=.6851410226      ↑b=.2464646465
 P=.5188332087      s=12.65456626
 df=6               r²=.0725595726
 ↓a=54.26767677     r=.2693688411
```

a. The scatterplot is given above at the left.

b. The screen at the right indicates r = 0.2694

c. $H_o: \rho = 0$
   $H_1: \rho \neq 0$
   $\alpha = 0.05$ and df = 6
   C.V. $t = \pm t_{\alpha/2} = \pm 2.447$
   calculations:
   $t_r = (r - \mu_r)/s_r = 0.685$
   P-value = $P(t_6 > 0.685) = 0.5188$

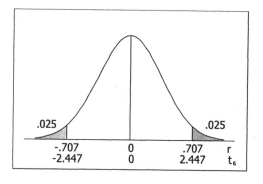

   conclusion:
       Do not reject $H_o$; there is not sufficient evidence to conclude that $\rho \neq 0$.
   No. In general, higher eruptions signify a greater release of pent up energy and are followed by longer intervals of "recovery" required to build up enough energy for another eruption – but the relationship is not strong enough to be statistically significant in a sample of size n = 8.

15. Place the budget amounts in list L1 and the gross amounts in list L2.
    Use STATPLOT ZOOM 9 and STAT TESTS LinRegTTest.

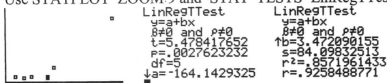

a. The scatterplot is given above at the left.

b. The screen at the right indicates r = 0.9258

c. $H_o$: $\rho = 0$
   $H_1$: $\rho \neq 0$
   $\alpha = 0.05$ and df = 5
   C.V.  t = $\pm t_{\alpha/2}$ = $\pm 2.571$
   calculations:
       $t_r = (r - \mu_r)/s_r$ = 5.478
       P-value = $2 \cdot P(t_5 > 5.478)$ = 0.0028
   conclusion:

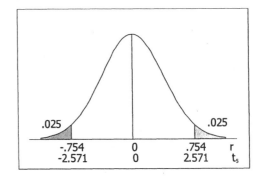

   Reject $H_o$; there is sufficient evidence to conclude that $\rho \neq 0$ (in fact, that $\rho > 0$).
Yes; there does appear to be a significant positive correlation between budget and gross – but the scatterplot indicates that this could be due to the effect of an outlier. Other important factors that affect the amount a movie earns are its quality, its publicity and its stars.

17. Place the chest sizes in list L1 and the weights in list L2.
    Use STATPLOT ZOOM 9 and STAT TESTS LinRegTTest.

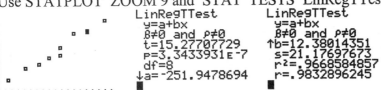

a. The scatterplot is given above at the left.

b. The screen at the right indicates r = 0.9833

c. $H_o$: $\rho = 0$
   $H_1$: $\rho \neq 0$
   $\alpha = 0.05$ and df = 8
   C.V.  t = $\pm t_{\alpha/2}$ = $\pm 2.306$
   calculations:
       $t_r = (r - \mu_r)/s_r$ = 15.277
       P-value = $2 \cdot P(t_8 > 15.277)$ = 3.34E-7 $\approx$ 0
   conclusion:

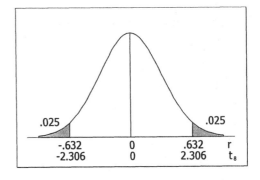

   Reject $H_o$; there is sufficient evidence to conclude that $\rho \neq 0$ (in fact, that $\rho > 0$).
Yes. Bear chest sizes may be used to estimate bear weights – without having to maneuver anesthetized bears or scales in the field.

190 CHAPTER 10 Correlation and Regression

19. Place the systolic readings in list L1 and the diastolic readings in list L2.
Use STATPLOT ZOOM 9 and STAT TESTS LinRegTTest.

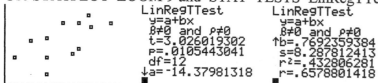

a. The scatterplot is given above at the left.

b. The screen at the right indicates r = 0.6579

c. $H_o$: $\rho = 0$
$H_1$: $\rho \neq 0$ and df = 12
$\alpha = 0.05$
C.V. $t = \pm t_{\alpha/2} = \pm 2.179$
calculations:
$t_r = (r - \mu_r)/s_r = 3.026$
P-value = $2 \cdot P(t_{12} > 3.026) = 0.0105$
conclusion:

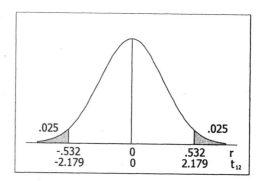

Reject $H_o$; there is sufficient evidence to conclude that $\rho \neq 0$ (in fact, that $\rho > 0$).
Yes; high systolic readings are associated with high diastolic readings. One important issue suggested by the data is the variability of the scores – is it due to differences among the students, differences among the equipment, or natural fluctuations within the patient?

21. Place the salaries in list L1 and the viewer numbers in list L2.
Use STATPLOT ZOOM 9 and STAT TESTS LinRegTTest.

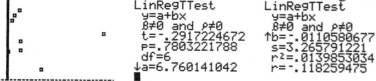

a. The scatterplot is given above at the left.

b. The screen at the right indicates r = -0.1183

c. $H_o$: $\rho = 0$
$H_1$: $\rho \neq 0$ and df = 6
$\alpha = 0.05$
C.V. $t = \pm t_{\alpha/2} = \pm 2.447$
calculations:
$t_r = (r - \mu_r)/s_r = -0.292$
P-value = $2 \cdot P(t_6 < -0.292) = 0.7803$
conclusion:

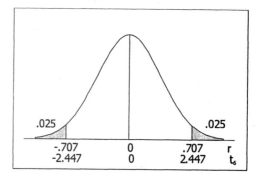

Do not reject $H_o$; there is not sufficient evidence to conclude that $\rho \neq 0$.
No; there is not a significant correlation between the salaries of stars and the number of viewers they attract. Susan Lucci has the lowest cost per viewer (23.81 cents), and Kelsey Grammer has the highest cost per viewer ($22.00).

23. Place the temperatures in list L1 and the times in list L2.
    Use STATPLOT ZOOM 9 and STAT TESTS LinRegTTest.

    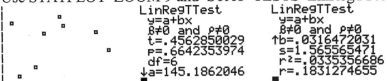

    a. The scatterplot is given above at the left.

    b. The screen at the right indicates r = 0.1831

    c. $H_o: \rho = 0$
       $H_1: \rho \neq 0$
       $\alpha = 0.05$ and df = 6
       C.V. $t = \pm t_{\alpha/2} = \pm 2.447$
       calculations:
       $t_r = (r - \mu_r)/s_r = 0.456$
       P-value = $2 \cdot P(t_6 > 0.456) = 0.6642$
       conclusion:
       Do not reject $H_o$; there is not sufficient evidence to conclude that $\rho \neq 0$.
    No; it does not appear that the winning time is affected by temperature. One factor complicating such a study is that winning times in most events tend to decrease over the years. Except for the point in the upper left corner of the scatterplot, the paired data seem to form a "∩" pattern. Intuitively, one might expect just the opposite if there were a temperature effect – i.e., a "" ∪ pattern, indicating high times on very cold or very hot days and an optimal middle range temperature at which the lowest times occur.

    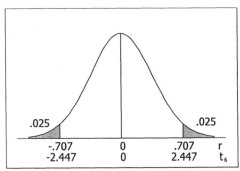

25. Place the chirp numbers in list L1 and the temperatures in list L2.
    Use STATPLOT ZOOM 9 and STAT TESTS LinRegTTest.

    a. The scatterplot is given above at the left.

    b. The screen at the right indicates r = 0.8737

    c. $H_o: \rho = 0$
       $H_1: \rho \neq 0$
       $\alpha = 0.05$ and df = 6
       C.V. $t = \pm t_{\alpha/2} = \pm 2.447$
       calculations:
       $t_r = (r - \mu_r)/s_r = 4.399$
       P-value = $2 \cdot P(t_6 > 4.399) = 0.0046$
       conclusion:
       Reject $H_o$; there is sufficient evidence to conclude that $\rho \neq 0$ (in fact, that $\rho > 0$).
    Yes; the higher the temperature, the more cricket chirps per minute.

    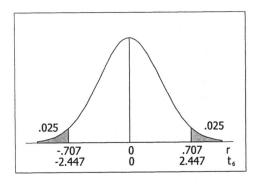

## 192  CHAPTER 10  Correlation and Regression

**27.** Place the heights in list L1 and the pulse rates in list L2.
Use STATPLOT ZOOM 9 and STAT TESTS LinRegTTest.

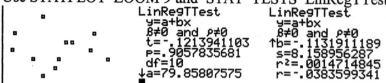

a. The scatterplot is given above at the left.

b. The screen at the right indicates r = -0.0384

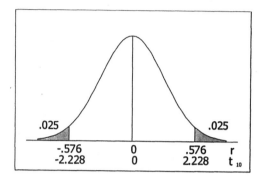

c. $H_o: \rho = 0$
$H_1: \rho \neq 0$
$\alpha = 0.05$ and df = 10
C.V. $t = \pm t_{\alpha/2} = \pm 2.228$
calculations:
$t_r = (r - \mu_r)/s_r = -0.121$
P-value = $2 \cdot P(t_{10} < -0.121) = 0.9058$
conclusion:
  Do not reject $H_o$; there is not sufficient evidence to conclude that $\rho \neq 0$.
No; the medical student's hypothesis does not appear to be correct.

**29.** Use STATPLOT ZOOM 9 and STAT TESTS LinRegTTest on lists HMLST and HMSP.

a. The scatterplot is given above at the left.

b. The screen at the right indicates r = 0.9949

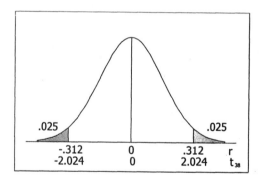

c. $H_o: \rho = 0$
$H_1: \rho \neq 0$
$\alpha = 0.05$ and df = 38
C.V. $t = \pm t_{\alpha/2} = \pm 2.024$
calculations:
$t_r = (r - \mu_r)/s_r = 60.804$
P-value = $2 \cdot P(t_{38} > 60.804) = 1.79\text{E-}39 \approx 0$
conclusion:
  Reject $H_o$; there is sufficient evidence to conclude that $\rho \neq 0$ (in fact, that $\rho > 0$).

31. a. Use STATPLOT ZOOM 9 and STAT TESTS LinRegTTest on lists TAR and NICOT.

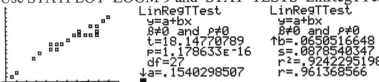

The scatterplot is given above at the left.
The screen at the right indicates r = 0.9614

$H_o: \rho = 0$
$H_1: \rho \neq 0$
$\alpha = 0.05$ and df = 27
C.V. $t = \pm t_{\alpha/2} = \pm 2.052$
calculations:
$\quad t_r = (r - \mu_r)/s_r = 18.148$
P-value = $2 \cdot P(t_{27} > 18.148) = 1.18\text{E-}16 \approx 0$
conclusion:
Reject $H_o$; there is sufficient evidence to conclude that $\rho \neq 0$ (in fact, that $\rho > 0$).
Yes; since the two variables are highly correlated, measuring both of them is unnecessary.

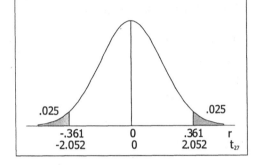

b. Use STATPLOT ZOOM 9 and STAT TESTS LinRegTTest on the lists CO and NICOT.

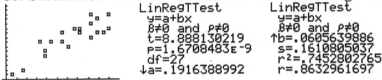

The scatterplot is given above at the left.
The screen at the right indicates r = 0.8633

$H_o: \rho = 0$
$H_1: \rho \neq 0$
$\alpha = 0.05$ and df = 27
C.V. $t = \pm t_{\alpha/2} = \pm 2.052$
calculations:
$\quad t_r = (r - \mu_r)/s_r = 8.888$
P-value = $2 \cdot P(t_{27} > 8.888) = 1.67\text{E-}9 \approx 0$
conclusion:
Reject $H_o$; there is sufficient evidence to conclude that $\rho \neq 0$ (in fact, that $\rho > 0$).

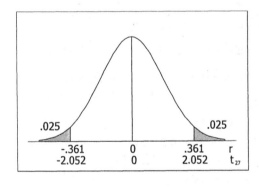

Yes; since the two variables are highly correlated, measuring both of them is unnecessary.

c. Tar, because it has a higher correlation with nicotine.

33. A linear correlation coefficient very close to zero indicates no significant linear correlation and no tendencies can be inferred.

35. A linear correlation coefficient very close to zero indicates no significant linear correlation, but there may be some other type of relationship between the variables.

**194** CHAPTER 10 Correlation and Regression

37. Place the x values in list L1 and the y values in list L2.

   a. Use STATPLOT ZOOM 9 and STAT TESTS LinRegTTest on L1 and L2.

   ```
   LinRegTTest              LinRegTTest
   y=a+bx                   y=a+bx
   β≠0 and ρ≠0              β≠0 and ρ≠0
   t=9.179769698            ↑b=.1759019741
   P=2.5730806E-4           s=.0392568821
   df=5                     r²=.9439889951
   ↓a=-.0631109598          r=.9715909608
   ```

   The scatterplot is given above at the left. The screen at the right indicates $r = 0.9716$.
   P-value = $0.0003 < 0.05$ indicates a significant linear correlation between x and y.

   b. $L1^2 \to L3$. Use STATPLOT ZOOM 9 and STAT TESTS LinRegTTest on L3 and L2.

   ```
   LinRegTTest              LinRegTTest
   y=a+bx                   y=a+bx
   β≠0 and ρ≠0              β≠0 and ρ≠0
   t=4.765532111            ↑b=.0295384511
   P=.0050346872            s=.070460146
   df=5                     r²=.8195616552
   ↓a=.1780211792           r=.905296446
   ```

   The scatterplot is given above at the left. The screen at the right indicates $r = 0.9053$.
   P-value = $0.0050 < 0.05$ indicates a significant linear correlation between $x^2$ and y.

   c. $\log(L1) \to L4$. Use STATPLOT ZOOM 9 and STAT TESTS LinRegTTest on L4 and L2.

   ```
   LinRegTTest              LinRegTTest
   y=a+bx                   y=a+bx
   β≠0 and ρ≠0              β≠0 and ρ≠0
   t=66.44115736            ↑b=1.004509551
   P=1.4623941E-8           s=.0055793181
   df=5                     r²=.9988686317
   ↓a=-.001306876           r=.9994341558
   ```

   The scatterplot is given above at the left. The screen at the right indicates $r = 0.9994$
   P-value = $0.00000001 < 0.05$ indicates a significant linear correlation between $\log(x)$ and y.

   d. $\sqrt{(L1)} \to L5$. Use STATPLOT ZOOM 9 and STAT TESTS LinRegTTest on L5 and L2.

   ```
   LinRegTTest              LinRegTTest
   y=a+bx                   y=a+bx
   β≠0 and ρ≠0              β≠0 and ρ≠0
   t=17.10597594            ↑b=.568087719
   P=1.2496623E-5           s=.0214999348
   df=5                     r²=.9831997384
   ↓a=-.5112469034          r=.9915642886
   ```

   The scatterplot is given above at the left. The screen at the right indicates $r = 0.9916$.
   P-value = $0.00001 < 0.05$ indicates a significant linear correlation between $\sqrt{x}$ and y.

   e. $L1^{-1} \to L6$. Use STATPLOT ZOOM 9 and STAT TESTS LinRegTTest on L6 and L2.

   ```
   LinRegTTest              LinRegTTest
   y=a+bx                   y=a+bx
   β≠0 and ρ≠0              β≠0 and ρ≠0
   t=-12.24561259           ↑b=-.8964633842
   P=6.4248224E-5           s=.0297962279
   df=5                     r²=.9677325733
   ↓a=.7784385538           r=-.9837339952
   ```

   The scatterplot is given above at the left. The screen at the right indicates $r = 0.9837$.
   P-value = $0.00006 < 0.05$ indicates a significant linear correlation between $1/x$ and y.

   While all the correlations are significant, the largest value for r occurs in part (c).

39. For r = 0.600, (1+r)/(1−r) = (1.6)/(0.4) = 4. Following the procedure in the text with $\alpha = 0.05$,
Step a. $z_{\alpha/2} = 1.960$
Step b. $W_L = \frac{1}{2} \cdot \ln(4) - 1.960/\sqrt{47} = 0.407$
$W_R = \frac{1}{2} \cdot \ln(4) + 1.960/\sqrt{47} = 0.979$
Step c. $(e^{.814} - 1)/(e^{.814} + 1) < \rho < (e^{1.958} - 1)/(e^{1.958} + 1)$
$(1.258)/(3.258) < \rho < (6.086)/(8.086)$
$0.386 < \rho < 0.753$

NOTE: While the distribution of r is not normal, the there is a transformation for which the transformed r's have a normal distribution. That transformation [which is actually the inverse hyperbolic tangent function] is $T(r) = \tanh^{-1}(r)$. Most mathematical/scientific calculators have the tanh(x) and $\tanh^{-1}$ functions.

$T(r)$ is normally distributed with $\mu_{T(r)} = T(\rho)$ and $\sigma_{T(r)} = \sqrt{1/(n-3)}$.

- The confidence interval for exercise #39 may be found by constructing a confidence interval for $T(\rho) = \tanh^{-1}(\rho)$ and then untransforming using the tanh(x) function.

$T(r) \pm z_{\alpha/2} \cdot \sigma_{T(r)}$
$\tanh^{-1}(0.600) \pm 1.96 \sqrt{1/47}$
$0.693 \pm 0.286$
$0.407 < T(\rho) < 0.979$
$0.386 < \rho < 0.753$

- The value of the power for exercise #40 may be found in the same manner as illustrated in the following solution.
- The hyperbolic tangent function is defined as $y = (e^x - e^{-x})/(e^x + e^{-x})$. The inverse hyperbolic tangent function is found by solving for x in terms of y to get $x = (\frac{1}{2}) \cdot \ln[(1+y)/(1-y)]$, which is the transformation for r given in the text. The formulas in the text are the usual normal formulas with $(\frac{1}{2}) \cdot \ln[(1+r)/(1-r)]$ substituted for r and with $\sqrt{1/(n-3)}$ substituted for the standard deviation.

## 10-3 Regression

1. In the model $y = \beta_0 + \beta_1 x$, the independent variable x is called the predictor variable and the dependent variable y is called the response variable. Conceptually, a given x value is used to predict a value for y and the y value is considered the resulting response for a given x value.

3. It is not unusual in an applied problem for one regression line to apply within a certain range of x values, and for a different regression line to apply within a different range of x values. When a regression line is found using a given range of x values, therefore, it might not be the appropriate regression line to predict y for x values outside of that given range.

5. For n=20, C.V. = ±0.444. Since r = 0.870 > 0.444, use the regression line for prediction.
$\hat{y} = -3.22 + 1.02x$
$\hat{y}_{110} = -3.22 + 1.02(110) = 109.0$

7. For n=16, C.V. = ±0.497. Since r = 0.765 > 0.497, use the regression line for prediction.
$\hat{y} = 3.46 + 1.01x$
$\hat{y}_{.40} = 3.46 + 1.01(0.40) = 3.86$ calories/gram

196  CHAPTER 10  Correlation and Regression

NOTE: Exercises #9 is worked by hand and using the TI-83/84 Plus calculator. Subsequent exercises are worked using only the calculator. The manual uses the following conventions
- Following the TI notation "a" and "b" are used for the y-intercept and the slope. Following the text, $\hat{y}$ is used to distinguish the "predicted y" from an actual y. And so the regression line is given by $\hat{y} = a + bx$ instead of the notation $\hat{y} = b_0 + b_1 x$ or the notation $y = a + bx$.
- When the correlation is not significant, 1-Var Stats is used on the list of the y's to find the $\bar{y}$ necessary for making predictions – but the TI screen uses the generic "x" label, even though the statistics are for what is being called "y" in the exercise. [See exercise #9 below.]

The calculator solution of exercise #9 gives the format used in the remainder of the manual. From the left screen, a = 0.857. From the right screen, b = 0.643. The ↓ arrow before the "a" and the ↑ arrow before the "b" are not part of the notation. The arrows indicate that more display information may be obtained by using keyboard to scroll down or up.

9. [by hand]  $\bar{x} = 2.4$                                                       $n = 5$
    $\bar{y} = 2.4$                                                                                 $\Sigma x = 12$
    $b_1 = [n(\Sigma xy) - (\Sigma x)(\Sigma y)]/[n(\Sigma x^2) - (\Sigma x)^2]$     $\Sigma y = 12$
    $= 36/56 = 0.643$                                                                     $\Sigma x^2 = 40$
    $b_0 = \bar{y} - b_1 \bar{x}$                                                          $\Sigma y^2 = 42$
    $= 2.4 - 0.643(2.4) = 0.857$                                                    $\Sigma xy = 36$
    $\hat{y} = b_0 + b_1 x$
    $\hat{y} = 0.857 + 0.643x$

9. [by calculator] Place the x values in list L1 and the y values in list L2.
   Use STAT TESTS LinRegTTest on L1 and L2. [and 1-Var Stats on L2]

   ```
   LinRegTTest            LinRegTTest            1-Var Stats
   y=a+bx                 y=a+bx                 x̄=6.6
   β≠0 and ρ≠0            β≠0 and ρ≠0            Σx=33
   t=1.272792206          ↑b=.6428571429         Σx²=335
   p=.292775452           s=1.690308509          Sx=5.412947441
   df=3                   r²=.3506493506         σx=4.841487375
   ↓a=.8571428571         r=.5921565255          ↓n=5
   ```

   $\hat{y} = a + bx = 0.857 + 0.643$

11. The following x and y values may be read from the display.
    x: 1 1 1 2 2 2 3 3 3 10
    y: 1 2 3 1 2 3 1 2 3 10
    Place the 10 x values in list L1 and the 10 y values in list L2.
    Omitting the (10,10) outlier, place the remaining 9 x and y values in lists L3 and L4.
    Use STAT TESTS LinRegTTest.

    part (a) screen results [all 10 points]              part (b) screen results [point (10,10) removed]
    ```
    LinRegTTest            LinRegTTest            LinRegTTest            LinRegTTest
    y=a+bx                 y=a+bx                 y=a+bx                 y=a+bx
    β≠0 and ρ≠0            β≠0 and ρ≠0            β≠0 and ρ≠0            β≠0 and ρ≠0
    t=6.041441045          ↑b=.9056603774         t=0                    t=0
    p=3.0883124E-4         s=1.19551047           p=1                    p=1
    df=8                   r²=.8202207191         df=7                   df=7
    ↓a=.2641509434         r=.9056603774          ↓a=2                   ↓a=2
    ```

    a. $\hat{y} = a + bx$
       $\hat{y} = 0.264 + 0.906x$
    b. $\hat{y} = a + bx$
       $\hat{y} = 2.0 + 0x$  [or simply $\hat{y} = 2.0$, for any x]
    c. The results are very different – without the outlier, x has no predictive value for y. A single outlier can have a dramatic effect on the regression equation.

13. Place the heights in list L1 and the intervals in list L2.
    Use STAT TESTS LinRegTTest.     [and 1-Var Stats on L2]
    ```
    LinRegTTest          LinRegTTest          1-Var Stats
     y=a+bx               y=a+bx               x̄=86
     β≠0 and ρ≠0          β≠0 and ρ≠0         Σx=688
     t=.6851410226        ↑b=.2464646465      Σx²=60204
     p=.5188332087        s=12.65456626       Sx=12.16552506
     df=6                 r²=.0725595726      σx=11.37980668
    ↓a=54.26767677        r=.2693688411       ↓n=8
    ```

    $\hat{y} = a + bx = 54.27 + 0.246x$

    $\hat{y}_{100} = \bar{y} = 86.00$ min [no significant correlation]

    Following an eruption with a height of 100 feet, the best predicted interval until the next eruption is 86 minutes.

15. Place the budget amounts in list L1 and the gross amounts in list L2.
    Use STAT TESTS LinRegTTest.
    ```
    LinRegTTest          LinRegTTest
     y=a+bx               y=a+bx
     β≠0 and ρ≠0          β≠0 and ρ≠0
     t=5.478417652        ↑b=3.472090155
     p=.0027623232        s=84.09832513
     df=5                 r²=.8571961433
    ↓a=-164.1429325       r=.9258488771
    ```

    $\hat{y} = a + bx = -164.14 + 3.472x$

    $\hat{y}_{40} = -164.14 + 3.472(40) = -25.26$ million dollars

    For a movie with a 40 million dollar budget, the best predicted gross amount is a 25 million dollar loss – assuming a linear relationship between the budget and the gross. Examination of the data points, however, suggests that -25 is not the appropriate predicted value for 40. The one movie with a large budget and large gross may be an outlier having a dramatic effect on the linear regression equation, or it could be that a non-linear model might better fit the data.

17. Place the chest sizes in list L1 and the weights in list L2.
    Use STAT TESTS LinRegTTest.
    ```
    LinRegTTest          LinRegTTest
     y=a+bx               y=a+bx
     β≠0 and ρ≠0          β≠0 and ρ≠0
     t=15.27707729        ↑b=12.38014351
     p=3.3433931E-7       s=21.17697673
     df=8                 r²=.9668584857
    ↓a=-251.9478694       r=.9832896245
    ```

    $\hat{y} = a + bx = -251.95 + 12.380x$

    $\hat{y}_{50} = -251.95 + 12.380(50) = 367.06$ lbs

    For a bear with a chest size of 50 inches, the best predicted weight is 367 pounds.

19. Place the systolic readings in list L1 and the diastolic readings in list L2.
    Use STAT TESTS LinRegTTest.
    ```
    LinRegTTest          LinRegTTest
     y=a+bx               y=a+bx
     β≠0 and ρ≠0          β≠0 and ρ≠0
     t=3.026019302        ↑b=.7692359384
     p=.0105443041        s=8.287812413
     df=12                r²=.432806281
    ↓a=-14.37981318       r=.6578801418
    ```

    $\hat{y} = a + bx = -14.38 + 0.769x$

    $\hat{y}_{140} = -14.38 + 0.769(140) = 93.31$ mmHg

    For a person with a systolic blood pressure reading of 140, the best predicted diastolic reading is 93 mmHg.

198   CHAPTER 10   Correlation and Regression

21. Place the salaries in list L1 and the viewer numbers in list L2.
    Use STAT TESTS LinRegTTest.   [and 1-Var Stats on L2]

    ```
    LinRegTTest           LinRegTTest           1-Var Stats
     y=a+bx                y=a+bx                x̄=6.5
     β≠0 and ρ≠0           β≠0 and ρ≠0           Σx=52
     t=-.2917224672        ↑b=-.0110580677       Σx²=402.9
     p=.7803221788         s=3.265791221         Sx=3.044902063
     df=6                  r²=.0139853034        σx=2.848245074
    ↓a=6.760141042         r=-.118259475        ↓n=8
    ```

    $\hat{y} = a + bx = 6.76 - 0.0111x$

    $\hat{y}_2 = \bar{y} = 6.50$ million viewers [no significant correlation]

    For a star with a salary of $2,000,000, the best predicted number of viewers is 6.5 million.

23. Place the temperatures in list L1 and the times in list L2.
    Use STAT TESTS LinRegTTest.   [and 1-Var Stats on L2]

    ```
    LinRegTTest           LinRegTTest           1-Var Stats
     y=a+bx                y=a+bx                x̄=147.077125
     β≠0 and ρ≠0           β≠0 and ρ≠0           Σx=1176.617
     t=.4562850029         ↑b=.0316472031        Σx²=173068.662
     p=.6642353974         s=1.565565471         Sx=1.4743647
     df=6                  r²=.0335356686        σx=1.379141892
    ↓a=145.1862046         r=.1831274655        ↓n=8
    ```

    $\hat{y} = a + bx = 145.19 + 0.0316x$

    $\hat{y}_{73} = \bar{y} = 147.077$  [no significant correlation]

    On a day when the temperature is 73°F, the predicted female winning time is 147.077 minutes. That compares reasonably well to the actual winning time of 150.750 minutes.

25. Place the chirp numbers in list L1 and the temperatures in list L2.
    Use STAT TESTS LinRegTTest.

    ```
    LinRegTTest           LinRegTTest
     y=a+bx                y=a+bx
     β≠0 and ρ≠0           β≠0 and ρ≠0
     t=4.398877495         ↑b=.0522722255
     p=.0045739116         s=4.269509134
     df=6                  r²=.7633147599
    ↓a=27.62835082         r=.8736788654
    ```

    $\hat{y} = a + bx = 27.63 + 0.0523x$

    $\hat{y}_{1000} = 27.63 + 0.0523(1000) = 79.90$ °F

    When a cricket is producing 1000 chirps per minute, the best predicted temperature is 79.9°F.

27. Place the heights in list L1 and the pulse rates in list L2.
    Use STAT TESTS LinRegTTest.   [and 1-Var Stats on L2]

    ```
    LinRegTTest           LinRegTTest           1-Var Stats
     y=a+bx                y=a+bx                x̄=72.66666667
     β≠0 and ρ≠0           β≠0 and ρ≠0           Σx=872
     t=-.1213941103        ↑b=-.1131911189       Σx²=64032
     p=.9057835681         s=8.158956287         Sx=7.784989442
     df=10                 r²=.0014714845        σx=7.453559925
    ↓a=79.85807575         r=-.0383599341       ↓n=12
    ```

    $\hat{y} = a + bx = 79.86 - 0.113x$

    $\hat{y}_{66} = \bar{y} = 72.67$ beats/minute [no significant correlation]

    For a woman who is 66 inches tall, the best predicted pulse rate is 73. Since all the pulse rates are multiples of 4 (as a result of counting for 15 seconds and multiplying by 4?), it appears that 73 might not be a possible rate in this situation – and one could argue that the best predicted rate in this context, following the practice of making predictions consistent with the accuracy of the original data, is 72.

29. Use STAT TESTS LinRegTTest on lists HMLST and HMSP.

```
LinRegTTest              LinRegTTest
 y=a+bx                   y=a+bx
 β≠0 and ρ≠0              β≠0 and ρ≠0
 t=60.80384559            ↑b=.9789557162
 p=1.792377E-39           s=8505.412608
 df=38                    r²=.9898262637
↓a=99.1840667             r=.9949001275
```

$\hat{y} = a + bx = 99.18 + 0.979x$

$\hat{y}_{400,000} = 99.18 + 0.979(400000) = 391{,}699.18$ dollars

For a house that lists for \$400,000, the best predicted selling price is \$391,699.

31. a. Use STAT TESTS LinRegTTest on lists TAR and NICOT.

```
LinRegTTest              LinRegTTest
 y=a+bx                   y=a+bx
 β≠0 and ρ≠0              β≠0 and ρ≠0
 t=18.14770789            ↑b=.0650516648
 p=1.178633E-16           s=.0878540347
 df=27                    r²=.9242295198
↓a=.1540298507            r=.961368566
```

$\hat{y} = a + bx = 0.154 + 0.0651x$

$\hat{y}_{15} = 0.154 + 0.0651(15) = 1.1305$ mg

For a cigarette with 15 mg of tar, the best predicted amount of nicotine is 1.1 mg.

b. Use STAT TESTS LinRegTTest on the lists CO and NICOT.

```
LinRegTTest              LinRegTTest
 y=a+bx                   y=a+bx
 β≠0 and ρ≠0              β≠0 and ρ≠0
 t=8.888130219            ↑b=.0605639886
 p=1.6708483E-9           s=.1610805037
 df=27                    r²=.7452802765
↓a=.1916388992            r=.8632961697
```

$\hat{y} = a + bx = 0.192 + 0.0606x$

$\hat{y}_{15} = 0.192 + 0.0606(15) = 1.101$ mg

For a cigarette with 15 mg of CO, the best predicted amount of nicotine is 1.1 mg.

33. If $H_o: \rho=0$ is true, there is no linear correlation between x and y and $\hat{y}=\bar{y}$ is the appropriate prediction for y for any x.

If $H_o: \beta_1=0$ is true, then the true regression line is $y = \beta_o + 0x = \beta_o$ and the best estimate for $\beta_o$ is $b_o = \bar{y} - 0\bar{x} = \bar{y}$, producing the line $\hat{y}=\bar{y}$.

Since both hypotheses imply precisely the same result, they are equivalent.

35. The scatterplot of the original data is given below in the screen at the left. While other scatterplots in this chapter appear to have a non-linear pattern, these data appear to lie almost exactly in a straight line and are not in need of being transformed. The analyses that follow confirm that a linear regression using the original x values (r = 0.9968) is a better fit than a linear regression using the ln(x) values (r = 0.9631).

Place the x values in list L1 and the y values in list L2. Use ln(L1)→L3.
Use STATPLOT ZOOM 9 and STAT TESTS LinRegTTest on lists L1 and L2.
Use STAT TESTS LinRegTTest on lists L3 and L2.

200  CHAPTER 10  Correlation and Regression

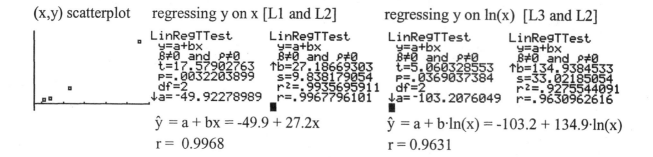

$\hat{y} = a + bx = -49.9 + 27.2x$    $\hat{y} = a + b \cdot \ln(x) = -103.2 + 134.9 \cdot \ln(x)$
$r = 0.9968$                          $r = 0.9631$

**10-4 Variation and Prediction Intervals**

1. A prediction interval is a confidence interval on a predicted value for y.

3. For each ordered pair (x,y), the explained deviation is the difference between the mean y value and the y value predicted by the regression equation. It has the original units of the problem and is equal to $\hat{y} - \bar{y}$. The explained variation is the sum of the squares of the explained deviations. It has the square of the units of the original problem and is equal to $\Sigma(\hat{y} - \bar{y})^2$.

5. The coefficient of determination is $r^2 = (0.3)^2 = 0.09$.
   The portion of the total variation in y explained by the regression is $r^2 = 0.09 = 9\%$.

7. The coefficient of determination is $r^2 = (-0.901)^2 = 0.812$.
   The portion of the total variation in y explained by the regression is $r^2 = 0.812 = 81.2\%$.

9. Since $b_1$ and r have the same sign, $b_1 = 0.0651$ implies that $r > 0$.
   Since $r^2 = 92.4\% = 0.924$ and $r > 0$, $r = +\sqrt{0.924} = 0.961$.
   From Table A-6 for n = 29, C.V. = ±0.361. Therefore r = 0.961 indicates a significant (positive) linear correlation.

11. $\hat{y}_{17} = 1.2599$, from the "fit" on the Minitab display. This also agrees with using the regression equation to obtain $\hat{y}_{17} = 0.154 + 0.0651(17) = 1.26$ mg.

13. Place the weights in list L1 and the consumptions in list L2.
    Use STAT TESTS LinRegTTest.    [L2 = y]

    Use 54.6950377 – 0.0079689183×L1 → L3 to place the $\hat{y}$ values in list L3.

    a. The explained variation is $\Sigma(\hat{y} - \bar{y})^2 = 147.397$
       Use L3 – 30.71428571 → L4 and the $\Sigma x^2$ from using 1-Var Stats on L4.

    b. The unexplained variation is $\Sigma(y - \hat{y})^2 = 18.032$
       Use L2 – L3 → L5 and the $\Sigma x^2$ from using 1-Var Stats on L5.

    c. The total variation is $\Sigma(y - \bar{y})^2 = 165.429$
       Use L2 – 30.71428571 → L6 and the $\Sigma x^2$ from using 1-Var Stats on L6
       check: explained variation + unexplained variation = total variation
       $147.397 + 18.032 = 165.429$

Variation and Prediction Intervals  SECTION 10-4   201

d. $r^2 = \Sigma(\hat{y}-\bar{y})^2 / \Sigma(y-\bar{y})^2 = 147.397/165.429 = 0.8910$         check: TI $r^2 = 0.8910$

e. $s_e^2 = \Sigma(y-\hat{y})^2/(n-2) = 18.032/5 = 3.6064$

   $s_e = \sqrt{3.6064} = 1.89905$         check: TI s = 1.89905

15. Place the systolic readings in list L1 and the diastolic readings in list L2.
    Use STAT TESTS LinRegTTest.    [L2 = y]

    ```
    LinRegTTest          LinRegTTest          1-Var Stats
    y=a+bx               y=a+bx               x̄=88.64285714
    β≠0 and ρ≠0          β≠0 and ρ≠0          Σx=1241
    t=3.026019302        ↑b=.7692359384       Σx²=111459
    p=.0105443041        s=8.287812413        Sx=10.57287635
    df=12                r²=.432806281        σx=10.18827858
    ↓a=-14.37981318      r=.6578801418        ↓n=14
    ```

    Use -14.37981318 + 0.7692359384×L1)→L3 to place the $\hat{y}$ values in list L3.

    a. The explained variation is $\Sigma(\hat{y}-\bar{y})^2 = 628.960$
       Use L3 − 88.64285714 → L4 and the $\Sigma x^2$ from using 1-Var Stats on L4.

    b. The unexplained variation is $\Sigma(y-\hat{y})^2 = 824.254$
       Use L2 − L3 → L5 and the $\Sigma x^2$ from using 1-Var Stats on L5.

    c. The total variation is $\Sigma(y-\bar{y})^2 = 1453.214$
       Use L2 − 88.64285714 → L6 and the $\Sigma x^2$ from using 1-Var Stats on L6
       check: explained variation + unexplained variation = total variation
                     628.960 + 828.254 = 1453.214

    d. $r^2 = \Sigma(\hat{y}-\bar{y})^2 / \Sigma(y-\bar{y})^2 = 628.960/1453.214 = 0.4328$         check: TI $r^2 = 0.4328$

    e. $s_e^2 = \Sigma(y-\hat{y})^2/(n-2) = 824.254/12 = 68.6878$

       $s_e = \sqrt{68.6878} = 8.28781$         check: TI s = 8.28781

17. Place the weights in list L1 and the consumptions in list L2.
    Use STAT TESTS LinRegTTest.    [L1 = x]

    ```
    LinRegTTest          LinRegTTest          1-Var Stats
    y=a+bx               y=a+bx               x̄=3009.285714
    β≠0 and ρ≠0          β≠0 and ρ≠0          Σx=21065
    t=-6.393021385       ↑b=-.0079689183      Σx²=65711675
    p=.001387292         s=1.89905468         Sx=621.9688401
    df=5                 r²=.8909980106       σx=575.8312536
    ↓a=54.6950377        r=-.9439269096       ↓n=7
    ```

    a. $\hat{y} = 54.69504 - 0.007968918x$

       $\hat{y}_{3700} = 54.69504 - 0.007968918(3700) = 25.210$, rounded to 25.2 mpg

    b. $n\Sigma x^2 - (\Sigma x)^2 = 7(65711765) - (21065)^2 = 16248130$
       $\alpha = 0.05$ and df = n−2 = 5

       $$\hat{y} \pm t_{\alpha/2} s_e \sqrt{1 + 1/n + n(x_o - \bar{x})^2/[n\Sigma x^2 - (\Sigma x)^2]}$$

       $\hat{y}_{3700} \pm t_{\alpha/2}(1.89905) \sqrt{1 + 1/7 + 7(3700-3009.29)^2/[16248130]}$

       $25.210 \pm (2.571)(1.89905) \sqrt{1.34839}$

       $25.210 \pm 5.670$

       $19.5 < y_{3700} < 30.9$ (mpg)

19. Place the systolic readings in list L1 and the diastolic readings in list L2.
Use STAT TESTS LinRegTTest.   [L1 = x]

```
LinRegTTest           LinRegTTest           1-Var Stats
y=a+bx                y=a+bx                x̄=133.9285714
β≠0 and ρ≠0           β≠0 and ρ≠0           Σx=1875
t=3.026019302         ↑b=.7692359384        Σx²=252179
P=.0105443041         s=8.287812413         Sx=9.042330245
df=12                 r²=.432806281         σx=8.71340745
↓a=-14.37981318       r=.6578801418         ↓n=14
```

a. ŷ = -14.37981 + 0.7692359x

   ŷ₁₂₀ = -14.37981 + 0.7692359(120) = 77.92, rounded to 78 mmHg

b. $n\Sigma x^2 - (\Sigma x)^2 = 14(252179) - (1875)^2 = 14881$

   α = 0.05 and df = n–2 = 12

   $\hat{y} \pm t_{\alpha/2} s_e \sqrt{1 + 1/n + n(x_o - \bar{x})^2/[n\Sigma x^2 - (\Sigma x)^2]}$

   $\hat{y}_{120} \pm t_{\alpha/2}(8.28781)\sqrt{1 + 1/14 + 14(120-133.929)^2/[14881]}$

   77.92 ± (2.179)(8.28781)√1.25395

   77.92 ± 20.22

   57.7 < y₁₂₀ < 98.2  (mmHg)

Exercises #21–#24 refer to the chapter problem of Table 10-1. Use the following, which are calculated and/or discussed in the text,

   n = 8   Σx = 1751   Σx² = 399451   ŷ = 34.770 + 0.23406   s_e = 4.97392

and the additional values

   x̄ = (Σx)/n = 1751/8 = 218.875    $n\Sigma x^2 - (\Sigma x)^2 = 8(399451) - (1751)^2 = 129607$

NOTE: Using a slightly different regression equation for ŷ or a slightly different value for s_e may result in slightly different values in exercises #21-#24.

21. ŷ₁₈₀ = 34.770 + 0.23406(180) = 76.901

    α = 0.01 and df = n–2 = 6

    $\hat{y} \pm t_{\alpha/2} s_e \sqrt{1 + 1/n + n(x_o - \bar{x})^2/[n\Sigma x^2 - (\Sigma x)^2]}$

    $\hat{y}_{180} \pm t_{\alpha/2}(4.97392)\sqrt{1 + 1/8 + 8(180-218.875)^2/[129607]}$

    76.901 ± (3.707)(4.97392)√1.21828

    76.901 ± 20.351

    56.5 < y₁₈₀ < 97.3  (min)

23. ŷ₂₀₀ = 34.770 + 0.23406(200) = 81.582

    α = 0.05 and df = n–2 = 6

    $\hat{y} \pm t_{\alpha/2} s_e \sqrt{1 + 1/n + n(x_o - \bar{x})^2/[n\Sigma x^2 - (\Sigma x)^2]}$

    $\hat{y}_{200} \pm t_{\alpha/2}(4.97392)\sqrt{1 + 1/8 + 8(200-218.875)^2/[129607]}$

    81.582 ± (2.447)(4.97392)√1.14699

    81.582 ± 13.035

    68.5 < y₂₀₀ < 94.6  (min)

25. Use the following, which are calculated and/or discussed in the text,
    $n = 8 \quad \Sigma x = 1751 \quad \Sigma x^2 = 399451 \quad \hat{y} = 34.770 + 0.23406 \quad s_e = 4.97392$
    and the additional values
    $\bar{x} = (\Sigma x)/n = 1751/8 = 218.875 \quad \Sigma x^2 - (\Sigma x)^2/n = 399451 - (1751)^2/8 = 16200.875$
    a. $\alpha = 0.05$ and $df = n-2 = 6$
    $$b_o \pm t_{\alpha/2} s_e \sqrt{1/n + (\bar{x})^2/[\Sigma x^2 - (\Sigma x)^2/n]}$$
    $34.770 \pm t_{\alpha/2}(4.97392)\sqrt{1/8 + (218.875)^2/[16200.875]}$
    $34.770 \pm (2.447)(4.97392)\sqrt{3.08202}$
    $34.770 \pm 21.367$
    $13.4 < \beta_o < 56.1$ (minutes)
    b. $\alpha = 0.05$ and $df = n-2 = 6$
    $$b_1 \pm t_{\alpha/2} s_e / \sqrt{\Sigma x^2 - (\Sigma x)^2/n}$$
    $0.23406 \pm t_{\alpha/2}(4.97392)/\sqrt{16200.875}$
    $0.23406 \pm (2.447)(4.97392)/\sqrt{16200.875}$
    $0.23406 \pm 0.09562$
    $0.1384 < \beta_1 < 0.3297$ (minutes/second)
    NOTE: The confidence interval for $\beta_o = y_o$ may also be found as the confidence interval [as distinguished from the prediction interval, see exercise #26] for $x = 0$.
    $\hat{y}_o = 34.770 + 0.23406(0) = 34.770$
    $\alpha = 0.05$ and $df = n-2 = 6$
    $$\hat{y} \pm t_{\alpha/2} s_e \sqrt{1/n + n(x_o - \bar{x})^2/[n\Sigma x^2 - (\Sigma x)^2]} \text{ modifies to become}$$
    $$\hat{y}_o \pm t_{\alpha/2} s_e \sqrt{1/n + n(\bar{x})^2/[n\Sigma x^2 - (\Sigma x)^2]}$$
    $34.770 \pm t_{\alpha/2}(4.97392)\sqrt{1/8 + 8(218.875)^2/[129607]}$
    $34.770 \pm (2.447)(4.97392)\sqrt{3.08202}$
    $34.770 \pm 21.367$
    $13.4 < \beta_o < 56.1$ (minutes)

## 10-5 Multiple Regression

1. Multiple regression is a method for exploring the relationship of one response variable to two or more predictor variables. It is a generalization of the regression equations in section 10-3, which used a single predictor variable.

3. No. In general, the requirements given at the beginning of the chapter for correlation and simple linear regression still apply – in particular, that the data are a random sample of independent quantitative measurements. Since eye color as typically measured is a categorical and not a quantitative variable, it is not a valid response variable in this context.

5. WEIGHT = -272 – 0.870(HEADLENGTH) + 0.554(LENGTH) + 12.2(CHEST)
   $\hat{y} = -272 - 0.870 x_1 + 0.554 x_2 + 12.2 x_3$
   NOTE: Each coefficient above is given to 3 significant digits. It may be desirable to use more accurate values in the remaining problems.

7. Yes, since the P-value is 0.000 <u>and</u> the adjusted $R^2$ is 0.9236. Although the P-value is less than 0.05, that alone in multiple regression problems does not necessarily indicate that the regression is of practical significance. Increasing the number of predictor variables, like increasing the sample size in the previous chapters, can create statistical significance that is not of practical value. The adjusted $R^2$ takes into account the number of x variables.

9. The best single predictor variable in a simple linear regression for weight is waist size, which explains about 79% of the variation in the men's weights. That regression has the highest $R^2$ (0.790) and adjusted $R^2$ (0.785), and the lowest P-value (0.000), among the three simple linear regressions.

11. Two of the given regression equations have the highest $R^2$ of 0.870 – but one of them achieves that level using only two variables. Since adding cholesterol level to height and waist size does not increase the adjusted $R^2$, the best and most efficient (statistically and financially) regression equation is $\hat{y}$ = -206 + 2.66(HEIGHT) + 2.15(WAIST).

13. Place lists NICOT,TAR,CO into a 29x3 matrix.

   a. Use PRGM A2MULREG with 1 independent variable from column 2.
```
       DF  SS           : B0=.1540298507 :
   RG  1   2.54194987   CL COEFF / T     P
   ER  27  .208394948   2  .0650516648
       F=329.34              18.15  0.000
       P=0.000
       R-SQ=.9242
       (ADJ).9214
       S=.0878540347
```
   NICOTINE = 0.154 + 0.0651(TAR)                adjusted $R^2$ = 0.9214

   b. Use PRGM A2MULREG with 1 independent variable from column 3.
```
       DF  SS           : B0=.1916388992 :
   RG  1   2.04977775   CL COEFF / T     P
   ER  27  .700567073   3  .0605639886
       F=79                  8.89   0.000
       P=0.000
       R-SQ=.7453
       (ADJ).7358
       S=.1610805037
```
   NICOTINE = 0.192 + 0.0606(CO)                 adjusted $R^2$ = 0.7358

   c. Use PRGM A2MULREG with 2 independent variables from columns 2,3.
```
       DF  SS           : B0=.1816453365 :
   RG  2   2.56691457   CL COEFF / T     P
   ER  26  .183430249   2  .0818371286
       F=181.92              8.56   0.000
       P=0.000          3  -.0186421919
       R-SQ=.9333           -1.88   .071
       (ADJ).9282
       S=.0839941045
```
   NICOTINE = 0.182 + 0.0818(TAR) – 0.0186(CO)   adjusted $R^2$ = 0.9282

   d. The best equation for predicting the amount of nicotine is the equation in (a). Although the adjusted $R^2$ is slightly higher in (c), the increase from adding another variable is negligible. In the interest of efficiency (both statistical and economic), the regression equation in (a) is the wisest choice.

   e. Yes, since the P-value = 0.000 in the "P=" line of the left screen indicates statistical significance and the adjusted $R^2$ = 0.9214 is high.

15. To find the best variables for a regression to predict selling price SELL, consider the correlations. Arranged in one table, the correlations are as follows.

```
       SELL   LIST   AREA   ACRE   AGE    TAX    ROOM   BDRM   BATH
SELL   1.000
LIST   0.995  1.000
AREA   0.802  0.808  1.000
ACRE   0.463  0.458  0.061  1.000
AGE   -0.488 -0.493 -0.382 -0.126  1.000
TAX    0.833  0.849  0.823  0.314 -0.572  1.000
ROOM   0.635  0.654  0.821 -0.125 -0.484  0.700  1.000
BDRM   0.552  0.555  0.610 -0.078 -0.333  0.479  0.726  1.000
BATH   0.586  0.580  0.653  0.100 -0.406  0.647  0.639  0.486  1.000
```

The variable having the highest correlation [0.995] with SELL is LIST. LIST is the single best variable for predicting SELL. The next variable to add would be one with a moderately high correlation with SELL and a low correlation with LIST (so that it is not merely duplicating information provided by LIST). This suggests that ACRE or AGE would be appropriate variables to consider next. But since the seller presumably considered the other variables when determining LIST, there is a sense in which their information is already contained in LIST. Set up a regression using only LIST and then decide what to do next.

Place lists HMSP,HMLST into a 40x2 matrix.
Use PRGM A2MULREG with 1 independent variable from column 2.

```
     DF  SS              B0=99.1840667
RG    1  2.67456322     CL COEFF / T     P
ER  138  2748997660     2  .9789557162
     F=3697.11                60.8   0.000
     P=0.000
     R-SQ=.9898
     (ADJ).9896
     S=8505.412611
```

SELL = 99.2 + 0.979(LIST)                                    adjusted $R^2$ = 0.9896

This is probably the best multiple regression equation with selling price HMSP as the dependent variable. The adjusted $R^2$ and overall P-value [0.000] indicate that this is a suitable regression equation for predicting selling price. Since the adjusted $R^2$ is so close to 1.00, there is virtually no room for improvement and no additional variable could be of any real value.

17. Use 2–BSEX→L1 to recode the males as 1 and the females as 0.
    Place lists BWGHT,L1,BAGE into a 54x3 matrix.
    Use PRGM A2MULREG with 2 independent variables from column 2,3.

```
     DF  SS              B0=3.062371221
RG    2  522288.422     CL COEFF / T     P
ER   51  263994.911     2  82.37937943
     F=50.45                 3.96   0.000
     P=0.000            3  2.905256883
     R-SQ=.6642             9.77   0.000
     (ADJ).6511
     S=71.94699996
```

WEIGHT = 3.06 + 82.4(SEX) + 2.91(AGE)                        adjusted $R^2$ = 0.6511
$\hat{y} = 3.1 + 82.4x_1 + 2.91x_2$

Yes, but not merely because the coefficient 82.4 is so large. The display indicates that for the test $H_o$: $\beta_1 = 0$, the sample value $b_1 = 82.4$ results in the test statistic $t_{51} = 3.96$ and P-value =0.000. As suggested by (a) and (b) below, sex does have a significant effect on the weight of a bear.

a. $\hat{y} = 3.1 + 82.4x_1 + 2.91x_2$

   $\hat{y}_{20,0} = 3.1 + 2.91(0) + 2.91(20) = 61.3$ lbs

b. $\hat{y} = 3.1 + 82.4x_1 + 82.4x_2$

   $\hat{y}_{20,1} = 3.1 + 2.91(1) + 82.4(20) = 143.7$ lbs

**10-6 Modeling**

1. A mathematical model is an equation, typically expressing one variable as a function of one or more other variables, that fits real world data.

3. The year 3000 is too far beyond the scope of the data used in the model to assert that the same conditions and assumptions that applied to the data used to build the model would still be relevant.

NOTE: The following format is used for exercises #2-#12, with the TI instructions given for each step: construct the scatterplot, evaluate the model(s) under consideration, choose the best model. The TI screen displays for all steps appear at the end in one row, in the order generated.

5. Place the x values in list L1 and the y values in list L2.
   Use STATPLOT ZOOM 9 to construct the scatterplot.
   The graph appears to be that of a straight line function.
   •Try a linear regression of the form y = ax + b.
   Use CATALOG DiagnosticOn STAT CALC LinReg on L1,L2.
   The display indicates y = 2x + 3 has $r^2$=1.00, a perfect fit.
   •Choose the linear model y = 2x + 3.

7. Place the x values in list L1 and the y values in list L2.
   Use STATPLOT ZOOM 9 to construct the scatterplot.
   The graph appears to be that of a quadratic function.
   •Try a quadratic regression of the form y = $ax^2$ + bx + c.
   Use CATALOG DiagnosticOn STAT CALC QuadReg on L1,L2.
   The display indicates y = $2x^2$ + 0x – 1 has $R^2$=1.000, a perfect fit.
   •Choose the quadratic model y = $2x^2$ – 1.

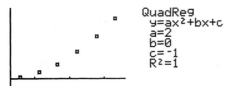

9. Place the years (x) in list L1 and the numbers of deaths (y) in list L2.
   NOTE: For simplicity, code the years with 1983=1. The values entered will be 1,2,3,…,22.
   Use STATPLOT ZOOM 9 to construct the scatterplot.
   The graph could be that of any of several functions.
   •Try a linear regression of the form y = ax + b.
   Use CATALOG DiagnosticOn STAT CALC LinReg on L1,L2.
   The display indicates y = 2.74x + 21.7 has $r^2$=0.7763, a reasonable fit.
   •Try a quadratic regression of the form y = $ax^2$ + bx + c.
   Use CATALOG DiagnosticOn STAT CALC QuadReg on L1,L2.
   The display indicates y = $0.0256x^2$ + 2.15x + 24.1 has $R^2$=0.7785, a reasonable fit.

- Try a power function regression of the form $y = a \cdot x^b$.
  Use CATALOG DiagnosticOn STAT CALC PwrReg on L1,L2.
  The display indicates $y = 18.1 \cdot x^{0.455}$ has $r^2 = 0.7892$, a reasonable fit.
- Even though the power function model has a slightly higher $R^2$, the difference is negligible and not worth the extra mathematical complication. Unless there are actuarial or other reasons to suspect that this phenomenon should follow a power function, choose the linear model $y = 21.7 + 2.74x$.
  NOTE: This is a judgment call – the quadratic and power function are also acceptable.

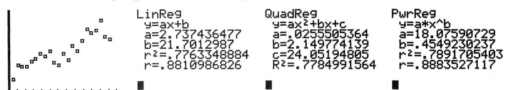

11. Place the times (x) in list L1 and the distances (y) in list L2.
    Use STATPLOT ZOOM 9 to construct the scatterplot.
    The graph appears to be that of a quadratic function.
    - Try a quadratic regression of the form $y = ax^2 + bx + c$.
      Use CATALOG DiagnosticOn STAT CALC QuadReg on L1,L2.
      The display indicates $y = 4.90x^2 - 0.0286x + 0.00476$ has $R^2 = 1.000$, a perfect fit.
    - Choose the quadratic model $y = 4.90x^2 - 0.0286x + 0.00476$.
      $\hat{y}_{12} = 4.90(12)^2 - 0.0286(12) + 0.00476 = 705.3$ meters.

    But if the building from which the ball is dropped is only 50 meters tall, the ball will hit the ground and stop falling long before 12 seconds elapse.

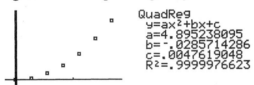

13. NOTE: The following analysis codes the years so that 1971 is x=1. In general, the coding of the years is arbitrary – and while a different equation may result, the individual predictions and other key characteristics will be identical.
    Consider the pattern in the box at the right.
    For a variable that starts with value $y = a$ at year 1 and doubles every 18 months,
    $y = a \cdot 2^{(x-1)/2}$
    $= a \cdot 2^{(x/2)-(1/2)} = a \cdot (2^{1/2})^x / 2^{1/2} = (a/\sqrt{2}) \cdot (\sqrt{2})^x$.

| year | doubles every 12 months | doubles every 18 months |
|---|---|---|
| 1 | $a = a \cdot 2^0$ | $a = a \cdot 2^0$ |
| 2 | $2a = a \cdot 2^1$ | |
| 3 | $4a = a \cdot 2^2$ | $2a = a \cdot 2^1$ |
| 4 | $8a = a \cdot 2^3$ | |
| 5 | $16a = a \cdot 2^4$ | $4a = a \cdot 2^2$ |
| … | … | … |
| x | $a \cdot 2^{x-1}$ | $a \cdot 2^{(x-1)/2}$ |

a. If Moore's law applies as indicated, and the years are coded with 1971 = 1, the data should be a good fit to the exponential model $y = (2.3/1.414) \cdot (1.414)^x = 1.626 \cdot (1.414)^x$.

b. Try an exponential regression of the form $y = a \cdot b^x$.
   Place the years (x) in list L1 and the numbers of deaths (y) in list L2.
   NOTE: For simplicity, code the years with 1971=1. The irregularly spaced values are 1,4,8,12,15,…,33 (i.e., subtract 1970 from the given year).

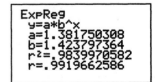

# 208 CHAPTER 10 Correlation and Regression

Use CATALOG DiagnosticOn STAT CALC ExpReg on L1,L2.
The display indicates y = 1.382·(1.423)$^x$ has r$^2$=0.9840, an excellent fit.
Choose the exponential model y = 1.382·(1.424)$^x$.

c. Yes. The 1.424≈1.414 indicates that the y value is doubling approximately every 18 months. In addition, the starting value for 1971 (x=1) is 1.382·(1.424) = 2.0, which is is close to the actual value of 2.3.

15. The manual follows the text and uses two decimal accuracy in the coefficients to obtain
    $\hat{y}_{lin}$ = 27.20x − 61.93
    $\hat{y}_{quad}$ = 2.77x$^2$ −6.00 x + 10.01
   Using more significant digits produces slightly different answers.
   Place the coded years in list L1 and the populations in list L2.

   a. Use L2 − (27.20×L1 − 61.93) → L3 and 1-Var Stats on L3 to get Σx$^2$ = 6641.7819
      Σ(y - $\hat{y}$)$^2$ = 6641.7819 for the linear model

   b. Use L2 − (2.77×L1$^2$ − 6.00×L1 +10.01) → L4 and 1-Var Stats on L4 to get Σx$^2$ = 73.6181
      Σ(y - $\hat{y}$)$^2$ = 73.6181

   c. Since 73.16 < 6641.78, the quadratic model is better − using the sum of squares criterion.

## Statistical Literacy and Critical Thinking

1. Correlation is a measure of the degree of association between two variables. Regression is a procedure to identify the formula that describes that association.

2. If the linear correlation coefficient is r = 0, then there is no <u>linear</u> relationship between the variables − but there might be a relationship defined by something other than a straight line graph.

3. The conclusion of cause-and-effect cannot be justified solely on the basis of the existence of a correlation. If there is a medical condition that happens to be associated with low cholesterol levels, for example, then people taking a drug to combat that condition would have low cholesterol − but it would not be because of the drug. If there is a carefully designed study with patients randomly assigned to a treatment group and a control group in such a way that the only thing different between the two groups is the drug, then such a conclusion could be justified.

4. Not necessarily. The accuracy of the prediction depends on s$_e$, the spread around the regression line. There can be a definite correlation, but little confidence in the prediction. For children between birth and 12 years, for example, there is an undeniable correlation between age and weight − you won't find any 12 year olds weighing 7 pounds, 8 ounces! But just knowing that a child is twelve years old gives you no information about his weight within the very large weight range of all 12 year olds.

**Review Exercises**

1. Place the watercraft (x) values in list L1 and the natural (y) values in list L2.
   Use STAT TESTS LinRegTTest on lists L1 and L2.
   Use STAT CALC 1-Var Stats on L2.  [L2 = y]

   ```
   LinRegTTest          LinRegTTest          1-Var Stats
   y=a+bx               y=a+bx               x̄=47
   β≠0 and ρ≠0          β≠0 and ρ≠0          Σx=517
   t=.2723943327        ↑b=.1664259406       Σx²=32827
   p=.7914651594        s=30.65628           Sx=29.2027396
   df=9                 r²=.0081768843       σx=27.84371971
   ↓a=35.65277677       r=.0904261261        ↓n=11
   ```

   a. $r = 0.0904$
      P-value = $0.7915 > 0.05$ does not indicate a significant linear correlation

   b. $\hat{y} = a + bx$
      $= 35.653 + 0.166x$

   c. $\hat{y}_{50} = \bar{y} = 47.00$ deaths [no significant correlation]
      For a year with 50 deaths from encounters with boats, the best predicted number of natural deaths is 47.

2. Place the duration (x) values in list L1 and the height (y) values in list L2.
   Use STAT TESTS LinRegTTest on lists L1 and L2.
   Use STAT CALC 1-Var Stats on L2.  [L2 = y]

   ```
   LinRegTTest          LinRegTTest          1-Var Stats
   y=a+bx               y=a+bx               x̄=128.75
   β≠0 and ρ≠0          β≠0 and ρ≠0          Σx=1030
   t=1.035700676        ↑b=.1076330754       Σx²=133850
   p=.3402684045        s=13.22759001        Sx=13.29607891
   df=6                 r²=.151664788        σx=12.43734296
   ↓a=105.1918106       r=.3894416362        ↓n=8
   ```

   a. $r = 0.3894$
      P-value = $0.3403 > 0.05$ does not indicate a significant linear correlation

   b. $\hat{y} = a + bx$
      $= 105.192 + 0.108x$

   c. $\hat{y}_{180} = \bar{y} = 128.750$ feet [no significant correlation]
      For an eruption that has a duration of 180 seconds, the best predicted height is 128.75 feet.

For exercises #3-#5, place the amounts of electricity consumed (kWh) in list L1.
the numbers of degree days in list L2.
the average daily temperatures in list L3.
the costs in list L4.

NOTE: As a review of hand calculations, the correlation in exercise #3 may be found as follows.

$n = 10$
$\Sigma x = 26481$
$\Sigma y = 2473.02$
$\Sigma xy = 6716083.75$
$\Sigma x^2 = 71605869$
$\Sigma y^2 = 634142.1730$

$n(\Sigma xy) - (\Sigma x)(\Sigma y) = 10(6716083.75) - (26481)(2473.02)$
$= 1672794.88$

$n(\Sigma x^2) - (\Sigma x)^2 = 10(71605869) - (26481)^2$
$= 14815329$

$n(\Sigma y^2) - (\Sigma y)^2 = 10(634142.1730) - (2473.02)^2$
$= 225593.8096$

$r = [n(\Sigma xy) - (\Sigma x)(\Sigma y)]/[\sqrt{n(\Sigma x^2) - (\Sigma x)^2} \sqrt{n(\Sigma y^2) - (\Sigma y)^2}]$

$= 1672794.88/[\sqrt{14815329} \sqrt{225593.8096}] = 0.9150$

3. Use STAT TESTS LinRegTTest on L1 and L4.

   ```
   LinRegTTest          LinRegTTest
   y=a+bx               y=a+bx
   β≠0 and ρ≠0          β≠0 and ρ≠0
   t=6.414853133        ↑b=.1129097356
   p=2.0585135E-4       s=21.42398831
   df=8                 r²=.8372340887
   ↓a=-51.6942708       r=.9150049665
   ```

   a. $H_0: \rho = 0$
      $H_1: \rho \neq 0$
      $\alpha = 0.05$ and df = 8
      C.V. $t = \pm t_{\alpha/2} = \pm 2.306$
      calculations:
      $t_r = (r - \mu_r)/s_r = 6.415$
      P-value = $2 \cdot P(t_8 > 6.415) = 0.0002$
      conclusion:
      Reject $H_0$; there is sufficient evidence to conclude that $\rho \neq 0$ (in fact, that $\rho > 0$).

   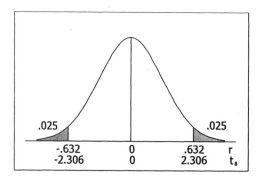

   b. $r^2 = 0.8372 = 83.72\%$

   c. $\hat{y} = a + bx$
      $= -51.694 + 0.1136x$

   d. $\hat{y}_{3000} = -51.694 + 0.1129(3000) = 287.03$ dollars
      For a time when 3000 kwh of electricity is used, the best predicted cost is $287.03.

4. Use STAT TESTS LinRegTTest on L3 and L4.
   Use STAT CALC 1-Var Stats on L4.   [L4 = y]

   ```
   LinRegTTest          LinRegTTest           1-Var Stats
   y=a+bx               y=a+bx                x̄=247.302
   β≠0 and ρ≠0          β≠0 and ρ≠0           Σx=2473.02
   t=.2416011221        ↑b=.2433060426        Σx²=634142.173
   p=.8151668571        s=52.91026752         Sx=50.06593537
   df=8                 r²=.0072435361        σx=47.49671669
   ↓a=234.7960694       r=.0851089662         ↓n=10
   ```

   a. $H_0: \rho = 0$
      $H_1: \rho \neq 0$
      $\alpha = 0.05$ and df = 8
      C.V. $t = \pm t_{\alpha/2} = \pm 2.306$
      calculations:
      $t_r = (r - \mu_r)/s_r = 0.242$
      P-value = $2 \cdot P(t_8 > 0.242) = 0.8152$
      conclusion:
      Do not reject $H_0$; there is not sufficient evidence to conclude that $\rho \neq 0$.

   b. $r^2 = 0.0072 = 0.72\%$

   c. $\hat{y} = a + bx$
      $= 234.796 + 0.243x$

   d. $\hat{y}_{40} = \bar{y} = 247.302$ (dollars) [no significant correlation]
      For a time when the average daily temperature is 40°F, the best predicted cost is $247.30.

5. Place lists L4,L1,L3 into a 10x3 matrix.
   Use PRGM A2MULREG with 2 independent variable from column 2,3.

   ```
   DF SS           B0=-128.0552374
   RG 2  21250.5811  CL COEFF / T    P
   ER 7  1308.79985  2 .1232002369
   F=56.83            10.62  0.000
   P=0.000          3 .955460897
   R-SQ=.942           3.56   .009
   (ADJ).9254
   S=13.67374883
   ```

   $\hat{y} = -128.06 + 0.123x_1 + 0.955x_2$     adjusted $R^2 = 0.9254$

   Yes; considering the adjusted $R^2$ and the overall P-value, this is an excellent regression for predicting cost. No; since the adjusted $R^2$ here (0.9254) is so much higher that the one in exercise #3, this is the best regression for predicting cost. NOTE: For exercise #3,

   adjusted $R^2 = 1 - \dfrac{n-1}{n-(k+1)}(1-R^2) = 1 - \dfrac{9}{10-2}(1-0.8372) = 1 - \dfrac{9}{8}(0.1628) = 0.8169$.

## Cumulative Review Exercises

Place the Super Bowls points (x) in list L1 and the DJIA numbers (y) in list L2.
The following display screens contain all the information necessary to work the exercises, and they were obtained using the following commands:
STAT TESTS LinRegTTest on L1 and L2 (#1)
STAT CALC 1-Var Stats on L2 (#2)
STAT TESTS TInterval on data L1 with C-Level 0.95 (#4 and #7)
STATPLOT for normal quantile plots on L1 (#6)

```
(#1)                  (#1)                  (#2)              (#4 and #7)       (#6)
LinRegTTest           LinRegTTest           1-Var Stats       TInterval
y=a+bx                y=a+bx                x̄=9950.25         (41.854,60.896)
β≠0 and ρ≠0           β≠0 and ρ≠0           Σx=79602          x̄=51.375
t=-1.169762415        ↑b=-67.2094176        Σx²=814142944     Sx=11.38843398
p=.2864561619         s=1731.19304          Sx=1776.157309    n=8
df=6                  r²=.1857057824        σx=1661.443029
↓a=13403.13383        r=-.4309359376        ↓n=8
```

1. $H_o: \rho = 0$
   $H_1: \rho \neq 0$
   $\alpha = 0.05$ and df = 6
   C.V. $t = \pm t_{\alpha/2} = \pm 2.447$
   calculations:
   $\quad t_r = (r - \mu_r)/s_r = -1.170$
   $\quad$P-value $= 2 \cdot P(t_6 < -1.170) = 0.2865$
   conclusion:
   $\quad$Do not reject $H_o$; there is not sufficient
   $\quad$evidence to conclude that $\rho \neq 0$.

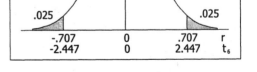

There is no significant linear correlation between Super Bowl points and the DJIA. Yes, this is the expected result.

2. $\hat{y} = a + bx$
   $\quad = 13403.134 - 67.209x$
   $\hat{y}_{50} = \bar{y} = 9950.25$ [no significant correlation]
   For a year in which there are 50 points scored in the Super Bowl, the best predicted high value for DJIA is 9950.

3. Yes. It is possible to carry out the mathematical calculations to make the test, but it would have no meaning. Because the two quantities have different units, differences would not be meaningful.

4. $41.9 < \mu < 60.9$ (points)

5. A confidence interval is based on all the data and applies to the population from which the data were selected. When there is a trend over time, the population is changing. Data collected from populations changing over time are relevant for making predictions for specific times only when the time factor is built into the model – as when "year" is one of the variables in a regression. Since ordinary confidence intervals use the data only from the variable of interest and do not consider the time factor, they are not appropriate for making predictions about populations that are changing over time.

6. The TI display given above may also ge generated by hand as follows/
   (1) Arrange the n=8 scores in order and place them in the x column.
   (2) For each $x_i$, calculate the cumulative probability using $cp_i = (2i-1)/2n$ for $i = 1,2,\ldots,n$.
   (3) For each $cp_i$, find the $z_i$ for which $P(z<z_i) = cp_i$ for $i = 1,2,\ldots,n$.
   The resulting normal quantile plot is relatively close to a straight line with no obvious patterns, indicating that the data appear to come from a population with a distribution that is approximately normal.

   | i | x | cp | z |
   |---|----|--------|-------|
   | 1 | 37 | 0.0625 | -1.53 |
   | 2 | 39 | 0.1875 | -0.89 |
   | 3 | 41 | 0.3125 | -0.49 |
   | 4 | 53 | 0.4375 | -0.16 |
   | 5 | 55 | 0.5625 | 0.16  |
   | 6 | 56 | 0.6875 | 0.49  |
   | 7 | 61 | 0.8125 | 0.89  |
   | 8 | 69 | 0.9375 | 1.53  |

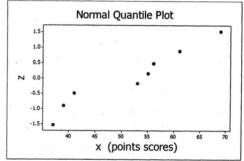

7. $\bar{x} = 51.375$, rounded to 51.4
   $s = 11.388$, rounded to 11.4

8. normal distribution
   $\mu = 51.375$
   $\sigma = 11.388$
   $P(x < 40)$
   $= P(z<-1.00)$
   $= 0.1587$

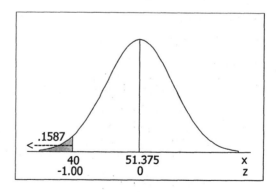

TI: DIST normalcdf(-999,40,51.375,11.388)
   = 0.1589
NOTE: One may wish to correct for the continuity of the normal, since only a whole number of points can be scored. In that case, use DIST normalcdf(-999,40,51.375,11.388) = 0.1485.

# Chapter 11

# Multinomial Experiments and Contingency Tables

## 11-2 Multinomial Experiments: Goodness of Fit

1. A test for "goodness-of-fit" examines the differences between values observed in the sample and values predicted by a specific mathematical model. If those differences are too large, then there is not a good fit between the observed data and the hypothesized model.

3. An observed frequency is the number of sample observations falling into a particular category. An expected frequency is the number of observations that the hypothesized model predicts will fall into a particular category.

5. a. $H_o: p_1 = p_2 = p_3$
   b. $\Sigma O = 30$. Under the hypothesis of equally likely categories, the expected frequency for each category is $\Sigma O/n = 30/3 = 10$.
   c. $\chi^2 = \Sigma[(O-E)^2/E] = [(5-10)^2/10] + [(5-10)^2/10] + [(20-10)^2/10]$
   $= 25/10 + 25/10 + 100/10$
   $= 2.5 + 2.5 + 10.0 = 15.0$
   d. For $\alpha = 0.05$ and df = 2, the critical value is $\chi^2_\alpha = 5.991$.
   e. Reject $H_o$; there is sufficient evidence to conclude that at least one proportion is not as claimed – i.e., that the categories are not equally likely.

7. a. For df=37 and $\alpha=0.10$, the critical value is $\chi^2_\alpha = 51.805$.
   b. For a $\chi^2$ distribution with df=37, 90% of the probability falls above 29.051 and 10% falls above 51.805. Since $29.051 < 38.232 < 51.805$, $0.10 < \text{P-value} < 0.90$.
   c. There is not sufficient evidence to reject the claim that the 38 results are equally likely.

9. $H_o: p_1 = p_2 = p_3 = \ldots = p_6 = 1/6$
   $H_1$: at least $p_i \neq 1/6$
   $\alpha = 0.05$ and df = 5
   C.V. $\chi^2 = \chi^2_\alpha = 11.071$
   calculations:

   | outcome | O | E | (O-E)²/E |
   |---|---|---|---|
   | 1 | 27 | 33.33 | 1.2033 |
   | 2 | 31 | 33.33 | 0.1633 |
   | 3 | 42 | 33.33 | 2.2533 |
   | 4 | 40 | 33.33 | 1.3333 |
   | 5 | 28 | 33.33 | 0.8533 |
   | 6 | 32 | 33.33 | 0.0533 |
   |   | 200 | 200.00 | 5.8600 |

   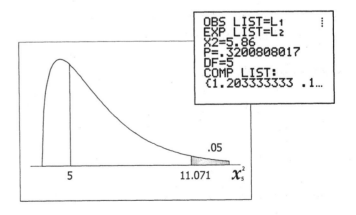

   $\chi^2 = \Sigma[(O-E)^2/E] = 5.860$
   P-value = $P(\chi^2_5 > 5.860) = 0.3201$

   conclusion:
   Do not reject Ho; there is not sufficient evidence to conclude that at least one the proportions is different from 1/6.
   No. This particular loaded die did not behave noticeably different from a fair die.

214   CHAPTER 11  Multinomial Experiments and Contingency Tables

NOTES FOR THE REMAINING EXERCISES IN THIS SECTION:
(1) In multinomial problems, <u>always verify</u> that $\Sigma E = \Sigma O$ before proceeding. If these sums are not equal, then an error has been made and further calculations have no meaning. The program for the TI-83/84 Plus calculator does not automatically perform this verification.
(2) As in the previous uses of the $\chi^2$ distribution, the accompanying illustrations follow the "usual" shape – even though that shape is not correct for df = 1 and df = 2.
(3) As in the previous sections, a subscript (the df) may be used to identify which $\chi^2$ distribution to use in the tables.
(4) Critical values are from Table A-4. As an aid to understanding, calculations are done by hand <u>and</u> using the $\chi^2$GOF program on the TI-83/84 Plus calculator as described in the text. To minimize round-off error, enter the E's as fractions – e.g., 200/6 instead of 33.3.

11. $H_0$: $p_{Sun} = p_{Mon} = p_{Tue} = \ldots = p_{Sat} = 1/7$
    $H_1$: at least one $p_i \neq 1/7$
    $\alpha = 0.05$ and df = 6
    C.V. $\chi^2 = \chi^2_\alpha = 12.592$
    calculations:

    | day | O | E | $(O-E)^2/E$ |
    |-----|-----|-----|-----|
    | Sun | 132 | 117 | 1.9231 |
    | Mon | 98  | 117 | 3.0855 |
    | Tue | 95  | 117 | 4.1368 |
    | Wed | 98  | 117 | 3.0855 |
    | Thu | 105 | 117 | 1.2308 |
    | Fri | 133 | 117 | 2.1880 |
    | Sat | 158 | 117 | 14.3675 |
    |     | 819 | 819 | 30.0171 |

    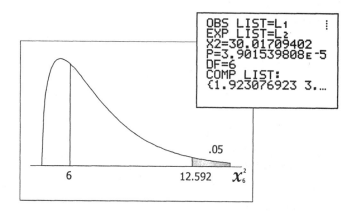

    $\chi^2 = \Sigma[(O-E)^2/E] = 30.017$
    P-value = $P(\chi^2_6 > 30.017) = 0.00004$

    conclusion:
        Reject Ho; there is sufficient evidence to conclude that at least one the proportions is different from 1/7 – i.e., that car crash fatalities occur with different frequencies on different days of the week.
    Drinking and/or increased recreational travel on the weekends could explain the differences – and the larger number on Saturday may be due to early morning crashes as a result of Friday night parties that last past midnight.

13. $H_0$: $p_{Jan} = p_{Feb} = p_{Mar} = \ldots = p_{Dec} = 1/12$
    $H_1$: at least one $p_i \neq 1/12$
    $\alpha = 0.05$ and df = 11
    C.V. $\chi^2 = \chi^2_\alpha = 19.675$

    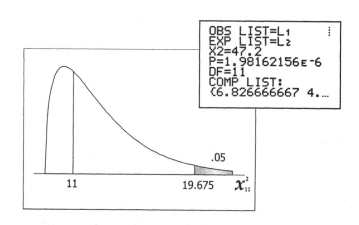

## Multinomial Experiments: Goodness of Fit  SECTION 11-2  215

calculations:

| month | O | E | $(O-E)^2/E$ |
|---|---|---|---|
| Jan | 6 | 16.67 | 6.82667 |
| Feb | 8 | 16.67 | 4.50667 |
| Mar | 10 | 16.67 | 2.66667 |
| Apr | 16 | 16.67 | 0.02667 |
| May | 22 | 16.67 | 1.70667 |
| Jun | 28 | 16.67 | 7.70667 |
| Jul | 24 | 16.67 | 3.22667 |
| Aug | 28 | 16.67 | 7.70667 |
| Sep | 26 | 16.67 | 5.22667 |
| Oct | 14 | 16.67 | 0.42667 |
| Nov | 10 | 16.67 | 2.66667 |
| Dec | 8 | 16.67 | 4.50667 |
|  | 200 | 200.00 | 47.20000 |

$\chi^2 = \Sigma[(O-E)^2/E] = 47.200$

P-value = $P(\chi^2_{11} > 47.200) = 0.000002$

conclusion:
  Reject Ho; there is sufficient evidence to reject the claim that motorcycle deaths occur with equal frequency in the different months.
There are probably fewer motorcycle deaths during the winter months because fewer persons ride their motorcycles when the weather is cold or the roads are icy.

15. $H_o$: $p_{Jan} = p_{Feb} = p_{Mar} = \ldots = p_{Dec} = 1/12$
    $H_1$: at least one $p_i \neq 1/12$
    $\alpha = 0.05$ and df = 11
    C.V. $\chi^2 = \chi^2_\alpha = 19.675$

calculations:

| month | O | E | $(O-E)^2/E$ |
|---|---|---|---|
| Jan | 7 | 6.33 | 0.07018 |
| Feb | 3 | 6.33 | 1.75439 |
| Mar | 7 | 6.33 | 0.07018 |
| Apr | 7 | 6.33 | 0.07018 |
| May | 8 | 6.33 | 0.43860 |
| Jun | 7 | 6.33 | 0.07018 |
| Jul | 6 | 6.33 | 0.01754 |
| Aug | 6 | 6.33 | 0.01754 |
| Sep | 5 | 6.33 | 0.28070 |
| Oct | 6 | 6.33 | 0.01754 |
| Nov | 9 | 6.33 | 1.12281 |
| Dec | 5 | 6.33 | 0.28070 |
|  | 76 | 76.00 | 4.21053 |

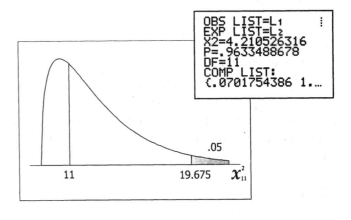

$\chi^2 = \Sigma[(O-E)^2/E] = 4.211$

P-value = $P(\chi^2_{11} > 4.211) = 0.9633$

conclusion:
  Do not reject Ho; there is not sufficient evidence to reject the claim that Oscar-winning actresses are born in the different months with the same frequency.
No; there is no reason why more Oscar-winning actresses would be born in some months more than others.

NOTE: This exercise has remarkable uniformity in the data – the chance a random sample of births has such little monthly variation is 1–0.9633 = 0.0367. While the critical region in goodness-of-fit tests is entirely in the upper tail to identify too much deviation from the null hypothesis, this calculated $\chi^2 = 4.211$ is significant in the lower tail. In some rare instances (e.g., to catch people suspected of fabricating data) a lower tail chi-square test is used to detect a fit that is "too good."

216  CHAPTER 11  Multinomial Experiments and Contingency Tables

17. $H_0$: $p_{Jan} = p_{Feb} = p_{Mar} = \ldots = p_{Dec} = 1/12$
    $H_1$: at least one $p_i \neq 1/12$
    $\alpha = 0.05$ and df = 11
    C.V. $\chi^2 = \chi^2_\alpha = 19.675$
    calculations:

    | month | O | E | (O-E)²/E |
    |---|---|---|---|
    | Jan | 5 | 9.17 | 1.89394 |
    | Feb | 8 | 9.17 | 0.14848 |
    | Mar | 6 | 9.17 | 1.09394 |
    | Apr | 8 | 9.17 | 0.14848 |
    | May | 11 | 9.17 | 0.36667 |
    | Jun | 14 | 9.17 | 2.54848 |
    | Jul | 10 | 9.17 | 0.07576 |
    | Aug | 9 | 9.17 | 0.00303 |
    | Sep | 10 | 9.17 | 0.07576 |
    | Oct | 12 | 9.17 | 0.87576 |
    | Nov | 8 | 9.17 | 0.14848 |
    | Dec | 9 | 9.17 | 0.00303 |
    |  | 110 | 110.00 | 7.38182 |

    $\chi^2 = \Sigma[(O-E)^2/E] = 7.382$
    P-value = $P(\chi^2_{11} > 7.382) = 0.7674$

    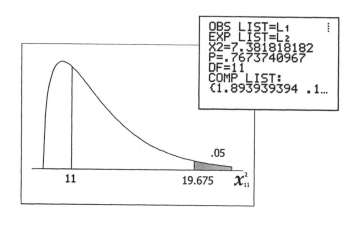

    conclusion:
    Do not reject $H_0$; there is not sufficient evidence to reject the claim that weddings are held in different months with the same frequency.
    The results refute the claim that "most" marriages occur in June – or even that June is significantly different from other months.

19. $H_0$: $p_4 = 2/16$, $p_5 = 4/16$, $p_6 = 5/16$, $p_7 = 5/16$
    $H_1$: at least one $p_i$ is not as claimed
    $\alpha = 0.05$ and df = 3
    C.V. $\chi^2 = \chi^2_\alpha = 7.815$
    calculations:

    | games | O | E | (O-E)²/E |
    |---|---|---|---|
    | 4 | 18 | 12.1250 | 2.84665 |
    | 5 | 20 | 24.2500 | 0.74485 |
    | 6 | 22 | 30.3125 | 2.27951 |
    | 7 | 37 | 30.3125 | 1.47539 |
    |  | 97 | 97.0000 | 7.34639 |

    $\chi^2 = \Sigma[(O-E)^2/E] = 7.346$
    P-value = $P(\chi^2_3 > 7.346) = 0.0616$

    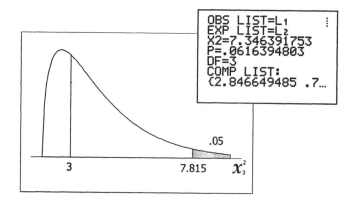

    conclusion:
    Do not reject $H_0$; there is not sufficient evidence to reject that claim that the observed frequencies agree with the theoretical proportions.
    The evidence does not support the claim that 7-game series occur more often than expected.

21. $H_0$: $p_{gr} = 0.16$, $p_{or} = 0.20$, $p_{ye} = 0.14$, $p_{bl} = 0.24$, $p_{re} = .13$, $p_{br} = 0.13$
$H_1$: at least one $p_i$ is not as claimed
$\alpha = 0.05$ and df = 5
C.V. $\chi^2 = \chi^2_\alpha = 11.071$

calculations:

| color | O | E | $(O-E)^2/E$ |
|---|---|---|---|
| green | 19 | 16 | 0.56250 |
| orange | 25 | 20 | 1.25000 |
| yellow | 8 | 14 | 2.57143 |
| blue | 27 | 24 | 0.37500 |
| red | 13 | 13 | 0.00000 |
| brown | 8 | 13 | 1.92308 |
|  | 100 | 100 | 6.68210 |

$\chi^2 = \Sigma[(O-E)^2/E] = 6.682$
P-value = $P(\chi^2_5 > 6.682) = 0.2454$

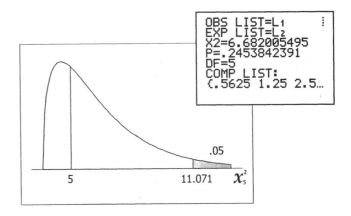

conclusion:
Do not reject $H_0$; there is not sufficient evidence to reject the claim that the color distribution is as claimed by Mars, Inc.

23. $H_0$: $p_w = 0.757$, $p_H = 0.091$, $p_b = 0.108$, $p_A = 0.038$, $p_N = 0.007$
$H_1$: at least one $p_i$ is not as claimed
$\alpha = 0.05$ and df = 4
C.V. $\chi^2 = \chi^2_\alpha = 9.488$

calculations:

| race | O | E | $(O-E)^2/E$ |
|---|---|---|---|
| white | 644 | 569.264 | 9.8117 |
| Hisp | 23 | 68.432 | 30.1623 |
| black | 69 | 81.216 | 1.8375 |
| Asian | 14 | 28.576 | 7.4349 |
| Native | 2 | 5.264 | 2.0239 |
|  | 752 | 752.000 | 51.2703 |

$\chi^2 = \Sigma[(O-E)^2/E] = 51.270$
P-value = $P(\chi^2_4 > 51.270) = 1.96\text{E}{-}10 \approx 0$

conclusion:
Reject $H_0$; there is sufficient evidence to reject the claim that the participants fit the same distribution as the US population.

If the participants are not racially representative of the US, and if there are racial differences in cancer occurrence and/or responses to treatments, then the unadjusted sample mean would not describe the US population. A correct national mean could still be obtained by keeping sample data by race and calculating a weighted sample mean – but if there are racial differences, then each race should be examined separately anyway and a national mean has no real value. The underlying concern here may be for representative inclusion in clinical trials more for appearances than for medical or statistical reasons.

25. $H_o$: the leading digit proportions conform to Benford's law
   $H_1$: at least one $p_i$ is not as claimed
   $\alpha = 0.05$ and df = 8
   C.V. $\chi^2 = \chi^2_\alpha = 15.507$
   calculations:

   | digit | O | p | E | (O-E)²/E |
   |---|---|---|---|---|
   | 1 | 72 | 0.301 | 60.2 | 2.3130 |
   | 2 | 23 | 0.176 | 35.2 | 4.2284 |
   | 3 | 26 | 0.125 | 25.0 | 0.0400 |
   | 4 | 20 | 0.097 | 19.4 | 0.0186 |
   | 5 | 21 | 0.079 | 15.8 | 1.7114 |
   | 6 | 18 | 0.067 | 13.4 | 1.5791 |
   | 7 | 8 | 0.058 | 11.6 | 1.1172 |
   | 8 | 8 | 0.051 | 10.2 | 0.4745 |
   | 9 | 4 | 0.046 | 9.2 | 2.9391 |
   |   | 200 | 1.000 | 200.0 | 14.4213 |

   $\chi^2 = \Sigma[(O - E)^2/E] = 14.421$
   P-value = $P(\chi^2_8 > 14.421) = 0.0714$

   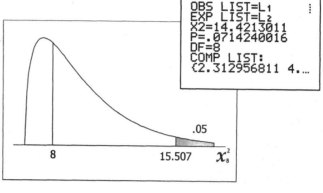

   conclusion:
   Do not reject Ho; there is not sufficient evidence to reject the claim that the author's leading digits conform to Benford's law.

27. Use the lower tail of the $\chi^2$ distribution to determine the critical region. If the null hypothesis is true and only random fluctuation exists, the expected value of $\chi^2 = \Sigma[(O - E)^2/E]$ is df – i.e., the degrees of freedom. If $\chi^2 = \Sigma[(O - E)^2/E]$ is significantly greater than df, then there is more variation than can be expected by chance and the data do not fit the null hypothesis. If $\chi^2 = \Sigma[(O - E)^2/E]$ is significantly less than df, then there is less deviation than can be expected by chance and the data fit the null hypothesis too well.

29. a. P(x=0) = binomialpdf(3,1/3,0) = 0.2963
   P(x=1) = binomialpdf(3,1/3,1) = 0.4444
   P(x=2) = binomialpdf(3,1/3,2) = 0.2222
   P(x=3) = binomialpdf(3,1/3,3) = 0.0370

   | x | O | E | (O-E)²/E |
   |---|---|---|---|
   | 0 | 89 | 88.89 | 0.000 |
   | 1 | 133 | 133.33 | 0.001 |
   | 2 | 52 | 66.67 | 3.227 |
   | 3 | 26 | 11.11 | 19.951 |
   |   | 300 | 300.00 | 23.179 |

   b. The requested values appear in the "E" column above at the right. Included with the expected frequencies are the usual $\chi^2$ calculations needed for part (c).

   c. $H_o$: there is goodness of fit to the binomial distribution with n=3 and p = 1/3
   $H_1$: there is not goodness of fit
   $\alpha = 0.05$ and df = 3
   C.V. $\chi^2 = \chi^2_\alpha = 7.815$
   calculations:
   $\chi^2 = \Sigma[(O - E)^2/E] = 23.179$
   P-value = $P(\chi^2_3 > 23.179) = 0.00004$

   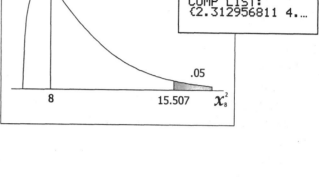

   conclusion:
   Reject Ho; there is sufficient evidence to reject the claim that the observed frequencies fit a binomial distribution with n=3 and p = 1/3.
   NOTE: While the phrase "goodness of fit" is the accepted statistical terminology in this context, the test could also be expressed as   $H_o$: $p_0 = 8/27$, $p_1 = 12/27$, $p_2 = 6/27$, $p_3 = 1/27$
   $H_1$: at least one $p_i$ is not as claimed

# 11-3 Contingency Tables: Independence and Homogeneity

1. In this section, the chi-square statistic measures the a $\chi_1^2$ mount of disagreement between the observed frequencies and the frequencies expected if the null hypothesis were true.

3. In general, a contingency is a related event or the relationship between two events. In this section, a contingency table is an array of frequencies of related events. The methods of this section seek to determine whether the relationship between two variables is one of independence, or whether the value of one of them is contingent upon the value of the other.

NOTES FOR THE REMAINING EXERCISES:
  (1) For each row and each column it must be true that $\Sigma O = \Sigma E$. After the marginal row and column totals are calculated, both the row totals and column totals must sum to produce the same grand total. If either of the preceding is not true, then an error has been made and further calculations have no meaning.
  (2) All contingency table analyses in this manual use the following conventions.
    • Even though they are not entered when using the TI-83/84 Plus calculator, the E values for each cell are given in parentheses below the O values.
    • The accompanying illustrations follow the "usual" chi-square shape, even though that shape is not correct for df=1 or df=2
  (3) As an added aid, the addends used to calculate the $\chi^2$ test statistic for exercise #5 are given following arrangement of the cells in the contingency table. For the remaining exercises the calculated $\chi^2$ statistic and the P-value are taken from the STAT TESTS $\chi^2$–test display screens.

5. $H_o$: ethnicity and being stopped are independent
   $H_1$: ethnicity and being stopped are related
   $\alpha = 0.05$ and df = 1
   C.V. $\chi^2 = \chi_\alpha^2 = 3.841$
   calculations:

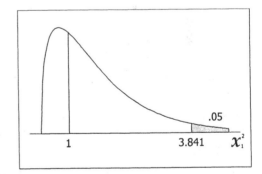

|  |  | ETHNICITY B | W |  |
|---|---|---|---|---|
| STOPPED? | Y | 24 (21.375) | 147 (149.625) | 171 |
|  | N | 176 (178.625) | 1253 (1250.375) | 1429 |
|  |  | 200 | 1400 | 1600 |

   $\chi^2 = \Sigma[(O - E)^2/E]$
       $= 0.3224 + 0.0461$
         $0.0386 + 0.0055 = 0.413$ [Minitab]
   P-value = $P(\chi_1^2 > 0.413) = 0.521$ [Minitab]

   conclusion:
     Do not reject Ho; there is not sufficient evidence to reject the claim that ethnicity and being stopped are independent. No. We cannot conclude that racial profiling is being used.

NOTE: While the table in exercise #5 claims to summarize "results for randomly selected drivers stopped by police," it appears that fixed numbers (200 black, 1400 white) of the ethnic groups were asked whether or not they had been stopped by the police. If so, the test should use "$H_o$: the proportion of drivers stopped is the same for both ethnic groups" to test homogeneity and not independence. The tests are equivalent mathematically, but there is a subtle difference in interpretation. This comment could apply to several other of the exercises in this section.

220   CHAPTER 11   Multinomial Experiments and Contingency Tables

7. Place the table values in a matrix and use STAT TESTS $\chi^2$–test.
   $H_0$: developing the flu and treatment used are independent
   $H_1$: developing the flu and treatment used are related
   $\alpha = 0.05$ and df = 1
   C.V. $\chi^2 = \chi^2_\alpha = 3.841$
   calculations:

   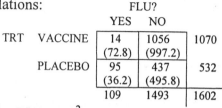

   |  |  | FLU? | |  |
   |---|---|---|---|---|
   |  |  | YES | NO |  |
   | TRT | VACCINE | 14 (72.8) | 1056 (997.2) | 1070 |
   |  | PLACEBO | 95 (36.2) | 437 (495.8) | 532 |
   |  |  | 109 | 1493 | 1602 |

   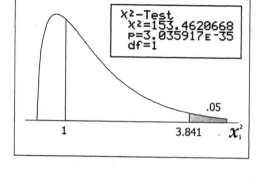

   $\chi^2 = \Sigma[(O - E)^2/E] = 153.462$
   P-value = $P(\chi^2_1 > 153.462)$ 3.04E-35 $\approx 0$

   conclusion:
   Reject $H_0$; there is sufficient evidence to reject the claim that developing the flu is independent of the treatment used.
   Yes. The vaccine appears to be effective.

9. Place the table values in a matrix and use STAT TESTS $\chi^2$–test.
   $H_0$: gender and left-handedness are independent
   $H_1$: gender and left-handedness are related
   $\alpha = 0.05$ and df = 1
   C.V. $\chi^2 = \chi^2_\alpha = 3.841$
   calculations:

   |  |  | LEFTY? | |  |
   |---|---|---|---|---|
   |  |  | Y | N |  |
   | GENDER | M | 83 (89) | 17 (11) | 100 |
   |  | F | 184 (178) | 16 (22) | 200 |
   |  |  | 267 | 33 | 300 |

   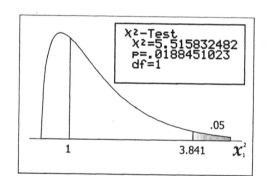

   $\chi^2 = \Sigma[(O - E)^2/E] = 5.516$
   P-value = $P(\chi^2_1 > 5.516) = 0.0188$

   conclusion:
   Reject $H_0$; there is sufficient evidence to reject the claim that gender and left-handedness are independent. It appears that the proportion of left-handers is higher for males than it is for females.

11. Place the table values in a matrix and use STAT TESTS $\chi^2$–test.
    $H_0$: subject truthfulness and polygraph reading are independent
    $H_1$: subject truthfulness and polygraph reading are related
    $\alpha = 0.05$ and df = 1
    C.V. $\chi^2 = \chi^2_\alpha = 3.841$
    calculations:

    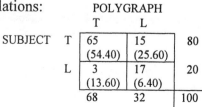

    |  |  | POLYGRAPH | |  |
    |---|---|---|---|---|
    |  |  | T | L |  |
    | SUBJECT | T | 65 (54.40) | 15 (25.60) | 80 |
    |  | L | 3 (13.60) | 17 (6.40) | 20 |
    |  |  | 68 | 32 | 100 |

    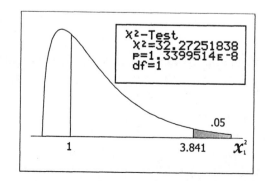

$\chi^2 = \Sigma[(O - E)^2/E] = 32.273$

P-value = $P(\chi_1^2 > 32.273) = 0.00000001$

conclusion:
   Reject Ho; there is sufficient evidence to reject the claim that a subject's truthfulness and polygraph reading are independent – and conclude that a person's truthfulness and the polygraph reading are related, so that the polygraph reading does indicate something about a person's truthfulness.

The results suggest that the two variables are related in that a person's truthfulness can be predicted from the polygraph reading – but the results do not suggest how accurate that prediction might be. Mathematically the chi square test addresses only the existence of a relationship and not the strength of a relationship – as an increase in sample size will increase the test statistic, even if all the proportions remain the same. Despite the extremely low P-value above, for example, the polygraph was incorrect (3+15)/100 = 18% of the time.

13. Place the table values in a matrix and use STAT TESTS $\chi^2$–test.
   Ho: plea and sentence are independent
   H1: plea and sentence are related
   α = 0.05 and df = 1
   C.V. $\chi^2 = \chi_\alpha^2 = 3.841$
   calculations:

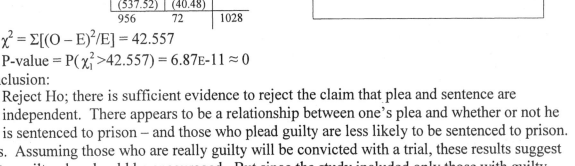

|  |  | PLEA G | PLEA NG |  |
|---|---|---|---|---|
| SENTENCE | P | 392 (418.48) | 58 (31.52) | 450 |
|  | NP | 564 (537.52) | 14 (40.48) | 578 |
|  |  | 956 | 72 | 1028 |

$\chi^2 = \Sigma[(O - E)^2/E] = 42.557$

P-value = $P(\chi_1^2 > 42.557) = 6.87\text{E-}11 \approx 0$

conclusion:
   Reject Ho; there is sufficient evidence to reject the claim that plea and sentence are independent. There appears to be a relationship between one's plea and whether or not he is sentenced to prison – and those who plead guilty are less likely to be sentenced to prison.

Yes. Assuming those who are really guilty will be convicted with a trial, these results suggest that a guilty plea should be encouraged. But since the study included only those with guilty pleas who were convicted in a trial, perhaps a guilty person who has a chance for acquittal should plead not guilty. If 50 guilty people with not guilty pleas were acquitted, as shown below, including them in the table in the no prison category changes the above conclusion.

|  |  | PLEA G | PLEA NG |  |
|---|---|---|---|---|
| SENTENCE | P | 392 (418.48) | 58 (31.52) | 450 |
|  | NP | 564 (537.52) | 64 (40.48) | 628 |
|  |  | 956 | 112 | 1078 |

$\chi^2 = \Sigma[(O - E)^2/E]$
$= 0.125 + 0.982$
$\phantom{=} 0.090 + 0.704 = 1.901$

15. Place the table values in a matrix and use STAT TESTS $\chi^2$-test.
    $H_o$: coin results and method used are independent
    $H_1$: coin results and method used are related
    $\alpha = 0.05$ and df = 1
    C.V. $\chi^2 = \chi_\alpha^2 = 3.841$
    calculations:

    |  |  | RESULT | | |
    |---|---|---|---|---|
    |  |  | H | T |  |
    | METHOD | FLIP | 14709 (14030.6) | 14306 (14984.4) | 29015 |
    |  | SPIN | 9197 (9875.4) | 11225 (10546.6) | 20422 |
    |  |  | 23906 | 25531 | 49437 |

    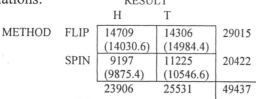

    $\chi^2 = \Sigma[(O - E)^2/E] = 153.739$
    P-value = $P(\chi_1^2 > 153.739) = 2.64\text{E-}35 \approx 0$

    conclusion:
       Reject Ho; there is sufficient evidence to reject the claim that coin results are independent of the method used – it appears that for a penny more heads result from flipping than from spinning.
    NOTE: In general, the data in the exercises in the text are real – i.e., not fabricated. As a protection against gross errors, if not for intellectual satisfaction, one should always stop to consider whether the conclusion in a test of hypothesis agrees with common sense. At first glance, the above coin results may seem surprising. But coins have different designs for their heads and tails, and they are not necessarily designed to be balanced. It appears that spinning, more than flipping, allows the coin enough time for the laws of physics and the unbalanced nature of the coin to affect the outcome.

17. Place the table values in a matrix and use STAT TESTS $\chi^2$-test.

    |  |  | OCCUPATION | | | | |
    |---|---|---|---|---|---|---|
    |  |  | POLICE | CASHIER | TAXI.D | GUARD |  |
    | DEATH BY HOMICIDE? | Y | 82 (112.92) | 107 (75.28) | 70 (64.25) | 59 (65.55) | 318 |
    |  | N | 92 (61.08) | 9 (40.72) | 29 (34.75) | 42 (35.45) | 172 |
    |  |  | 174 | 116 | 99 | 101 | 490 |

    $H_o$: occupation and manner of death are independent
    $H_1$: occupation and manner of death are related
    $\alpha = 0.05$ [assumed] and df = 3
    C.V. $\chi^2 = \chi_\alpha^2 = 7.815$
    calculations:
       $\chi^2 = \Sigma[(O - E)^2/E] = 65.524$
       P-value = $P(\chi_3^2 > 65.524)$ 3.87E-14 $\approx 0$

    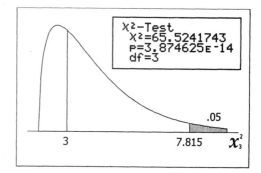

    conclusion:
       Reject Ho; there is sufficient evidence to reject the claim that occupation and manner of death are independent.
    Yes. The occupation cashier appears to be significantly more prone to death by homicide.

19. Place the table values in a matrix and use STAT TESTS $\chi^2$-test.

|  |  | NUMBER OF CIGARETTES | | | | |
|---|---|---|---|---|---|---|
|  |  | 0 | 1-14 | 15-34 | 35+ |  |
| USE BELT? | Y | 175 (171.53) | 20 (19.59) | 42 (43.94) | 6 (7.94) | 243 |
|  | N | 149 (152.47) | 17 (17.41) | 41 (39.06) | 9 (7.06) | 216 |
|  |  | 324 | 37 | 83 | 15 | 459 |

$H_0$: amount of smoking and seat belt use are independent
$H_1$: amount of smoking and seat belt use are related
$\alpha = 0.05$ [assumed] and df = 3
C.V. $\chi^2 = \chi^2_\alpha = 7.815$
calculations:
$\chi^2 = \Sigma[(O - E)^2/E] = 1.358$
P-value = $P(\chi^2_3 > 1.358) = 0.7154$

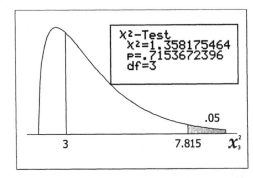

conclusion:
Do not reject $H_0$; there is not sufficient evidence to reject the claim that amount of smoking and seat belt use are independent.
No. The theory is not supported by the sample data. While the deviation from the null hypothesis is in the direction of the theory, it is even less deviation from the null hypothesis than could be expected by chance alone.

21. The following table is used for the calculations in exercise #21.

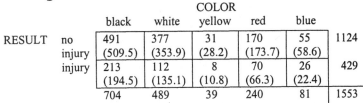

|  |  | COLOR | | | | |  |
|---|---|---|---|---|---|---|---|
|  |  | black | white | yellow | red | blue |  |
| RESULT | no injury | 491 (509.5) | 377 (353.9) | 31 (28.2) | 170 (173.7) | 55 (58.6) | 1124 |
|  | injury | 213 (194.5) | 112 (135.1) | 8 (10.8) | 70 (66.3) | 26 (22.4) | 429 |
|  |  | 704 | 489 | 39 | 240 | 81 | 1553 |

Place the table values in a matrix and use STAT TESTS $\chi^2$-test.
$H_0$: injuries and helmet color are independent
$H_1$: injuries and helmet color are related
$\alpha = 0.05$ [assumed] and df = 4
C.V. $\chi^2 = \chi^2_\alpha = 9.488$
calculations:
$\chi^2 = \Sigma[(O - E)^2/E] = 9.971$
P-value = $P(\chi^2_4 > 9.971) = 0.0409$

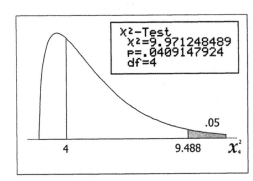

conclusion:
Reject $H_0$; there is sufficient evidence to reject the claim that injuries are independent of helmet color.
Yes. These data lead to the same conclusion reached in the example of this section.

224   CHAPTER 11   Multinomial Experiments and Contingency Tables

23. This exercise cannot use STAT TESTS $\chi^2$–test, and it must be worked by hand.
    $H_o$: ethnicity and being stopped are independent
    $H_1$: ethnicity and being stopped are related
    $\alpha = 0.05$ and df = 1
    C.V. $\chi^2 = \chi^2_\alpha = 3.841$
    calculations:

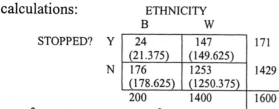

|  | | ETHNICITY | | |
|---|---|---|---|---|
| | | B | W | |
| STOPPED? | Y | 24 (21.375) | 147 (149.625) | 171 |
| | N | 176 (178.625) | 1253 (1250.375) | 1429 |
| | | 200 | 1400 | 1600 |

$\chi^2 = \Sigma[(|O - E| - 0.5)^2/E]$
   $= 0.2113 + 0.0302$
   $0.0253 + 0.0036 = 0.270$

P-value = $P(\chi^2_1 > 0.270) = \chi^2\text{cdf}(0.270,99,1) = 0.6033$

conclusion:
   Do not reject Ho; there is not sufficient evidence to reject the claim that ethnicity and being stopped are independent.
   No. We cannot conclude that racial profiling is being used. Without the correction for continuity [see the solution for exercise #5 for the details] the calculated test statistic is 0.413. Since $(|O-E|-0.5)^2 < (O-E)^2$ whenever $|O-E| > 0.25$, Yates' correction for continuity generally lowers the calculated test statistic.

**11-4 McNemar's Test for Matched Pairs**

1. In section 11-3 the row and column variables/responses are different (but possibly related) and apply to the same subject. In this section the row and column variables/responses are the same and apply to two different (but matched) subjects. McNemar's test is appropriate for data in a 2x2 table when
   a. the variables/responses for the rows and columns are the same.
   b. the data consist of matched pairs.
   c. the row variable/responses apply to one of the pair, and the column variable/responses apply to its match.
   d. the frequency in each cell is the number of matched pairs that are described by that cell.

3. Discordant pairs of results are pairs which do not share the same result for the binomial variable being measured.

Exercises #5-#12 refer to the following 2x2 table, given both in general and with the specific frequencies observed in the experiment.

| | | BEFORE | |
|---|---|---|---|
| | | smoker | non-smoker |
| AFTER | smoker | a | b |
| | non-smoker | c | d |

| | | BEFORE | |
|---|---|---|---|
| | | smoker | non-smoker |
| AFTER | smoker | 50 | 6 |
| | non-smoker | 8 | 80 |

5. The total number of subjects in the experiment is $50 + 6 + 8 + 80 = 144$.

7. The number of subjects who appear unaffected by the experiment is $50 + 80 = 130$.

9. Discordant pairs are pairs for which the BEFORE and AFTER states do not agree.
   a. No. The data from the a=50 subjects who were smokers BEFORE and smokers AFTER do not represent discordant pairs of results.
   b. Yes. The data from the b=8 subjects who were smokers BEFORE and non-smokers AFTER represent discordant pairs of results.
   c. Yes. The data from the c=6 subjects who were non-smokers BEFORE and smokers AFTER represent discordant pairs of results.
   d. No. The data from the d=80 subjects who were non-smokers BEFORE and non-smokers AFTER do not represent discordant pairs of results.

11. For $\alpha = 0.01$ and df = 1, the critical value is $\chi^2 = 6.635$.

13. $H_o$: there is no difference between the treatment and the placebo
    $H_1$: there is a difference between the treatment and the placebo
    $\alpha = 0.05$ and df = 1
    C.V. $\chi^2 = \chi^2_\alpha = 3.841$
    calculations:
        b = 12, c = 22
        $\chi^2 = (|b-c| - 1)^2/(b+c)$
           $= (10-1)^2/(12+22)$
           $= 81/34 = 2.382$
        P-value = $\chi^2$cdf (2.382,99,1) = 0.1227

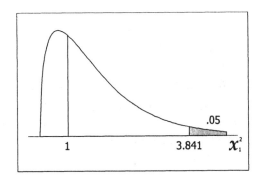

conclusion: Do not reject Ho; there is not sufficient evidence to reject the claim that there is no difference between the treatment and the placebo.

15. $H_o$: there is no difference between the PET-CT and MRI technologies
    $H_1$: there is a difference between the PET-CT and MRI technologies
    $\alpha = 0.05$ [assumed] and df = 1
    C.V. $\chi^2 = \chi^2_\alpha = 3.841$
    calculations:
        b = 1, c = 11
        $\chi^2 = (|b-c| - 1)^2/(b+c)$
           $= (10-1)^2/(1+11)$
           $= 81/12 = 6.750$
        P-value = $\chi^2$cdf (6.750,99,1) = 0.0094

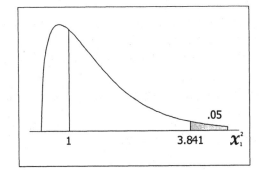

conclusion: Reject Ho; there is sufficient evidence to reject the claim that there is no difference between the PET-CT and MRI technologies. Since the majority of the discordant pairs are within the PET-CT=correct/MRI=incorrect cell, conclude that the PET-CT technology is more accurate.

17. b=8, c=40, $\chi^2 = (b-c)^2/(b+c) = (8-40)^2/(8+40) = 1024/48 = 21.333$
    The uncorrected test statistic of 21.333 is slightly larger than the corrected for continuity test statistic of 20.021. Since $(|b-c|-1)^2 < (b-c)^2$ whenever $|b-c| > 0.5$, the correction for continuity generally lowers the calculated test statistic.

19. This exercise follows the usual format for testing hypotheses with two modifications.
    * In order to compare two different techniques, the decision is based on comparing the α level and the P-value. This approach is valid for any hypothesis test and any test statistic, and it makes comparisons easier.
    * There are two "calculations" sections – one for each of the two techniques being compared.

$H_o$: there is no difference between the two treatments
$H_1$: there is a difference between the two treatments
α = 0.05

calculations using the exact binomial:
  x = 6 = largest frequency in cells representing discordant pairs
  P-value = 2·P(x≥6)
    = 2·[P(x=6) + P(x=7) + P(x=8)]
    = 2·[28(0.5)$^6$(0.5)$^2$ + 8(0.5)$^7$(0.5)$^1$ + 1(0.5)$^8$(0.5)$^0$]
    = 2·[0.109375 + 0.03125 + 0.00390625]
    = 2·[0.14453125]
    = 0.2891

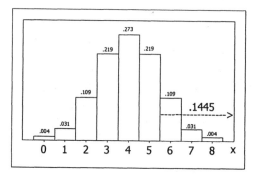

calculations using the $\chi^2$ approximation:
  b = 2, c = 6
  $\chi^2$ = (|b-c| – 1)$^2$/(b+c)
    = (4-1)$^2$/(2+6)
    = 9/8 = 1.125
  P-value = $\chi^2$cdf (1.125,99,1) = 0.2888

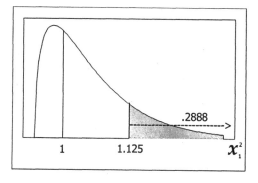

conclusion: Do not reject Ho; there is not sufficient evidence to reject the claim that there is no difference between the two treatments.

## Statistical Literacy and Critical Thinking

1. Categorical data are responses that can be placed in groups identified according to a non-numeric characteristic.

2. Nothing is wrong with the project – unless she tries to generalize the results to an inappropriate population. Without knowing how the sample was selected and how she worded her conclusion, it's difficult to make a precise critique. If she selected the "classmates" at random from a large university, for example, the results apply to the students of that university – but not necessarily to another school, and certainly not to the US population in general. If "classmates" means those students who are in one of her classes, however, the sample is a convenience sample and the results apply only to the students selected.

3. a. True. Chi-square formulas involve squaring a difference, and squares are never negative.
   b. False. The chi-square distribution is positively skewed – there is a lower bound at 0, but there is no upper bound.
   c. True. Each chi-square distribution "bunches up" around its degrees of freedom.

d. False. In contingency tables, for example, the degrees of freedom is (r-1)(c-1) – regardless of the number of subjects that were surveyed to construct the table.

e. False. Failure to have a random sample, in any situation, can create a bias that cannot be overcome merely by taking a larger sample.

4. Yes. The requirement is that all of the <u>expected</u> frequencies be at least 5, not that all of the <u>observed</u> frequencies be at least 5. No requirement is attached to the numbers of observed frequencies.

## Review Exercises

1. Place the O's in list Li and the E's in list L2. Use PRGM $\chi^2$GOF.
   $H_0$: $p_{Sun} = p_{Mon} = p_{Tue} = \ldots = p_{Sat} = 1/7$
   $H_1$: at least one $p_i \neq 1/7$
   $\alpha = 0.05$ and df = 6
   C.V. $\chi^2 = \chi^2_\alpha = 12.592$
   calculations:

   | day | O | E | (O-E)²/E |
   |-----|-----|--------|-------|
   | Sun | 40 | 30.857 | 2.709 |
   | Mon | 24 | 30.857 | 1.524 |
   | Tue | 25 | 30.857 | 1.112 |
   | Wed | 28 | 30.857 | 0.265 |
   | Thu | 29 | 30.857 | 0.112 |
   | Fri | 32 | 30.857 | 0.042 |
   | Sat | 38 | 30.857 | 1.653 |
   |     | 216 | 216.000 | 7.417 |

   $\chi^2 = \Sigma[(O-E)^2/E] = 7.417$
   P-value = $P(\chi^2_6 > 7.417) = 0.2840$

   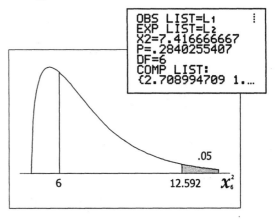

   conclusion:
   Do not Reject Ho; there is not sufficient evidence to reject the claim that fatal crashes occur on the different days with equal frequency.
   There were more crashes on Friday, Saturday and Sunday – which supports the Friday night and Saturday night (extending into Sunday morning) binge theory. But the increases were not large enough to be 95% sure that they were not random fluctuations from an equal frequency for each day. While the daily drinker (i.e., equal frequency) theory cannot be rejected, the results do not provide confirming support for either theory.

2. Place the table values in a matrix and use STAT TESTS $\chi^2$-test.
   $H_0$: response and company status are independent
   $H_1$: response and company status are related
   $\alpha = 0.05$ and df = 1
   C.V. $\chi^2 = \chi^2_\alpha = 3.841$
   calculations:

   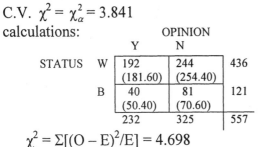

   $\chi^2 = \Sigma[(O-E)^2/E] = 4.698$

   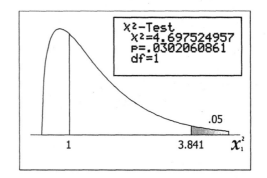

P-value = $P(\chi_1^2 > 4.698) = 0.0302$

conclusion:
Reject Ho; there is sufficient evidence to reject the claim that response and company status are independent. It appears that workers are more likely than bosses to think that it is unethical to monitor employee e-mail.
Yes; the decision changes if the 0.01 level is used – since P-value = 0.0302 > 0.01.
No; workers and bosses do not appear to agree on this issue.

3. Place the table values in a matrix and use STAT TESTS $\chi^2$–test.
   $H_o$: type of crime and criminal/victim connection are independent
   $H_1$: type of crime and criminal/victim connection are related
   $\alpha = 0.05$ and df = 2
   C.V. $\chi^2 = \chi_\alpha^2 = 5.991$
   calculations:

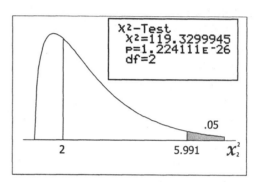

|  |  | CRIME | | | |
|---|---|---|---|---|---|
|  |  | homicide | robbery | assault | |
| C/V | S | 12 (29.93) | 379 (284.64) | 727 (803.43) | 1118 |
|  | A | 39 (21.07) | 106 (200.36) | 642 (565.57) | 787 |
|  |  | 51 | 485 | 1369 | 1905 |

$\chi^2 = \Sigma[(O - E)^2/E] = 119.330$
P-value = $P(\chi_2^2 > 119.330) = 1.22\text{E-}26 \approx 0$

conclusion:
Reject Ho; there is sufficient evidence to reject the claim that the type of crime and the criminal/victim connection are independent and conclude that the variables are related.
Police should not overlook acquaintances and relatives when investigating homicides.

4. $H_o$: there is no difference between the treatment and the placebo
   $H_1$: there is a difference between the treatment and the placebo
   $\alpha = 0.05$ and df = 1
   C.V. $\chi^2 = \chi_\alpha^2 = 3.841$
   calculations:
   b = 8, c = 32
   $\chi^2 = (|b-c| - 1)^2/(b+c)$
       $= (24-1)^2/(8+32)$
       $= 528/40 = 13.225$
   P-value = $\chi^2\text{cdf}(13.225, 99, 1) = 0.0003$

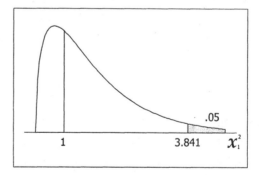

conclusion: Reject Ho; there is sufficient evidence to reject the claim that there is no difference between the treatment and the placebo. Since the majority of the discordant pairs are within the treatment=relief/placebo=norelief cell, conclude that the treatment is effective.

**Cumulative Review Exercises**

Place the 4 x scores in list L1, the 4 y scores in list L2, and all 8 scores in list L3.
The following commands and screen displays are used to answer the exercises as indicated.
Use STAT CALC 1-Var Stats on list L3. (#1)
Use STAT TESTS LinRegTTest on lists L1 and L2. (#4)

display for L3

display for L1 (x) and L2 (y)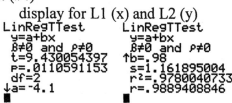

1. The following answers may be read from the 1-Var Stats display for L3.
   $\bar{x}$ = 79.625, rounded to 79.6
   $\tilde{x}$ = 80.0
   R = 90 − 70 = 20
   $s^2$ = (6.696214282)$^2$ = 44.839, rounded to 44.8
   s = 6.7
   five number summary: min = 70, $Q_1$ = 74.0, $\tilde{x}$ = 80.0, $Q_3$ = 84.5, max = 90

2. Consider the table at the right.
   a. P(C) = 153/637 = 0.240
   b. P(M) = 330/637 = 0.518
   c. P(C or M) = P(C) + P(M) − P(C and M)
      = 153/637 + 330/637 − 80/637
      = 403/637 = 0.633
   d. P($F_1$ and $F_2$) = P($F_1$)·P($F_2|F_1$)
      = (307/637)(306/636) = 0.232
   e. P(B|F) = 84/307 = 0.274

|   | A | B | C | D |   |
|---|---|---|---|---|---|
| M | 85 | 90 | 80 | 75 | 330 |
| F | 80 | 84 | 73 | 70 | 307 |
|   | 165 | 174 | 153 | 145 | 637 |

3. The following table is used for the calculations in exercise #3.

|   |   | SELECTION | | | | |
|---|---|---|---|---|---|---|
|   |   | A | B | C | D |   |
| GENDER | M | 85 | 90 | 80 | 75 | 330 |
|   |   | (85.48) | (90.14) | (79.26) | (75.12) |   |
|   | F | 80 | 84 | 73 | 70 | 307 |
|   |   | (79.52) | (83.86) | (73.74) | (69.88) |   |
|   |   | 165 | 174 | 153 | 145 | 637 |

This test follows the pattern established in section 11-3.
Place the table values in a matrix and use STAT TESTS $\chi^2$–test.
$H_o$: gender and selection are independent
$H_1$: gender and selection are related
$\alpha$ = 0.05 [assumed] and df = 3
C.V. $\chi^2 = \chi^2_\alpha$ = 7.815
calculations:
   $\chi^2 = \Sigma[(O − E)^2/E]$ = 0.0207
   P-value = P($\chi^2_3$ > 0.0207) = 0.9992

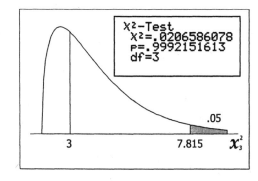

conclusion:
   Do not reject $H_o$; there is not sufficient evidence to reject the claim that men and women choose the different answers in the same proportions.

230 CHAPTER 11 Multinomial Experiments and Contingency Tables

4. This test follows the pattern established in section 10-2.
   Use the display screens above for LinRegTTest on L1 and L2.
   r = 0.9989
   $H_o$: $\rho = 0$
   $H_1$: $\rho \neq 0$
   $\alpha = 0.05$ and df = 2
   C.V. t = $\pm t_{\alpha/2}$ = $\pm 4.303$
   calculations:
   $t_r = (r - \mu_r)/s_r = 9.430$
   P-value = $2 \cdot P(t_2 > 9.430) = 0.0111$

   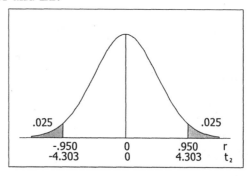

   conclusion:
   Reject $H_o$; there is sufficient evidence to conclude that $\rho \neq 0$ (in fact, that $\rho > 0$).

5. This test follows the pattern established in section 9-4.
   Use L1–L2→L4 to place d = x–y = pre– post in L4. Use STAT TESTS T-Test on list L4.
   original claim: $\mu_d = 0$
   $H_o$: $\mu_d = 0$
   $H_1$: $\mu_d \neq 0$
   $\alpha = 0.05$ [assumed]
   and df = 3
   C.V. t = $-t_{\alpha/2}$ = $\pm 3.182$
   calculations
   $t_{\bar{d}} = (\bar{d} - \mu_{\bar{d}})/s_{\bar{d}} = 12.011$
   P-value = $2 \cdot P(t_3 > 12.011) = 0.0012$

   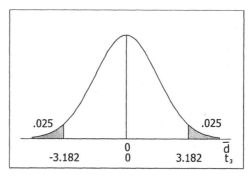

   conclusion:
   Reject $H_o$; there is sufficient evidence to reject the claim that $\mu_d = 0$ and conclude that $\mu_d \neq 0$ (in fact, that $\mu_d > 0$ – i.e., the pre-training scores are significantly higher than the post-training scores).

6. Let the men be group 1. This test follows the pattern established in section 9-3.
   Use STAT TESTS 2-SampTTest on lists L1 and L2
   original claim: $\mu_1 - \mu_2 = 0$
   $H_o$: $\mu_1 - \mu_2 = 0$
   $H_1$: $\mu_1 - \mu_2 \neq 0$
   $\alpha = 0.05$ [assumed]
   and df = 3
   C.V. t = $\pm t_{\alpha/2}$ = $\pm 3.182$
   calculations:
   $t_{\bar{x}_1-\bar{x}_2} = (\bar{x}_1-\bar{x}_2 - \mu_{\bar{x}_1-\bar{x}_2})/s_{\bar{x}_1-\bar{x}_2}$
   = 1.265
   P-value = $2 \cdot P(t_6 > 1.265) = 0.2526$

   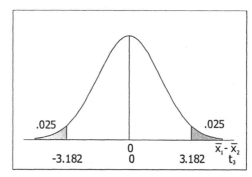

   conclusion:
   Do not reject $H_o$; there is not sufficient evidence to reject the claim that $\mu_1 - \mu_2 = 0$.
   NOTE: Recall that the text uses the conservative df = min(df$_1$,df$_2$) = 3. This is the df used to set up the test. The TI-83/84 Plus calculated uses a more sophisticated formula to determine df = 5.9995 ≈ 6. This is df used to calculate the P-value.

# Chapter 12

## Analysis of Variance

### 12-2 One-Way ANOVA

1. One-way analysis of variance tests the equality of two or more population means by analyzing sample variances. It finds a difference in the population means if the variance between the sample means is larger than can expected from the variance within the samples.

3. The variance between samples is an adjusted variance of the sample means – obtained by treating the sample means as individual scores. The variance within samples is a weighted average of the individual sample variances.

5. a. The null hypothesis is that $\mu_1 = \mu_2 = \mu_3 = \mu_4$, where the subscripts identify the four treatments.
   b. The alternative hypothesis is that at least one $\mu_i$ is different – i.e., that at least one of the treatments has a population mean that is different from the others.
   c. The test statistic is F = 8.448.
   d. From Table A-5, the critical value for $\alpha = 0.05$ and $df_{num} = 3$, $df_{den} = 15$ is $F_\alpha = 3.2874$.
   e. The P-value is $P(F^3_{15} > 8.448) = 0.0016$.
   f. The conclusion is as follows. Reject Ho; there is sufficient evidence to reject the claim that $\mu_1 = \mu_2 = \mu_3 = \mu_4$ and to conclude that at least one $\mu_i$ is different.
   g. Removing the outlier changes the df for the variance within samples from 16 to 15, the calculated F statistic from 5.731 to 8.448, and the P-value from 0.0073 to 0.0016. The conclusion does not change. Removing the outlier made the lack of equality among the means slightly easier to identify, but it did not have much of an effect on the results.

NOTE: Solutions for exercises #7 and #9 are read from the software displays given in the text. Intermediate steps are given to show how the various component of the display fit together with each other and with the notation given on the following page. Beginning with exercise #11, only the final solution is given (with the appropriate display screens from STAT TESTS ANOVA).

7. $H_o$: $\mu_1 = \mu_2 = \mu_3 = \mu_4$
   $H_1$: at least one $\mu_i$ is different
   $\alpha = 0.05$ and $df_{num} = 3$, $df_{den} = 156$
   C.V. $F = F_\alpha = 2.6626$ [Excel]
   calculations:
   $F = s_B^2 / s_p^2$
   $= 11.99995/34.08497 = 0.35206$
   P-value = $P(F^3_{156} > 0.35206) = 0.7877$

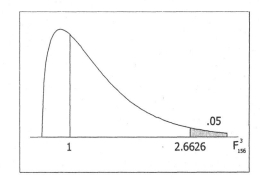

conclusion:
   Do not reject Ho; there is not sufficient evidence to reject the claim that $\mu_1 = \mu_2 = \mu_3 = \mu_4$. Given the four mean losses that range from -2.1 lbs to 3.3 lbs, it appears that one year of following the diet does not result in a weight loss worth the effort. The effort may be justified, however, when other potential benefits are considered – viz., maintaining one's weight (i.e., not gaining weight), enjoying a healthy lifestyle, etc.

9. $H_0$: $\mu_1 = \mu_2 = \mu_3 = \mu_4 = \mu_5 = \mu_6$
   $H_1$: at least one $\mu_i$ is different
   $\alpha = 0.05$ and $df_{num} = 5$, $df_{den} = 94$
   C.V. $F = F_\alpha = 2.3113$ [Statdisk]
   calculations:
   $F = ns_{\bar{x}}^2/s_p^2$
   $= 0.001/0.003 = 0.443$
   P-value $= P(F_{94}^5 > 0.443) = 0.8173$
   conclusion:
   Do not reject $H_0$; there is not sufficient evidence to reject the claim that
   $\mu_1 = \mu_2 = \mu_3 = \mu_4 = \mu_5 = \mu_6$.
   No. The results do not suggest that there is any problem requiring corrective action.

NOTE: This section is calculation-oriented. If you are asked to do any calculations by hand, do not get so involved with the formulas that you miss the concepts. In general, the manual uses the STAT TESTS ANOVA procedure on the TI-83/84 Plus calculator. When necessary or appropriate, the following notation is used.

$k$ = the number of groups
$n_i$ = the number of scores in group i (where i = 1,2,…,k)
$\bar{x}_i$ = the mean of group i
$s_i^2$ = the variance of group i
$\bar{\bar{x}}$ = the overall mean of all the scores in all the groups
$\quad = (\Sigma n_i \bar{x}_i)/\Sigma n_i$ = the (weighted) mean of the group means
$\quad = \Sigma \bar{x}_i /k$ = simplified form when each group has equal size n
$s_B^2$ = the variance between the groups [also called MS(treatment) or MS(factor)]
$\quad = \Sigma n_i(\bar{x}_i - \bar{\bar{x}})^2/(k-1)$
$\quad = n\Sigma(\bar{x}_i - \bar{\bar{x}})^2/(k-1) = ns_{\bar{x}}^2$ = simplified form when each group has equal size n
$s_p^2$ = the variance within the groups [also called MS(error)]
$\quad = (\Sigma df_i s_i^2)/\Sigma df_i$
$\quad = \Sigma s_i^2/k$ = simplified form when each group has equal size n
numerator df = k–1
denominator df = $\Sigma df_i$
$\quad = k(n-1)$ = simplified form when each group has equal size n
$F = s_B^2/s_p^2$ = (variance between groups)/(variance within groups)
P-value = Fcdf(F, 99, numerator df, denominator df)

As a crude check against errors, and to help get a feeling for the problem, always verify that the following "overall" values are realistic in that
$\quad \bar{\bar{x}}$ is a value between the lowest and the highest of the $\bar{x}_i$ values.
$\quad s_p^2$ is a value between the lowest and the highest of the $s_i^2$ values.

As in previous chapters, the superscripts and subscripts (the numerator df and denominator df) may be used to identify which F distribution to look up in the tables.

One-Way ANOVA SECTION 12-2   233

11. Place the 5 subcompact values in list L1, the 5 compact values in list L2, the 5 midsize values in list L3, and the 5 full-size values in list L4.
Use STAT TESTS ANOVA(L1,L2,L3,L4).
$H_o$: $\mu_1 = \mu_2 = \mu_3 = \mu_4$
$H_1$: at least one $\mu_i$ is different
$\alpha = 0.05$ and $df_{num} = 3$, $df_{den} = 16$
C.V.  $F = F_\alpha = 3.2389$
calculations:

$F = s_B^2/s_p^2 = 0.9922$

P-value = $P(F_{16}^3 > 0.9922) = 0.4216$

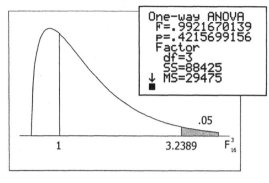

conclusion:
Do not reject Ho; there is not sufficient evidence to reject the claim that $\mu_1 = \mu_2 = \mu_3 = \mu_4$.
No. The head injury data do not suggest that larger cars are safer.

13. Place the 6 female/black values in list L1, the 6 male/black values in list L2, and the 6 female/white values in list L3, and the 6 male/white values in list L4.
Use STAT TESTS ANOVA(L1,L2,L3,L4).
$H_o$: $\mu_1 = \mu_2 = \mu_3 = \mu_4$
$H_1$: at least one $\mu_i$ is different
$\alpha = 0.05$ and $df_{num} = 3$, $df_{den} = 20$
C.V.  $F = F_\alpha = 3.0984$
calculations:

$F = s_B^2/s_p^2 = 2.4749$

P-value = $P(F_{20}^3 > 2.4750) = 0.0911$

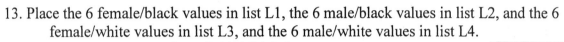

conclusion:
Do not reject Ho; there is not sufficient evidence to reject the claim that $\mu_1 = \mu_2 = \mu_3 = \mu_4$.
Yes. By this measure, these groups can be considered to be samples from the same population.

15. There are 5 CPIND values, 35 CPWHT values, 35 CPPRE values, and 37 CPPST values.
Use STAT TESTS ANOVA(CPIND,CPWHT,CPPRE,CPPST).
$H_o$: $\mu_1 = \mu_2 = \mu_3 = \mu_4$
$H_1$: at least one $\mu_i$ is different
$\alpha = 0.05$ and $df_{num} = 3$, $df_{den} = 108$
C.V.  $F = F_\alpha = 2.6802$
calculations:

$F = s_B^2/s_p^2 = 1713.725$

P-value = $P(F_{108}^3 > 1713.747) = 6.87\text{E}{-}91 \approx 0$

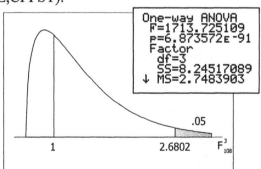

conclusion:
Reject Ho; there is sufficient evidence to reject the claim that $\mu_1 = \mu_2 = \mu_3 = \mu_4$ and conclude that at least one $\mu_i$ is different.
No. It appears that machines cannot treat the weights of all pennies the same way.

234   CHAPTER 12   Analysis of Variance

17. Place the 5 "None" values in list L1, and the 5 "Fertilizer" values in list L2.
    For part (a), use STAT TESTS 2-SampTTest on lists L1 and L2 with pooled = yes.
    For part (b), use STAT TESTS ANOVA(L1,L2).
    The 2 resulting screen displays for each part are given below. The relevant values may be read
    from these screens or calculated by hand using the summary statistics given in the text.

           for part (a)                    for part (b)

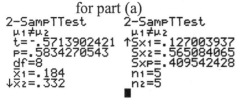 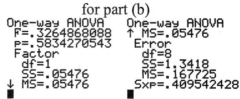

a. Testing $H_o$: $\mu_1 = \mu_2$ using the pooled variance t test of section 9-3.
   $\bar{x}_1 - \bar{x}_2 = 0.184 - 0.332 = -0.148$     df = $df_1 + df_2 = 4 + 4 = 8$
   $s_p^2 = [(df_1)s_1^2 + (df_2)s_2^2]/[df_1+df_2] = [(4)(0.127)^2 + (4)(0.565)^2]/[4+4] = 1.341416/8 = 0.1677$
   $H_o$: $\mu_1 - \mu_2 = 0$
   $H_1$: $\mu_1 - \mu_2 \neq 0$
   $\alpha = 0.05$ and df = 8
   C.V.  $t = \pm t_{\alpha/2} = \pm 2.306$
   calculations:
   $t_{\bar{x}_1-\bar{x}_2} = (\bar{x}_1-\bar{x}_2 - \mu_{\bar{x}_1-\bar{x}_2})/s_{\bar{x}_1-\bar{x}_2}$
   $= (-0.148 - 0)/\sqrt{0.167677/5 + 0.167677/5}$
   $= -0.148/0.25898 = -0.5714$
   P-value = $2 \cdot P(t_8 < -0.57147) = 0.5834$

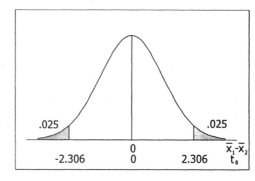

   conclusion:
      Do not reject $H_o$; there is not sufficient evidence to reject the claim that $\mu_1 - \mu_2 = 0$.

b. Testing $H_o$: $\mu_1 = \mu_2$ using the analysis of variance test of this section.
   Since each of the k=2 groups has equal size n=5, use the simplified form of the calculations.
   $\bar{\bar{x}} = \Sigma \bar{x}_i/k = (0.184 + 0.332)/2 = 0.258$
   $s_{\bar{x}}^2 = \Sigma(\bar{x}_i - \bar{\bar{x}})^2/(k-1) = [(0.184-0.258)^2 + (0.332-0.258)^2] = 0.010952$   $s_B^2 = n \cdot s_{\bar{x}}^2 = 5 \cdot 0.0010952 = 0.05476$
   $s_p^2 = \Sigma s_i^2/k = [(0.127)^2 + (0.565)^2]/2 = 0.335354/2 = 0.1677$
   $H_o$: $\mu_1 = \mu_2$
   $H_1$: at least one $\mu_i$ is different
   $\alpha = 0.05$ and $df_{num} = 1$, $df_{den} = 8$
   C.V.  $F = F_\alpha = 5.3177$
   calculations:
   $F = s_B^2/s_p^2$
   $= 0.05476/0.1677 = 0.3265$
   P-value = $P(F_8^1 > 0.3266) = 0.5834$

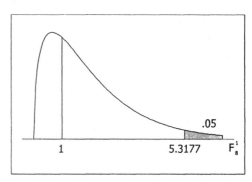

   conclusion:
      Do not reject $H_o$; there is not sufficient evidence to reject the claim that $\mu_1 = \mu_2$.

c. The tests are equivalent, as demonstrated by the P-values. In general $(t_{df})^2 = F_{df}^1$.
   In this particular case, for $\alpha = 0.05$,
      critical values: $(\pm t_{8,\alpha/2})^2 = (\pm 2.306)^2 = 5.3176 \approx 5.3177 = F_{8,\alpha}^1$ [except for round-off]
      calculated statistic: $t^2 = (-0.5714)^2 = 0.3265 = F$

## 12-3 Two-Way ANOVA

1. Analysis of variance is a test to discover differences in population means by examining population variances. The test is two-way when the experimental units can be classified according to levels of two different factors so that the data can be arranged in a two-dimensional table – with the rows and columns representing the two different factors.

3. When there is interaction, it is not possible to make an overall statement about the results for each factor – because the results for each factor vary according to the various levels of the other factor.

NOTE: The formulas and principles in this section are logical extensions of those in the previous section. In particular,

$SS_{Row} = \Sigma n_i (\bar{x}_i - \bar{\bar{x}})^2$ for i = 1,2,3... [for each row]

$SS_{Col} = \Sigma n_j (\bar{x}_j - \bar{\bar{x}})^2$ for j = 1,2,3... [for each column]

$SS_{Tot} = \Sigma (x - \bar{\bar{x}})^2$ for all the x's

When there is only one observation per cell, the unexplained variation is

$SS_{Err} = SS_{Tot} - SS_{Row} - SS_{Col}$, and there is not enough data to measure interaction.

When there is more than one observation per cell, the unexplained variation (i.e., the failure of experimental units in the same cell to respond the same) is

$SS_{Err} = \Sigma (x - \bar{x}_{ij})^2 = \Sigma df_{ij} s_{ij}^2$ [for each cell – i.e., for each i,j (row,col) combination]

and the interaction sum of squares is

$SS_{Int} = SS_{Tot} - SS_{Row} - SS_{Col} - SS_{Err}$.

Since the data will be analyzed with statistical software, however, the above formulas need not be used by hand.

5. $H_o$: there is no site-treatment interaction
$H_1$: there is site-treatment interaction
$\alpha = 0.05$ [assumed] and $df_{num} = 3$, $df_{den} = 32$
C.V. $F = F_\alpha = 2.9233$
calculations:
$F = MS_{Int}/MS_{Err}$
$= (1.9060/3)/(7.2755/32)$
$= 2.79$
P-value = $P(F_{32}^3 > 2.79) = 0.056$

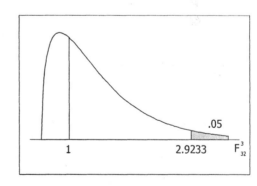

conclusion:
Do not reject $H_o$; there is not sufficient evidence to reject the claim that there is no interaction between site and treatment.

NOTE: In exercise #5, and in the remaining exercises of this section, intermediate steps are included to indicate how the final calculated F is derived from some of the other given values. The intermediate steps are not a necessary part of the solution and are included only to give added insight into the exercises and the TI displays.

7. $H_o$: $\mu_1 = \mu_2 = \mu_3 = \mu_4$ [there is no treatment effect]
$H_1$: at least one treatment $\mu_i$ is different
$\alpha = 0.05$ [assumed] and $df_{num} = 3$, $df_{den} = 32$
C.V. $F = F_\alpha = 2.9233$
calculations:
$F = MS_{Trt}/MS_{Err}$
$= (1.7518/3)/(7.2755/32)$
$= 2.57$
P-value $= P(F_{32}^{3} > 2.57) = 0.072$

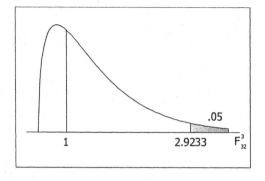

conclusion:
Do not reject $H_o$; there is not sufficient evidence to support the claim that treatment has an effect on weight.

9. $H_o$: $\mu_1 = \mu_2$ [there is no gender effect]
$H_1$: at least one gender $\mu_i$ is different
$\alpha = 0.05$ [assumed] and $df_{num} = 1$, $df_{den} = 36$
C.V. $F = F_\alpha = 4.0847$
calculations:
$F = MS_{Gen}/MS_{Err}$
$= (52635/1)/(376748/36)$
$= 5.03$
P-value $= P(F_{36}^{1} > 5.03) = 0.031$

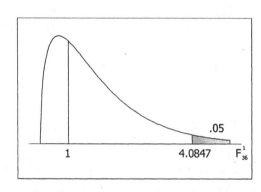

conclusion:
Reject $H_o$; there is sufficient evidence to support the claim that gender has an effect on SAT scores.

11. $H_o$: $\mu_1 = \mu_2 \ldots = \mu_{24}$ [there is no subject effect]
$H_1$: at least one subject $\mu_i$ is different
$\alpha = 0.05$ [assumed] and $df_{num} = 23$, $df_{den} = 69$
C.V. $F = F_\alpha = 1.7001$
calculations:
$F = MS_{Subj}/MS_{Err}$
$= (3231.6/23)/(2506.5/69)$
$= 3.87$
P-value $= P(F_{69}^{23} > 3.87) = 0.000$

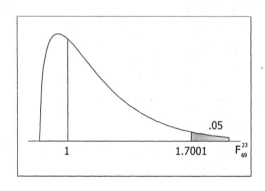

conclusion:
Reject $H_o$; there is sufficient evidence to support the claim that the choice of subject has an effect on the hearing test score.
This makes practical sense, because all people do not have the same level of hearing.

13. Enter the 24 pulse rates in list L1, moving across row 1 (male) and then across row 2 (female).
Enter the 24 row (gender) values in list L2: 12 1's (male) followed by 12 2's (female).
Enter the 24 column (age) values in list L3: 4 1's followed by 4 2's followed by 4 3's for the males, and then 4 1's followed by 4 2's followed by 4 3's for the females.
Use MATRIX MATH List▶Matr(L1,L2,L3,[D]). NOTE: The [D] is from MATRIX NAMES.
Use PRGM A1ANOVA and the option RAN BLOCK DESI as directed in the text to produce the screen on the following page.

```
   DF    SS                    P=.701
A  1   10.6666666       F(AB)=.36
B  2   81.3333333              P=.701
AB 2   81.3333333
ER 18  2024             S=10.60398248
   F(A)=.09
       P=.762
   F(B)=.36
```

The exercise can be worked by reading values directly from the above screens, or one may proceed to construct a formal ANOVA table for the tests of hypotheses in this exercise.

| Source | df | SS | MS | F | P-value |
|---|---|---|---|---|---|
| Gender | 1 | 10.667 | 10.667 | 0.09 | 0.762 |
| Age | 2 | 81.333 | 40.667 | 0.36 | 0.701 |
| Interaction | 2 | 81.333 | 40.667 | 0.36 | 0.701 |
| Error | 18 | 2024.000 | 112.444 | | |
| Total | 23 | 2197.333 | | | |

$H_o$: there is no gender-age interaction
$H_1$: there is gender-age interaction
$\alpha = 0.05$ [assumed] and $df_{num} = 2$, $df_{den} = 18$
C.V.  $F = F_\alpha = 3.5546$
calculations:
  $F = MS_{Int}/MS_{Err}$
    $= (81.33/2)/(2024/18) = 0.36$
  P-value $= P(F^2_{18} > 0.36) = 0.701$

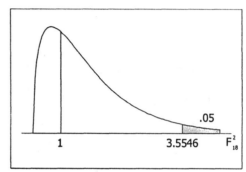

conclusion:
  Do not reject $H_o$; there is not sufficient evidence to reject the claim that there is no interaction between gender and age.

$H_o$: $\mu_1 = \mu_2$ [there is no gender effect]
$H_1$: at least one gender $\mu_i$ is different
$\alpha = 0.05$ [assumed] and $df_{num} = 1$, $df_{den} = 18$
C.V.  $F = F_\alpha = 4.4139$
calculations:
  $F = MS_{Gen}/MS_{Err}$
    $= (10.67/1)/(2024//18) = 0.09$
  P-value $= P(F^1_{18} > 0.09) = 0.762$

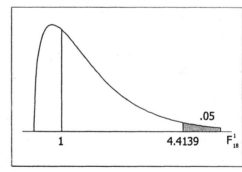

conclusion:
  Do not reject $H_o$; there is not sufficient evidence to support the claim that gender has an effect on pulse rate.

$H_o$: $\mu_1 = \mu_2 = \mu_3$ [there is no age effect]
$H_1$: at least one age $\mu_i$ is different
$\alpha = 0.05$ [assumed] and $df_{num} = 2$, $df_{den} = 18$
C.V.  $F = F_\alpha = 3.5546$
calculations:
  $F = MS_{Age}/MS_{Err}$
    $= (81.33/2)/(2024/18) = 0.36$
  P-value $= (F^2_{18} > 0.36) = 0.701$

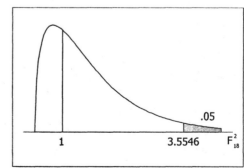

conclusion:
  Do not reject $H_o$; there is not sufficient evidence to support the claim that age has an effect on pulse rate.

238   CHAPTER 12   Analysis of Variance

15. a. No change. The ANOVA calculated F statistics are ratios of variances. Since adding the same constant to each score does not affect the variances, the ANOVA table will not change.
   b. No change. The ANOVA calculated F statistics are ratios of variances. Since multiplying each score by the same nonzero constant will multiply all the variances by the square of that constant, the numerators and denominators of each ANOVA F ratio will be multiplied by the same constant – and the ratios will not change.
   c. No change. The same values will appear in different positions, but each value will refer to the same factor it did before the transposition.
   d. Depends. The change will affect the mean for row 1, the mean for column 1, and the variability within the first cell. In general, all calculated ANOVA statistics will change.

## Statistical Literacy and Critical Thinking

1. The analysis of variance procedure tests the equality of two or more population means by analyzing sample variances. It finds a difference in the population means if the variance between the sample means is larger than can expected from the variance within the samples. And so even though the procedure is used to make decisions about means, it is called "analysis of variance."

2. The clinical trial described should be analyzed using a one-way ANOVA of section 12-2.

3. The clinical trial as modified should be analyzed using the t-test of section 9-3 for two independent means. The one-way ANOVA test of section 12-2 could be used with k=2, but the test of section 9-3 is preferred because
   (1) a one-tailed test is possible.
   (2) there is a possibility of using the test (and constructing straightforward confidence intervals) with or without the assumption of equal variances.

4. No. Eye color as commonly reported is not a quantitative variable. One of the analysis of variance requirements is that the response variable be continuous and quantitative – in particular, that the sample values in each cell come from a population that is approximately normal.

## Review Exercises

1. $H_o: \mu_A = \mu_B = \mu_C$
   $H_1$: at least one $\mu_i$ is different
   $\alpha = 0.05$ and $df_{num} = 2$, $df_{den} = 14$
   C.V.  $F = F_\alpha = 3.7389$
   calculations:
   $F = s_B^2/s_p^2$
   $\phantom{F} = 46.9$
   P-value = $P(F_{14}^2 > 46.9) = 6.23\text{E-}7 \approx 0$

   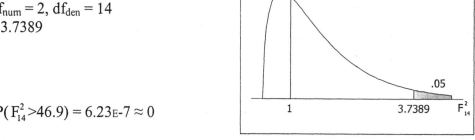

   conclusion:
       Reject Ho; there is sufficient evidence to reject the claim that the three groups have the same mean blood alcohol level.

2. Place the 12 Clancy values in list L1, the 12 Rowling values in list L2, and the 12 Tolstoy values in list L3.
   Use STAT TESTS ANOVA(L1,L2,L3).
   $H_0$: $\mu_1 = \mu_2 = \mu_3$
   $H_1$: at least one $\mu_i$ is different
   $\alpha = 0.05$ and $df_{num} = 2$, $df_{den} = 33$
   C.V. $F = F_\alpha = 3.3158$
   calculations:
   $F = s_B^2/s_p^2 = 9.4695$
   P-value = $P(F_{33}^2 > 9.4695) = 0.0006$

   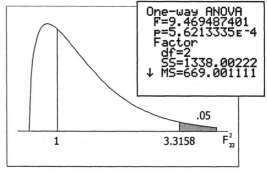

   conclusion:
   Reject $H_0$; there is sufficient evidence to reject the claim that $\mu_1 = \mu_2 = \mu_3$ and to conclude that at least one $\mu_i$ is different.
   No. Based on these results, the books do not appear to have the same reading level.

3. $H_0$: there is no gender-major interaction
   $H_1$: there is gender-major interaction
   $\alpha = 0.05$ [assumed] and $df_{num} = 2$, $df_{den} = 12$
   C.V. $F = F_\alpha = 3.8853$
   calculations:
   $F = MS_{Int}/MS_{Err}$
   $= (14.11/2)/(453.33/12)$
   $= 0.19$
   P-value = $P(F_{12}^2 > 0.19) = 0.832$

   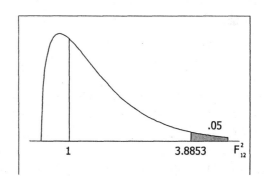

   conclusion:
   Do not reject $H_0$; there is not sufficient evidence to reject the claim that there is no interaction between gender and major.

4. $H_0$: $\mu_1 = \mu_2$ [there is no gender effect]
   $H_1$: at least one gender $\mu_i$ is different
   $\alpha = 0.05$ [assumed] and $df_{num} = 1$, $df_{den} = 12$
   C.V. $F = F_\alpha = 4.7472$
   calculations:
   $F = MS_{Gen}/MS_{Err}$
   $= (29.38/1)/(453.33/12)$
   $= 0.78$
   P-value = $P(F_{12}^1 > 0.78) = 0.395$

   conclusion:
   Do not reject $H_0$; there is not sufficient evidence to support the claim that estimated length is affected by gender.

240 CHAPTER 12 Analysis of Variance

5. $H_o$: $\mu_1 = \mu_2 = \mu_3$ [there is no major effect]
$H_1$: at least one major $\mu_i$ is different
$\alpha = 0.05$ [assumed] and $df_{num} = 2$, $df_{den} = 12$
C.V. $F = F_\alpha = 3.8853$
calculations:
$F = MS_{Maj}/MS_{Err}$
$= (10.11/2)/(453.33/12)$
$= 0.13$
P-value $= P(F_{12}^2 > 0.13) = 0.876$

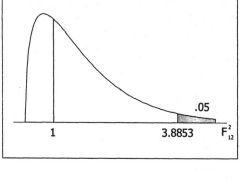

conclusion:
Do not reject $H_o$; there is not sufficient evidence to support the claim that estimated length is affected by major.

6. Enter the 16 temperatures in list L1, moving across the 8 males in row 1 followed by the 8 females in row 2.
Enter the 16 row (gender) values in list L2: 8 1's (male) followed by 8 2's (female).
Enter the 16 column (smoker) values in list L3: 4 1's followed by 4 2's for the males, and then 4 1's followed by 4 2's for the females.
Use MATRIX MATH List▶Matr(L1,L2,L3,[D]). NOTE: The [D] is from MATRIX NAMES.
Use PRGM A1ANOVA and the option RAN BLOCK DESI as directed in the text to produce the following screens.

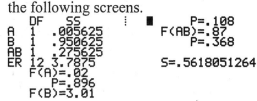

The exercise can be worked by reading values directly from the above screens, or one may proceed to construct a formal ANOVA table for the tests of hypotheses in this exercise.

| Source | df | SS | MS | F | P-value |
|---|---|---|---|---|---|
| Gender | 1 | 0.005625 | 0.005625 | 0.02 | 0.896 |
| Smoking | 1 | 0.950625 | 0.950625 | 3.01 | 0.108 |
| Interaction | 1 | 0.275625 | 0.275625 | 0.87 | 0.368 |
| Error | 12 | 3.787500 | 0.315625 | | |
| Total | 15 | 5.019375 | | | |

$H_o$: there is no gender-smoking interaction
$H_1$: there is gender-smoking interaction
$\alpha = 0.05$ and $df_{num} = 1$, $df_{den} = 12$
C.V. $F = F_\alpha = 4.7472$
calculations:
$F = MS_{Int}/MS_{Err}$
$= (0.275625/1)/(3.7875/12) = 0.87$
P-value $= P(F_{12}^1 > 0.87) = 0.368$

conclusion:
Do not reject $H_o$; there is not sufficient evidence to reject the claim that there is no body temperature interaction between gender and whether a person smokes.

$H_o$: $\mu_1 = \mu_2$ [there is no gender effect]
$H_1$: at least one gender $\mu_i$ is different
$\alpha = 0.05$ and $df_{num} = 1$, $df_{den} = 12$
C.V.  $F = F_\alpha = 4.7472$
calculations:
   $F = MS_{Gen}/MS_{Err}$
     $= (0..005625/1)/(3.7875/12) = 0.02$

   P-value $= P(F_{12}^1 > 0.02) = 0.896$

conclusion:
   Do not reject $H_o$; there is not sufficient evidence to support the claim that gender has an effect on body temperature.

$H_o$: $\mu_1 = \mu_2$ [there is no smoking effect]
$H_1$: at least one smoking $\mu_i$ is different
$\alpha = 0.05$ and $df_{num} = 1$, $df_{den} = 12$
C.V.  $F = F_\alpha = 4.7472$
calculations:
   $F = MS_{Smo}/MS_{Err}$
     $= (0.950625/1)/(3.7875/12) = 3.01$
   P-value $= P(F_{12}^1 > 3.01) = 0.108$

conclusion:
   Do not reject $H_o$; there is not sufficient evidence to support the claim that whether a person smokes has an effect on body temperature.

7. Enter the 6 emission values in list L1, moving across the 3 values in row 1 followed by the 3 values in row 2.
   Enter the 6 row (transmission) values in list L2: 3 1's (automatic) followed by 3 2's (manual).
   Enter the 6 column (cylinder) values in list L3: 1 2 3 followed by 1 2 3.
   Use MATRIX MATH List▶Matr(L1,L2,L3,[D]).  NOTE: The [D] is from MATRIX NAMES.
   Use PRGM A1ANOVA and the option RAN BLOCK DESI as directed in the text to produce the following screens.

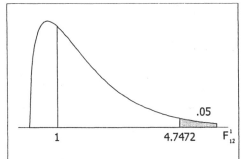

The exercise can be worked by reading values directly from the above screens, or one may proceed to construct a formal ANOVA table for the tests of hypotheses in this exercise.

| Source       | df | SS       | MS      | F    | P-value |
|--------------|----|----------|---------|------|---------|
| Transmission | 1  | 0.66667  | 0.66667 | 1.00 | 0.423   |
| Cylinder     | 2  | 9.33333  | 4.66667 | 7.00 | 0.125   |
| Error        | 2  | 1.33333  | 0.66667 |      |         |
| Total        | 5  | 11.33333 |         |      |         |

a. $H_o$: $\mu_1 = \mu_2$ [there is no transmission effect]
$H_1$: at least one transmission $\mu_i$ is different
$\alpha = 0.05$ [assumed] and $df_{num} = 1$, $df_{den} = 2$
C.V. $F = F_\alpha = 18.513$
calculations:
$F = MS_{Tra}/MS_{Err}$
$= (0.667/1)/(1.333/2)$
$= 1.0000$
P-value = $P(F_2^1 > 1.00) = 0.423$

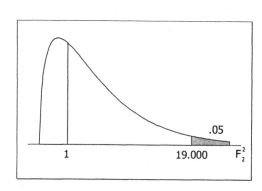

conclusion:
Do not reject $H_o$; there is not sufficient evidence to support the claim that the amounts of emitted greenhouse gases are affected by the type of transmission.

b. $H_o$: $\mu_1 = \mu_2 = \mu_3$ [there is no cylinder effect]
$H_1$: at least one cylinder $\mu_i$ is different
$\alpha = 0.05$ [assumed] and $df_{num} = 2$, $df_{den} = 2$
C.V. $F = F_\alpha = 19.000$
calculations:
$F = MS_{Cyl}/MS_{Err}$
$= (9.333/2)/(1.333/2)$
$= 7.0000$
P-value = $P(F_2^2 > 7.00) = 0.125$

conclusion:
Do not reject $H_o$; there is not sufficient evidence to support the claim that the amounts of emitted greenhouse gases are affected by the numbers of cylinders.

c. No. We cannot conclude there is no effect, but only that there is not enough evidence to be 95% certain that there is an effect. The very small sample sizes make it unlikely that the data would provide enough evidence to reject any fairly reasonable null hypothesis.

8. Enter the 38 President values in list L1, the 24 Pope values in list L2, and the 14 Monarch values in list L3.
Use STAT TESTS ANOVA(L1,L2,L3).
$H_o$: $\mu_1 = \mu_2 = \mu_3$
$H_1$: at least one $\mu_i$ is different
$\alpha = 0.05$ [assumed] and $df_{num} = 2$, $df_{den} = 73$
C.V. $F = F_\alpha = 3.1504$
calculations:
$F = s_B^2/s_p^2 = 3.1095$
P-value = $P(F_{73}^2 > 3.2422) = 0.0506$

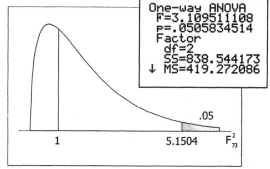

conclusion:
Do not reject Ho; there is not sufficient evidence to reject the claim that $\mu_1 = \mu_2 = \mu_3$.
There is not sufficient evidence to conclude that the survival times for the three groups differ.

# Cumulative Review Exercises

1. Enter the 38 President values in list L1, the 24 Pope values in list L2, and the 14 Monarch values in list L3.
   Use STAT CALC 1-Var Stats on each of L1, L2 and L3 to produces the following screens.

   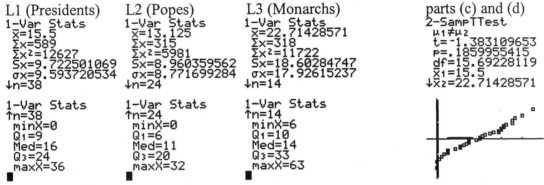

   a. $\bar{x}_{pres} = 15.5$ yrs

      $\bar{x}_{pope} = 13.1$ yrs

      $\bar{x}_{mon} = 22.7$ yrs

      Since there are essentially no minimum age requirements for monarchs, one expects that their lengths of life after assuming office could be longer.

   b. $s_{pres} = 9.7$ yrs

      $s_{popes} = 9.0$ yrs

      $s_{mon} = 18.6$ yrs

   c. Let the presidents be group 1.
      Use STAT TESTS 2-SampTTest on L1 and L3.
      See the screen above for part (c).
      original claim: $\mu_1 - \mu_2 \neq 0$ yrs
      $H_o$: $\mu_1 - \mu_2 = 0$ yrs
      $H_1$: $\mu_1 - \mu_2 \neq 0$ yrs
      $\alpha = 0.05$ [assumed] and df = 13
      C.V. $t = \pm t_{\alpha/2} = \pm 2.160$
      calculations:
         $t_{\bar{x}_1-\bar{x}_2} = (\bar{x}_1-\bar{x}_2 - \mu_{\bar{x}_1-\bar{x}_2})/s_{\bar{x}_1-\bar{x}_2}$
         $= -1.383$
      P-value = $2 \cdot P(t_{15.7} < -1.383) = 0.1860$
      conclusion:
         Do not reject $H_o$; there is not sufficient evidence to conclude that $\mu_1 - \mu_2 \neq 0$.

   d. Answers will vary. One approach is to use STATPLOT and ZOOM to produce a normal quantile plot for L1. That screen display is given above. Since points can be approximated by a straight line, the distribution appears to be approximately normal.

   e. $\alpha = 0.05$ and df = 37       OR       Use STAT TESTS TInterval

      $\bar{x} \pm t_{\alpha/2} \cdot s/\sqrt{n}$                      List: L1

      $15.500 \pm 2.028(9.7225)/\sqrt{38}$           C-Level: 0.95

      $15.500 \pm 3.198$                             $12.3 < \mu < 18.7$ (years)

      $12.3 < \mu < 18.7$ (years)

244  CHAPTER 12  Analysis of Variance

2. Enter the red values in list L1, the green values in list L2, and the blue values in list L3.
   Use STAT CALC 1-Var Stats on each of L1, L2 and L3 to produces the following screens.

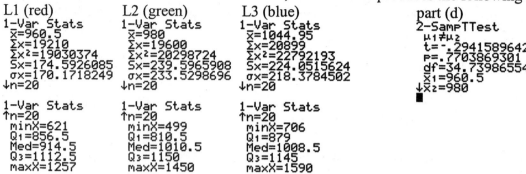

a. $\bar{x}_R = 960.5$, $\bar{x}_G = 980.0$, $\bar{x}_B = 1044.95$

   No. While the means are fairly close, the one for the blue M&M's appears to be significantly higher – a statistical test is needed to decide for sure.

b. $\tilde{x}_R = 914.5$, $\tilde{x}_G = 1010.5$, $\tilde{x}_B = 1008.5$

   No. While the medians are fairly close, the one for the red M&M's appears to be significantly lower – a statistical test is needed to decide for sure.

c. $s_R = 174.6$, $s_G = 239.6$, $s_B = 224.1$

   No. While the standard deviations are fairly close, the one for the red M&M's appears to be significantly lower – a statistical test is needed to decide for sure.

d. Let the red M&M's be group 1.
   Use STAT TESTS 2-SampTTest on L1 and L2.
   See the screen above for part (d).
   original claim: $\mu_1 - \mu_2 = 0$
   $H_o$: $\mu_1 - \mu_2 = 0$
   $H_1$: $\mu_1 - \mu_2 \neq 0$
   $\alpha = 0.05$ [assumed] and df = 19
   C.V.  $t = \pm t_{\alpha/2} = \pm 2.093$
   calculations:
   $t_{\bar{x}_1 - \bar{x}_2} = (\bar{x}_1 - \bar{x}_2 - \mu_{\bar{x}_1 - \bar{x}_2})/s_{\bar{x}_1 - \bar{x}_2}$
   $= -0.294$
   P-value = $2 \cdot P(t_{34.7} < -0.294) = 0.7704$
   conclusion:
   Do not reject $H_o$; there is not sufficient evidence to reject the claim that $\mu_1 - \mu_2 = 0$.

e. $\alpha = 0.05$ and df = 19         OR        Use STAT TESTS TInterval.
   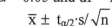                          List: L1
   $960.5 \pm 2.093(174.593)/\sqrt{20}$          C-Level: 0.95
   $960.5 \pm 81.7$                              $878.8 < \mu < 1042.2$
   $878.8 < \mu < 1042.2$

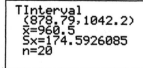

f. Use STAT TESTS ANOVA(L1,L2,L3).
   $H_o: \mu_1 = \mu_2 = \mu_3$
   $H_1:$ at least one $\mu_i$ is different
   $\alpha = 0.05$ and $df_{num} = 2$, $df_{den} = 57$
   C.V. $F = F_\alpha = 3.1504$
   calculations:
   $F = s_B^2/s_p^2 = 0.8495$
   P-value = $P(F_{57}^2 > 0.8495) = 0.4330$
   conclusion:
   Do not reject Ho; there is not sufficient evidence to reject the claim that $\mu_1 = \mu_2 = \mu_3$.

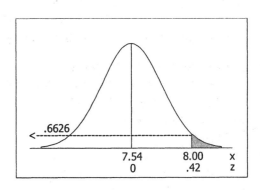

3. Let x = the number of offspring with blue eyes.
   binomial: n=100 and p=0.25
   Use DIST binomcdf.
   $P(x \le 19) = $ binomcdf(100,0.25,19)
   $= 0.0995$
   No. Since $0.0995 > 0.05$, getting 19 or fewer offspring with blue eyes is not an unusual occurrence – but rather one that could occur fairly easily when the one-quarter rate is correct.

4. a. normal distribution
   $\mu = 7.54$
   $\sigma = 1.09$
   $P(x>8.00)$
   $= P(z>0.42)$
   $= 1 - P(z<0.42)$
   $= 1 - 0.6628$
   $= 0.3372$
   TI: normalcdf(8.00,999,7.54,1.09) = 0.3364

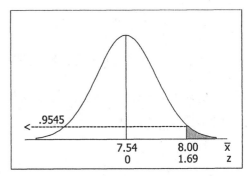

b. normal distribution, since the original distribution is so
   $\mu_{\bar{x}} = \mu = 7.54$
   $\sigma_{\bar{x}} = \sigma/\sqrt{n} = 1.09/\sqrt{16} = 0.2725$
   $P(\bar{x} > 8.00)$
   $= P(z>1.69)$
   $= 1 - P(z<1.69)$
   $= 1 - 0.9545$
   $= 0.0455$
   TI: normalcdf(8.00,999,7.54,1.06/$\sqrt{16}$) = 0.0457

c. Let B = a baby weighs more than 7.54 lbs.
   $P(B) = 0.5000$, for each birth
   $P(B_1$ and $B_2$ and $B_3) = P(B_1) \cdot P(B_2) \cdot P(B_3)$
   $= (0.5000)(0.5000)(0.5000)$
   $= 0.125$

# Chapter 13

# Nonparametric Statistics

## 13-2 Sign Test

1. The sign test is "nonparametric" or "distribution-free" because it does not require the data to come from a particular distribution defined in terms of certain parameters.

3. The conclusion must be that there is not enough evidence to support the claim. In order for the data to support the claim the number of girl births would have to be significantly greater than (½)·(80) = 40.

5. 15 +'s and 4 –'s
   n = 19 ≤ 25; use C.V. = 4 from Table A-7
   Since min(15,4) = 4 ≤ 4 = C.V., reject the hypothesis of no difference.
   TI: P-value = 2·binomcdf(19,0.5,4) = 0.0192
   Since there were more +'s, conclude that the first variable has the larger scores.

7. 30 +'s and 35 –'s
   n = 65 > 25; use C.V. = -1.96 from the z table
   $z = [(x+0.5) - n/2]/[\sqrt{n}/2] = [30.5 - 32.5]/[\sqrt{65}/2] = -2.0/4.031 = -0.496$
   TI: P-value = 2·binomcdf(65,0.5,30) = 0.6201
   Since -0.496 > -1.96, do not reject the hypothesis of no difference.

NOTES FOR THE REMAINING EXERCISES IN THIS SECTION.
(1) FOR n ≤ 25. Table A-7 gives only $x_L$, the <u>lower</u> critical value for the sign test. And so the text lets x be the <u>smaller</u> of the number of +'s or the number of –'s, and warns the reader to use common sense to avoid concluding the reverse of what the data indicates. But the problem's symmetry means that the upper critical value is $x_U = n - x_L$ and that $\mu_x = n/2$, the natural expected value of x when $H_o$ is true. For completeness, this manual indicates those values whenever using the sign test – and uses a normal curve for illustration, even though the distribution of x is discrete. P-values are calculated using DIST binomcdf as given in the text.

Letting x always be the number of +'s is an alternative approach that maintains the natural agreement between the alternative hypothesis and the critical region – and is consistent with the logic and notation of parametric tests.

(2) FOR n > 25. Just as in the case for n ≤25, the problem is worked directly in terms of x and the exact P-value determined using DIST binomcdf as given in the text. With the TI-83/84 Plus calculator, there is no need to use the normal approximation to the binomial. For those choosing to use that approximation, the C.V. is given in terms of z – but no z calculations are shown. Note that the formula given in the text is the usual one for converting a score into its standard score, using the correction for continuity: $z_x = \dfrac{x - \mu_x}{\sigma_x} = \dfrac{(x \pm 0.5) - (n/2)}{\sqrt{n}/2}$.

9. Let the Friday the 6$^{th}$ values be group 1.

| month | 1 | 2 | 3 | 4 | 5 | 6 |
|---|---|---|---|---|---|---|
| 6$^{th}$–13$^{th}$ | – | – | – | + | – | – |

n = 6: 1 +'s and 5 –'s
$H_o$: median difference = 0 admissions
$H_1$: median difference ≠ 0 admissions
α = 0.05
C.V.  $x = x_{L,α/2} = 0$
  $x = x_{U,α/2} = 6 - 0 = 6$

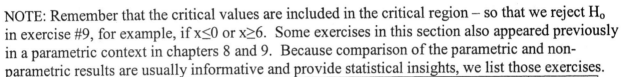

calculations:
  x = 1
  P-value = 2·P(x≤1) = 2·binomcdf(6,0.5,1) = 0.21875
conclusion:
  Do not reject $H_o$; there is not sufficient evidence to reject the claim that the number of admissions are not affected by it being Friday the 13$^{th}$.

NOTE: Remember that the critical values are included in the critical region – so that we reject $H_o$ in exercise #9, for example, if x≤0 or x≥6. Some exercises in this section also appeared previously in a parametric context in chapters 8 and 9. Because comparison of the parametric and non-parametric results are usually informative and provide statistical insights, we list those exercises.

| this section | #9 | #10 | #11 | #12 | #15 | #16 | #17 | #18 |
|---|---|---|---|---|---|---|---|---|
| previously | 9-4 #12 | 9-rev #8 | 9-4 #16 | 9-4 #14 | 8-3 #10 | 8-3 #11 | 8-3 #12 | 8-3 #18 |

11. Let the reported heights be group 1.

| subject# | 1 | 2 | 3 | 4 | 5 | 6 | 7 | 8 | 9 | 10 | 11 | 12 |
|---|---|---|---|---|---|---|---|---|---|---|---|---|
| rep-mea | + | + | – | + | + | – | + | – | – | – | + | + |

n = 12: 7 +'s and 5 –'s
$H_o$: median difference = 0 in
$H_1$: median difference ≠ 0 in
α = 0.05
C.V.  $x = x_{L,α/2} = 2$
  $x = x_{U,α/2} = 12 - 2 = 10$

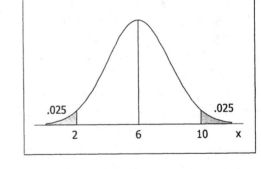

calculations:
  x = 5 (using less frequent count)
  x = 7 (using + count)
  P-value = 2·P(x≤3)   [or 2·P(x≥7)]
    = 2·binomcdf(12,0.5,5) = 0.7744
conclusion:
  Do not reject $H_o$; there is not sufficient evidence to support the claim that there is a difference between the self-reported heights and the measured heights.

248  CHAPTER 13  Nonparametric Statistics

13. claim: median < 98.6

| temp | 1 | 2 | 3 | 4 | 5 | 6 | 7 | 8 | 9 | 10 | 11 | 12 |
|---|---|---|---|---|---|---|---|---|---|---|---|---|
| t–98.6 | – | – | 0 | – | – | + | – | – | – | – | – | – |

n = 11: 1 + and 10 –'s
$H_0$: median = 98.6 °F
$H_1$: median < 98.6 °F
$\alpha = 0.05$
C.V. $x = x_{L,\alpha} = 2$
calculations:
    x = 1
    P-value = $P(x \leq 2)$
        = binomcdf(11,0.5,1) = 0.0059

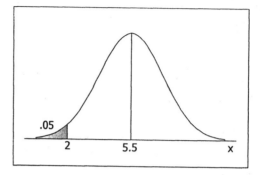

conclusion:
    Reject $H_0$; there is sufficient evidence to support the claim that these body temperatures come from a population with a median that is less than 98.6°F.

15. Let x = the number of boy births.
claim: p > 0.5
n = 51: 39 +'s and 12 –'s
[the data are in the proper direction to proceed]
$H_0$: p = 0.5
$H_1$: p > 0.5
$\alpha = 0.01$
C.V. $z = -z_\alpha = -2.326$
calculations:
    x = 12
    P-value = $P(x \leq 12)$ = binomcdf(51,0.5,12) = 9.90E-5 = 0.0000990
conclusion:
    Reject $H_0$; there is sufficient evidence to support the claim that with this method the probability of a boy is greater than 0.5.
Yes. The method appears to work.

NOTE: Working the exercise using the alternate method preserves the natural agreement between the alternative hypothesis and the critical region as follows.
$H_0$: p = 0.5
$H_1$: p > 0.5
$\alpha = 0.01$
C.V. $z = z_\alpha = 2.326$
calculations:
    x = 39
    P-value = $P(x \geq 39) = 1 -$ binomcdf(51,0.5,38)
        = 9.90E-5 = 0.0000990

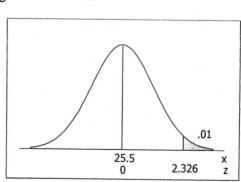

conclusion:
    Reject $H_0$; conclude the same as above.

17. Let x = the number of Internet users who use the Internet for making travel plans.
claim: p < 0.5
n = 734: 360 +'s and 374 −'s
[the data are in the proper direction to proceed]
$H_o$: p = 0.5
$H_1$: p < 0.5
α = 0.01
C.V. z = $-z_\alpha$ = −2.326
calculations:
   x = 360
   P-value = P(x≤360)
        = binomcdf(734,0.5,360) = 0.3157

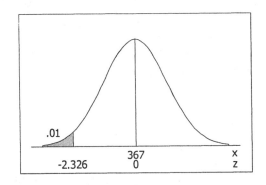

conclusion:
   Do not reject $H_o$; there is not sufficient evidence to support the claim that less than 50% of Internet users use the Internet for making travel plans.
Yes. Since each person who makes a reservation on the Internet is one less person who needs the services of a travel agent, it is important for travel agents to keep abreast of the trends in that area.

19. claim: median < 100
results: 40 +'s and 60 −'s and 21 0's
The three requested tests are given on the following page. As seen by the increasing P-values, and as suggested by common sense, the tests are increasingly conservative. In the third method of part (c), counting all the zeros to favor the null hypothesis actually renders the test unnecessary – and so the alternative method is used so that a comparison can be made.

  a. Use the usual method, discarding the 0's
     n = 100: 40 +'s and 60 −'s, so that x = 40
     $H_o$: median = 100
     $H_1$: median < 100
     α = 0.05
     C.V. z = $-z_{\alpha/2}$ = −1.645
     calculations:
        x = 40
        P-value = P(x≤40)
           = binomcdf(100,0.5,40) = 0.0284

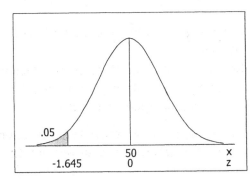

     conclusion:
       Reject $H_o$; there is sufficient evidence to support the claim that median < 100.

  b. Use the second method, counting half the 0's each way (and discarding the odd zero).
     n = 120: 50 +'s and 70 −'s, so that x = 50
     $H_o$: median = 100
     $H_1$: median < 100
     α = 0.05
     C.V. z = $-z_\alpha$ = −1.645
     calculations:
        x = 50
        P-value = P(x≤50)
           = binomcdf(120,0.5,50) = 0.0412

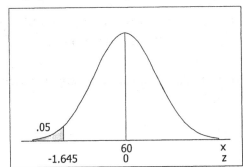

250 CHAPTER 13 Nonparametric Statistics

conclusion:
Reject $H_o$; there is sufficient evidence to support the claim that median < 100.

c. Use the third method, counting all the 0's to favor $H_o$ (i.e., against $H_1$).
n = 121: 61 +'s and 60 −'s, so that x=61
[the data are not in the proper direction to proceed; do not reject $H_o$, use the alternate method to obtain the P-value]
$H_o$: median = 100
$H_1$: median < 100
α = 0.05
C.V. $z = -z_\alpha = -1.645$
calculations:
  x = 61
  P-value = $P(x \leq 61)$
       = binomcdf (121,0.5,61) = 0.5721

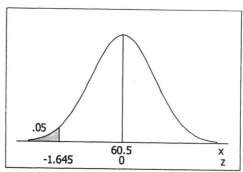

conclusion:
Do not reject $H_o$; there is not sufficient evidence to support the claim that median < 100.

21. Let p = the proportion of those hired that are female.
  a. Use the normal approximation.
  claim: p < 0.5
  n = 54: 18 +'s and 36 −'s
  Since n>25, use z with
    $\mu_x = n/2 = 54/2 = 27$
    $\sigma_x = \sqrt{n}/2 = \sqrt{54}/2 = 3.674$
  $H_o$: p = 0.5
  $H_1$: p < 0.5
  α = 0.01
  C.V. $z = -z_\alpha = -2.326$
  calculations:
    x = 18

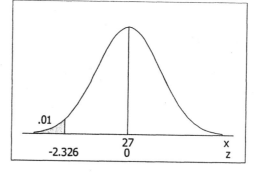

$$z_x = \frac{x - \mu_x}{\sigma_x} = \frac{(x+0.5) - (n/2)}{\sqrt{n}/2} = \frac{18.5 - 27}{3.674} = \frac{-8.5}{3.674} = -2.313$$

  P-value = $P(z<-2.31) = 0.0104$
conclusion:
  Do not reject $H_o$; there is not sufficient evidence to support the claim that p < 0.5.

b. Use the exact binomial distribution.
  claim: p < 0.5
  n = 54: 18 +'s and 36 −'s
  $H_o$: p = 0.5
  $H_1$: p < 0.5
  α = 0.01
  calculations:
    x = 18
    P-value = $P(x \leq 18)$
         = binomcdf (54,0.5,18) = 0.00992
conclusion:
  Reject $H_o$; there is sufficient evidence to support the claim that p < 0.5

Wilcoxon Signed-Ranks Test for Matched Pairs  SECTION 13-3  251

## 13-3 Wilcoxon Signed-Ranks Test for Matched Pairs

1. The parametric methods of section 9-4 require that the differences obtained from the matched pairs come from a population that is approximately normal. The nonparametric Wilcoxon signed-ranks test requires only that the differences obtained from the matched pairs come from a population that is approximately symmetric. When the differences come from a population that is approximately normal, the parametric methods of section 9-4 are to be preferred.

3. When the differences come from a population that is approximately symmetric, the Wilcoxon signed-ranks test is more efficient than the sign test – i.e., it can reach the same conclusion with a smaller sample size. Table 13-2 indicates that when working with normal populations the Wilcoxon signed-ranks test and sign test have efficiency ratings of 0.95 and 0.63 respectively.

5. Put the x values in list L1 and the y values in list L2.
   Use L1 – L2 → L3.  Use PRGM SRTEST on list L3.
   Let the x values be group 1.

   | x-y | 25 | 28 | 42 | 48 | 39 | 36 | 42 | 33 |
   |-----|----|----|----|----|----|----|----|----|
   | R   | 1  | 2  | 6.5| 8  | 5  | 4  | 6.5| 3  |

   $\Sigma R- = 0$          n = 8 non-zero ranks
   $\Sigma R+ = 36$         check: $\Sigma R = n(n+1)/2 = 8(9)/2 = 36$
   $\Sigma R = 36$

   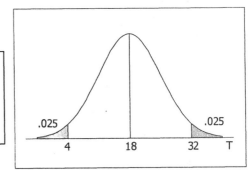

   $H_0$: median difference = 0
   $H_1$: median difference ≠ 0
   $\alpha = 0.05$
   C.V.  $T = T_{L,\alpha/2} = 4$
         $T = T_{U,\alpha/2} = 36-4 = 32$
   calculations:
         T = 0 (using smaller ranks)
         T = 36 (using positive ranks)
   conclusion:
         Reject $H_0$; there is sufficient evidence to conclude that median difference ≠ 0 (in fact, that the x scores are greater).

NOTE FOR THE REMAINING EXERCISES IN THIS SECTION.

(1) For n≤30, use the critical values in Table A-8 and the test statistic T. No P-values are given. T may be calculated by hand, or using the downloaded PRGM SRTEST. The manual works each exercise both ways for given data, and only on the calculator for data sets from the appendices.

Table A-8 gives only $T_L$, the lower critical value for the signed-ranks test. And so the text lets T be the smaller of the sum of the positive ranks or the sum of the negative ranks, and warns the reader to use common sense to avoid concluding the reverse of what the data indicates. But the problem's symmetry means that the upper critical value is $T_U = \Sigma R - T_L$ and that $\mu_T = \Sigma R/2$, the natural expected value of T when $H_0$ is true. For completeness, this manual indicates those values whenever using the signed-ranks test – and uses a normal curve for illustration, even though the distribution of T is discrete. And remember that the critical values are included in the critical region – so that we reject $H_0$ in exercise #5, for example, if T≤4 or T≥32.

Letting T always be the sum of the positive ranks is an alternative approach [used by PRGM SRTEST to calculate z] that maintains the natural agreement between the alternative hypothesis and the critical region – and is consistent with the logic and notation of parametric tests.

## 252 CHAPTER 13 Nonparametric Statistics

Finally, this manual uses a minus sign preceding ranks associated with negative differences. The ranks themselves are not negative, but use of the minus sign helps to organize the information.

(2) For n>30, use the normal approximation. The critical values come from Table A-2, and the T statistic is transformed to z by the formula in the text. The manual works these problems only using the PRGM SRTEST program downloaded to the TI-83/84 Plus calculator. As usual, appropriate set up comments and display screens are given. P-values are calculated using the DIST normalcdf function.

NOTE: The following exercises in this section were worked in chapter 9 using the parametric test for the difference between two means and in section 13-2 using the sign test. Because comparison of the results are usually informative and provide statistical insights, we list those exercises.

| this section | #7 | #8 | #9 | #10 |
|---|---|---|---|---|
| chapter 9 | 9-4 #12 | review #8 | 9-4 #16 | 9-4 #14 |
| section 13-2 | #9 | #10 | #11 | #12 |

7. Put the Friday the $6^{th}$ values in list L1 and the Friday the $13^{th}$ values in list L2.
   Use L1 – L2 → L3. Use PRGM SRTEST on list L3.
   Let the Friday the $6^{th}$ values be group 1.

   | $6^{th}$–$13^{th}$ | -4 | -6 | -3 | 1 | -1 | -7 |
   |---|---|---|---|---|---|---|
   | R | -4 | -5 | -3 | 1.5 | -1.5 | -6 |

   $\Sigma R- = 19.5$    n = 6 non-zero ranks
   $\Sigma R+ = 1.5$     check: $\Sigma R = n(n+1)/2 = 6(7)/2 = 21$
   $\Sigma R = 21$
   $H_o$: median difference = 0 admissions
   $H_1$: median difference ≠ 0 admissions
   $\alpha = 0.05$
   C.V. $T = T_{L,\alpha/2} = 1$
   $T = T_{U,\alpha/2} = 21-1 = 20$
   calculations:
   $T = 1.5$
   conclusion:
   Do not reject $H_o$; there is not sufficient evidence to conclude that median difference ≠ 0.

   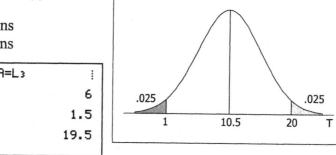

9. Put the reported heights in list L1 and the measured heights in list L2.
   Use L1 – L2 → L3. Use PRGM SRTEST on list L3.
   Let the reported values be group 1.

   | rep-mea | .1 | 1.1 | -1.9 | 1.7 | .7 | -0.6 | .5 | -3.0 | -1.6 | -11.2 | 1.0 | 1.2 |
   |---|---|---|---|---|---|---|---|---|---|---|---|---|
   | R | 1 | 6 | -10 | 9 | 4 | -3 | 2 | -11 | -8 | -12 | 5 | 7 |

   $\Sigma R- = 44$     n = 12 non-zero ranks
   $\Sigma R+ = 34$     check: $\Sigma R = n(n+1)/2 = 12(13)/2 = 78$
   $\Sigma R = 78$

$H_o$: median difference = 0 in
$H_1$: median difference ≠ 0 in
α = 0.05
C.V. $T = T_{L,α/2} = 14$
$T = T_{U,α/2} = 78 – 14$
$= 64$
calculations:
T = 34

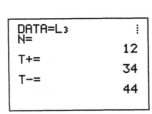
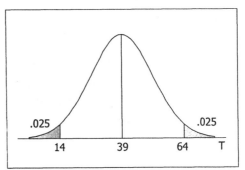

conclusion:
Do not reject $H_o$; there is not sufficient evidence to conclude that median difference ≠ 0.

11. Use ACTHI – PHI3 → L3. Use PRGM SRTEST on list L3.
Since there are only 29 non-zero differences,
use Table A-8 with ΣR = n(n+1)/2 = (29)(30)/2 = 435
and $μ_T$ = 435/2 = 217.5.
$H_o$: median difference = 0 °F
$H_1$: median difference ≠ 0 °F
α = 0.05
C.V. $T = T_{L,α/2} = 127$
$T = T_{U,α/2} = 435 – 127$
$= 308$
calculations:
T = 158.5

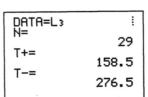
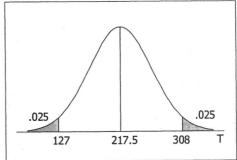

conclusion:
Do not reject $H_o$; there is not sufficient evidence to conclude that median difference ≠ 0.
No, there does not appear to be a difference.

13. Use BTEMP – 98.6 → L3. Use PRGM SRTEST on list L3.

```
DATA=L3        MEAN=
N=                   2093
        91     STD DEV=
T+=              252.6588609
        661    Z=
T-=              5.66772127
        3525          Done
```

For n=91 non-zero ranks, use the z approximation
with $μ_T$ = n(n+1)/4 = 91(92)/4 = 2093
$H_o$: median = 98.6 °F
$H_1$: median ≠ 98.6 °F
α = 0.05
C.V. $z = ±z_{α/2} = ±1.960$
calculations:
T = 661
$z = (T – μ_T)/σ_T = -5.668$
P-value = 2·P(z<-5.668 = 2·normalcdf(-999,-5.667721) = 1.45E-8 ≈ 0
conclusion:
Reject $H_o$; there is sufficient evidence to reject the claim that the median is 98.6 and conclude that median ≠ 98.6 (in fact, that the median body temperature of healthy adults is less than 98.6°F).

254  CHAPTER 13  Nonparametric Statistics

## 13-4 Wilcoxon Rank-Sum Test for Two Independent Samples

1. The Wilcoxon signed-ranks test applies to data consisting of matched pairs – which means that there must be an equal number of observations from each population, and that the observations are paired. The Wilcoxon rank-sum test applies to independent samples – which means that there may or may not be the same number of observations from each population.

3. The tests in section 9-3 require that the original populations be approximately normal. Since the Wilcoxon rank-sum test places no restrictions on the original populations, it may be used in a wider variety of situations.

NOTE FOR THE REMAINING EXERCISES IN THIS SECTION.

(1) When working an exercise by hand, it is strongly recommended that you calculate both $\Sigma R_1$ and $\Sigma R_2$ (even though only the former is necessary for the test) and that you verify that $\Sigma R_1 + \Sigma R_2 = n(n+1)/2$. This check can identify the presence of errors that must be corrected before proceeding. P-values are calculated using the DIST normalcdf function.

Always letting $R = \Sigma R_1$ (i.e., the sum of the ranks from the scores designated as group 1) guarantees that there will be agreement between $H_1$ and the C.R. (regardless of whether the test is two-tailed or one-tailed in either direction) and that the sign of the test statistic indicates in the natural manner which group has the larger median. Note that sometimes PRGM RSTEST uses $R = \Sigma R_2$, and the calculated z will be the negative of its usual value.

Exercises with both $n_1 < 15$ and $n_2 < 15$ (#5, #6, #7 and #10) are worked by hand <u>and</u> using PRGM RSTEST. The ordered scores and accompanying ranks for these exercises are given at the bottom of this page. This is all the information necessary to do each of those exercises.

(2) When using the downloaded PRGM RSTEST, remember that the output sometimes uses $R = \Sigma R_2$, and the calculated z will be the negative of its usual value. This manual works all the exercises (including #5, #6, #7 and #10) using PRGM RSTEST. As usual, appropriate set up comments and display screens are given. P-values are calculated using the DIST normalcdf function.

| | exercise #5 | | | | exercise #6 | | | | exercise #7 | | | | exercise #10 | | | |
|---|---|---|---|---|---|---|---|---|---|---|---|---|---|---|---|---|
| row | s1 | R | s2 | R | s1 | R | s2 | R | o-c | R | ctrl | R | m | R | f | R |
| 1 | 2 | 1 | 3 | 2 | 8 | 3 | 3 | 1 | 0.210 | 1.0 | 0.334 | 7.5 | 56 | 1.0 | 60 | 3.0 |
| 2 | 7 | 4 | 4 | 3 | 15 | 7 | 5 | 2 | 0.287 | 2.0 | 0.349 | 11.0 | 60 | 3.0 | 64 | 6.0 |
| 3 | 10 | 5 | 11 | 6 | 27 | 9 | 9 | 4 | 0.288 | 3.0 | 0.402 | 12.0 | 60 | 3.0 | 68 | 9.5 |
| 4 | 16 | 8 | 14 | 7 | 39 | 11 | 11 | 5 | 0.304 | 4.0 | 0.413 | 14.0 | 64 | 6.0 | 68 | 9.5 |
| 5 | 20 | 9 | 28 | 14 | 45 | 13 | 14 | 6 | 0.305 | 5.0 | 0.429 | 15.0 | 64 | 6.0 | 68 | 9.5 |
| 6 | 22 | 10 | 35 | 17 | 62 | 15 | 21 | 8 | 0.308 | 6.0 | 0.445 | 16.0 | 68 | 9.5 | 72 | 14.0 |
| 7 | 23 | 11 | 40 | 18 | 68 | 16 | 33 | 10 | 0.334 | 7.5 | 0.460 | 18.0 | 72 | 14.0 | 72 | 14.0 |
| 8 | 26 | 12 | 46 | 19 | 72 | 18 | 44 | 12 | 0.340 | 9.0 | 0.476 | 20.0 | 72 | 14.0 | 76 | 18.0 |
| 9 | 27 | 13 | 47 | 20 | 77 | 19 | 61 | 14 | 0.344 | 10.0 | 0.483 | 21.0 | 72 | 14.0 | 76 | 18.0 |
| 10 | 30 | 15 | 52 | 21 | 80 | 20 | 70 | 17 | 0.307 | 13.0 | 0.501 | 22.0 | 76 | 18.0 | 80 | 20.5 |
| 11 | 33 | <u>16</u> | 53 | 22 | 87 | <u>22</u> | 85 | <u>21</u> | 0.455 | 17.0 | 0.519 | 23.0 | 88 | 23.0 | 80 | 20.5 |
| 12 | | 104 | 60 | <u>23</u> | | 153 | | 100 | 0.463 | <u>19.0</u> | 0.594 | <u>24.0</u> | 88 | 23.0 | 88 | <u>23.0</u> |
| 13 | | | | 172 | | | | | | 96.5 | | 203.5 | 96 | <u>25.0</u> | | 165.5 |
| | | | | | | | | | | | | | | 159.5 | | |

5. Place the 11 Sample 1 values in list L1 and the 12 Sample 2 values in list L2.
   Use PRGM RSTEST on lists L1 and L2.
   ```
   GROUP A USED              16.24807681
   SUM OF RANKS R=    Z SCORE R
                104          -1.723280874
   MEAN R=            TEST STAT U=
                132                    38
   STD DEV R=         MEAN U=
        16.24807681                  66
   ```

   Let the sample 1 values be group 1.

   | | | | | | |
   |---|---|---|---|---|---|
   | $n_1 = 11$ | $R = \Sigma R_1 = 104$ | chk: $\Sigma R_i = n(n+1)/2$ | $\mu_R = n_1(n+1)/2$ | $\sigma^2_R = n_1 n_2(n+1)/12$ |
   | $n_2 = 12$ | $\Sigma R_2 = 172$ | $= 23(24)/2$ | $= 11(24)/2$ | $= (11)(12)(24)/12$ |
   | $n = \Sigma n = 23$ | $\Sigma R_i = 276$ | $= 276$ | $= 132$ | $= 264$ |
   | | | | | $\sigma_R = 16.248$ |

   $H_0$: median$_1$ = median$_2$
   $H_1$: median$_1 \neq$ median$_2$
   $\alpha = 0.05$
   C.V. $z = \pm z_{\alpha/2} = \pm 1.960$
   calculations:
   $z = (R - \mu_R)/\sigma_R$
   $= (104 - 132)/\sqrt{264}$
   $= -28/16.248 = -1.723$
   P-value $= 2 \cdot P(z < -1.72328) = 0.0849$

   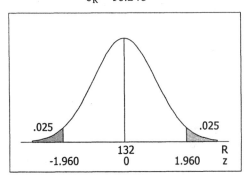

   conclusion:
   Do not reject $H_0$; there is not sufficient evidence to reject the claim that median$_1$ = median$_2$.

7. Place the 12 O-C values in list L1 and the 12 control values in list L2.
   Use PRGM RSTEST on lists L1 and L2.
   ```
   GROUP A USED              17.32050808
   SUM OF RANKS R=    Z SCORE R
                 96.5         -3.08882394
   MEAN R=            TEST STAT U=
                150                  18.5
   STD DEV R=         MEAN U=
        17.32050808                  72
   ```

   Let the obsessive-compulsive values be group 1.

   | | | | | |
   |---|---|---|---|---|
   | $n_1 = 12$ | $R = \Sigma R_1 = 96.5$ | chk: $\Sigma R_i = n(n+1)/2$ | $\mu_R = n_1(n+1)/2$ | $\sigma^2_R = n_1 n_2(n+1)/12$ |
   | $n_2 = 12$ | $\Sigma R_2 = 203.5$ | $= 24(25)/2$ | $= 12(25)/2$ | $= (12)(12)(25)/12$ |
   | $n = \Sigma n = 24$ | $\Sigma R_i = 300.0$ | $= 300$ | $= 150$ | $= 300$ |
   | | | | | $\sigma_R = 17.321$ |

   $H_0$: median$_1$ = median$_2$
   $H_1$: median$_1 \neq$ median$_2$
   $\alpha = 0.01$
   C.V. $z = \pm z_{\alpha/2} = \pm 2.575$
   calculations:
   $z = (R - \mu_R)/\sigma_R$
   $= (96.5 - 150)/\sqrt{300}$
   $= -53.5/17.321 = -3.089$
   P-value $= 2 \cdot P(z < -3.08882) = 0.0020$

   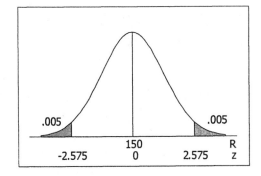

   conclusion:
   Reject $H_0$; there is sufficient evidence to reject the claim that median$_1$ = median$_2$ and conclude that median$_1 \neq$ median$_2$ (in fact, that obsessive compulsive patients have a smaller median brain volume).
   Yes. Based on this result, it appears that obsessive-compulsive disorders have a biological basis – but not necessarily that the smaller brain volume *causes* the disorders.

9. Place the 27 car values in list L1 and the 20 taxi values in list L2.
   Use PRGM RSTEST on lists L1 and L2.
   ```
   GROUP B USED         46.47580015
   SUM OF RANKS R=      Z SCORE R
         511.5             .6777720856
   MEAN R=              TEST STAT U=
         480                 301.5
   STD DEV R=           MEAN U=
         46.47580015         270
   ```
   NOTE: TI used group 2, for group 1 z = -0.677
   $H_o$: $median_1 = median_2$
   $H_1$: $median_1 \neq median_2$
   $\alpha = 0.05$
   C.V. $z = \pm z_{\alpha/2} = \pm 1.960$
   calculations:
   $\quad z = (R - \mu_R)/\sigma_R = -0.678$
   $\quad$ P-value = $2 \cdot P(z<-0.67777) = 0.4979$
   conclusion:
   $\quad$ Do not reject $H_o$; there is not sufficient evidence to support the claim that $median_1 \neq median_2$.
   The results do not suggest that taxis are newer – in fact in this sample data, the median age for the cars was lower.

11. Use PRGM RSTEST on lists MPULS and FPULS.
    ```
    GROUP A USED         103.9230485
    SUM OF RANKS R=      Z SCORE R
          1339.5             -2.699112508
    MEAN R=              TEST STAT U=
          1620                519.5
    STD DEV R=           MEAN U=
          103.9230485         800
    ```
    $H_o$: $median_1 = median_2$
    $H_1$: $median_1 \neq median_2$
    $\alpha = 0.05$ [assumed]
    C.V. $z = \pm z_{\alpha/2} = \pm 1.960$
    calculations:
    $\quad z = (R - \mu_R)/\sigma_R = -2.699$
    $\quad$ P-value = $2 \cdot P(z<-2.69911) = 0.0070$
    conclusion:
    $\quad$ Reject $H_o$; there is sufficient evidence to reject the claim that $median_1 = median_2$ and to conclude that $median_1 \neq median_2$ (in fact, that men have a lower median pulse rate than women).

13. The denominator of the Mann-Whitney and Wilcoxon statistics are the same.
    Assuming $R = \Sigma R_1$, the numerator of the Wilcoxon statistic is
    $\quad R - \mu_R = R - n_1(n+1)/2$
    Assuming $R = \Sigma R_1$, the numerator of the Mann-Whitney statistic is
    $\quad U - n_1n_2/2 = [n_1n_2 + n_1(n_1+1)/2 - R] - n_1n_2/2$
    $\quad\quad\quad\quad\quad = n_1n_2/2 + n_1(n_1+1)/2 - R$
    $\quad\quad\quad\quad\quad = (n_1/2)(n_2 + n_1 + 1) - R$
    $\quad\quad\quad\quad\quad = n_1(n+1)/2 - R = \mu_R - R = $ -[numerator of the Wilcoxon statistic]

For the BMI measures listed in Table 13-5,
$$U = n_1n_2 + n_1(n_1+1)/2 - R = 13(12) + 13(14)/2 - 187 = 60$$
$$\mu_U = n_1n_2/2 = 13(12)/2 = 78$$
$$\sigma^2_U = n_1n_2(n_1+n_2+1)/12 = 13(12)(26)/12 = 338$$
$$z_U = (\mu_U - U)/\sqrt{\sigma^2_U} = (60 - 78)/\sqrt{338} = -18/18.385 = -0.98$$

As predicted, $z_U$ has the same absolute value but the opposite sign of $z_R$.
NOTE: This exercise may also be worked on the TI-83/84 Plus calculator as follows.
Place the male BMI values in list L1 and the female BMI values in list L2.
Use PRGM RSTEST on lists L1 and L2.

```
GROUP B USED              18.38477631   MEAN U=
SUM OF RANKS R=      Z SCORE R                        78
             138          -.9790709278  STD DEV U=
MEAN R=              TEST STAT U=              18.38477631
             156                    60   Z SCORE U=
STD DEV R=           MEAN U=                  -.9790709278
     18.38477631              78                      Done
```

TI used group 2. For group 1, $z_R = +0.98$ — which agrees with the value in the text.
For the Mann-Whitney U statistic, $z_U = 0.98$ — which, as expected, is the negative of $z_R$.

## 13-5 Kruskal-Wallis Test

1. The one-way analysis of variance test requires that the original populations be approximately normal. Since the Kruskal-Wallis test places no restrictions on the original populations, it may be used in a wider variety of situations.

3. The efficiency of the Kruskal-Wallis test is 0.95. This means that for normal populations, the Kruskal-Wallis test requires a sample of size n=100 to identify departures from the null hypothesis that the one-way analysis of variance test could identify with a sample of size n=95.

NOTE: The critical value and accompanying graph for the exercises in this section are given in terms of $\chi^2$, which approximates the distribution of the calculated statistic H. P-values are calculated using DIST $\chi^2$cdf. Exercises #5 and #6 are worked both by hand and using the downloaded PRGM KWTEST. The remainder of the exercises are worked using only PRGM KWTEST.

The exercises in this section also appeared previously in a parametric context in section 12-2. Because comparison of the parametric and non-parametric results are usually informative and provide statistical insights, we list those pairings.

| this section | #5  | #6  | #7  | #8  | #9  | #10 | #11 | #12 |
|--------------|-----|-----|-----|-----|-----|-----|-----|-----|
| section 12-2 | #11 | #12 | #10 | #14 | #13 | #8  | #15 | #9  |

5. Place the data in 5x4 matrix [A], with the 5 subcompact values in column1, the 5 compact values in column 2, etc. NOTE: This is the transpose of the arrangement in the exercise.
   Use PROG KWTEST.
   Below are the ordered scores and ranks for each group. The group listed first is group 1, etc.

```
   sub  R     com  R      mid  R     full  R        n₁ = 5     R₁ = 64
   420  4     442  6      259  1     360   2        n₂ = 5     R₂ = 52.5
   428  5     514  9      454  7     384   3        n₃ = 5     R₃ = 44.5
   681  16    525  10.5   469  8     602   12       n₄ = 5     R₄ = 49
   898  19    643  13     525  10.5  656   15        n = Σn = 20   ΣR= 210
   917  20    655  14     727  18    687   17
        64         52.5        44.5        49       chk: ΣR = n(n+1)/2 = 20(21)/2 = 210
```

$H_0$: the populations have the same median
$H_1$: at least one median is different
$\alpha = 0.05$ and df $= 3$
C.V. $H = \chi_\alpha^2 = 7.815$
calculations:
$$H = \{12/[n(n+1)]\}\cdot[\Sigma(R_i^2/n_i)] - 3(n+1)$$
$$= \{12/[20(21)]\}\cdot[(64)^2/5 + (52.5)^2/5 + (44.5)^2/5 + (49)^2/5] - 3(21)$$
$$= \{0.0286\}\cdot[2246.7] - 63 = 1.191$$
P-value $= \chi^2\text{cdf}(1.19143,99,3) = 0.7551$
conclusion:

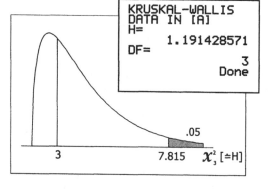

Do not reject $H_0$; there is not sufficient evidence to support the claim that head injury measurements for the four car weight categories are not all the same.

No. By this measurement, the data do not show that heavier cars are safer in a crash.

7. Place the data in 6x3 matrix [A], with the 6 sunny values in column1, the 6 cloudy values in column 2, and the 6 rainy values in column 3.
Use PRG KWTEST.
$H_0$: the populations have the same median
$H_1$: at least one median is different
$\alpha = 0.05$ and df $= 2$
C.V. $H = \chi_\alpha^2 = 5.991$
calculations:
$$H = \{12/[n(n+1)]\}\cdot[\Sigma(R_i^2/n_i)] - 3(n+1)$$
$$= 14.749$$
P-value $= \chi^2\text{cdf}(14.74853,99,2) = 0.0006$
conclusion:

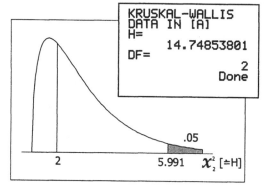

Reject $H_0$; there is sufficient evidence to reject the claim that voltage readings have the same median for the three different kinds of days.

Sunny days appear to produce greater amounts of electrical energy, but subsequent analysis would be required to determine exactly which pairs of medians are significantly different.

9. Place the data in 6x4 matrix [A], with the 6 female/black values in column1, the 6 male/black values in column 2, etc.
Use PRG KWTEST.
$H_0$: the populations have the same median
$H_1$: at least one median is different
$\alpha = 0.05$ and df $= 3$
C.V. $H = \chi_\alpha^2 = 7.815$
calculations:
$$H = \{12/[n(n+1)]\}\cdot[\Sigma(R_i^2/n_i)] - 3(n+1)$$
$$= 6.032$$
P-value $= \chi^2\text{cdf}(6.03167,99,3) = 0.1101$
conclusion:

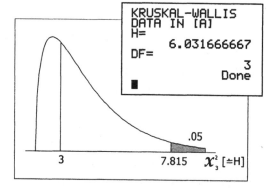

Do not reject $H_0$; there is not sufficient evidence to reject the claim that the different groups of people have the same blood pressure.

Yes, the groups can be considered samples from the same blood pressure population.

11. Place the data in 37x4 matrix [A] as follows.
    MATRX MATH List▶matr(CPIND,CPWHT,CPPRE,CPPST,[A]).
    NOTE: Enter [A] from the MATRX NAMES menu. The dimensions are set automatically.
    Use PRG KWTEST.
    $H_0$: the populations have the same median
    $H_1$: at least one median is different
    $\alpha = 0.05$ [assumed] and df = 3
    C.V. $H = \chi^2_\alpha = 7.815$

    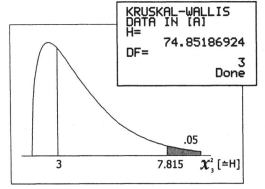

    calculations:
    $H = \{12/[n(n+1)]\} \cdot [\Sigma(R_i^2/n_i)] - 3(n+1)$
    $= 74.852$
    P-value $= \chi^2$cdf (74.85187,999,3)
    $= 3.90\text{E-}16 \approx 0$
    conclusion:
    Reject $H_0$; there is sufficient evidence to reject the claim that the median weight is the same for the four different penny categories.
    No. Machines that consider a coin's weight cannot treat all pennies the same.

13. This requires working exercise #7 by hand. The calculations for H are as follows.

    | sunny | R | cloudy | R | rainy | R |
    |---|---|---|---|---|---|
    | 13.0 | 12.0 | 12.5 | 7 | 11.6 | 1 |
    | 13.2 | 14 | 12.6 | 8 | 11.9 | 2 |
    | 13.5 | 15 | 12.7 | 9.5 | 12.1 | 3 |
    | 13.8 | 16 | 12.7 | 9.5 | 12.2 | 4.5 |
    | 13.9 | 17 | 13.0 | 12.0 | 12.2 | 4.5 |
    | 14.0 | 18 | 13.0 | 12.0 | 12.3 | 6 |
    |  | 92 |  | 58 |  | 21 |

    $n_1 = 6$  $R_1 = 92$
    $n_2 = 6$  $R_2 = 58$
    $n_3 = 6$  $R_3 = 21$
    $n = \Sigma n = 18$  $\Sigma R = 171$

    chk: $\Sigma R = n(n+1)/2 = 18(19)/2 = 171$

    $H = \{12/[n(n+1)]\} \cdot [\Sigma(R_i^2/n_i)] - 3(n+1)$
    $= \{12/[18(19)]\} \cdot [(92)^2/6 + (58)^2/6 + (21)^2/6] - 3(19)$
    $= \{0.0351\} \cdot [2044.8] - 57$
    $= 14.74853801$

    For the n=18 observations in exercise #7, the correction is determined as follows.

    | rank | t | $T = t^3 - t$ |
    |---|---|---|
    | 4.5 | 2 | 6 |
    | 9.5 | 2 | 6 |
    | 12.0 | 3 | 24 |
    |  |  | 36 |

    correction factor:
    $1 - \Sigma T/(n^3 - n) = 1 - 36(18^3 - 18)$
    $= 1 - 13/5418$
    $= 0.99380805$

    The original calculated test statistic in exercise #7 is H = 14.74853801
    The corrected calculated test statistic is H = 14.74853801/0.99380805 = 14.8404
    No, the corrected value of H does not differ substantially from the original one.
    NOTE: Be careful when identifying the tied ranks. In addition to the easily recognized ".5's" that result from an even number of ties, there may be multiple whole number ranks resulting from an odd number of ties. The notation used in exercise #7 helps to identify tied ranks and to bring some order to the process of ranking across the different groups: let only the whole number represent ranks that are not tied (e.g., 3 and 14), and let a decimal place represent ranks that are tied (e.g., 9.5 and 12.0).

# 260 CHAPTER 13 Nonparametric Statistics

## 13-6 Rank Correlation

1. The linear correlation method of section 10-2 requires that the original populations be approximately normal. Since the rank correlation places no restrictions on the original populations, it may be used in a wider variety of situations.

3. The subscript "s" is necessary to distinguish rank correlation from the linear correlation of section 10-2. The "s" stands for "Spearman" – and rank correlation is sometimes referred to as Spearman's rank correlation. The subscript "s" used to designate Spearman's rank correlation is not related to the lower case "s" used to designate standard deviation.

NOTE: The rank correlation is correctly calculated using the ranks in the Pearson product moment correlation formula 10-1 to produce

$$r_s = [n\Sigma R_x R_y - (\Sigma R_x)(\Sigma R_y)] / [\sqrt{n(\Sigma R_x^2) - (\Sigma R_x)^2} \sqrt{n(\Sigma R_y^2) - (\Sigma R_y)^2}]$$

Since $\Sigma R_x = \Sigma R_y = 1 + 2 + \ldots + n = n(n+1)/2$ [always], and

$$\Sigma R_x^2 = \Sigma R_y^2 = 1^2 + 2^2 + \ldots + n^2 = n(n+1)(2n+1)/6 \text{ [when there are no ties in the ranks],}$$

it can be shown by algebra that, for $d = R_x - R_y$, the above formula can be shortened to

$$r_s = 1 - [6(\Sigma d^2)] / [n(n^2 - 1)] \text{ when there are no ties in the ranks.}$$

Using the TI-84/84 Plus calculator eliminates the need for this shortened formula.

This manual converts the original scores to ranks by hand, as there is no RANK function or other convenient way to accomplish this on the calculator. When the original scores are not ranks, the ranks are stated as part of the solution. STAT TESTS LinRegTTest is used on the lists of ranks to calculate $r_s$, the test statistic t, and the P-value.

In addition, some exercises in this section appeared previously in a parametric context in section 10-2. Because comparison of the parametric and non-parametric results are usually informative and provide statistical insights, we list those pairings.

| this section | #13 | #15 | #16 | #17 | #18 |
|---|---|---|---|---|---|
| section 10-2 | #25 | #14 | #15 | #31 | #32 |

5. The original scores are already ranks.
   Place the $R_x$ values in list L1 and the $R_y$ values in list L2.
   Use STAT TESTS LinRegTTest on lists L1 and L2.
   Use STATPLOT and ZOOM to produce the scatterplot.

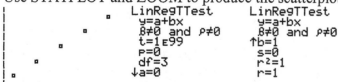

   a. The left screen above gives the scatterplot.
   b. The right screen above indicates that $r_s = 1.00$.
   c. The center screen above indicates that P-value = 0.00 < 0.05. Yes. There appears to be a positive correlation between $R_x$ and $R_y$ – and, therefore, between x and y.

7. a. Since n≤30, use Table A-9.
      CV: $r_s = \pm 0.587$
   b. Since n≤30, use Table A-9.
      CV: $r_s = \pm 0.570$
   c. Since n>30, use Formula 13-1.
      CV: $r_s = \pm 1.960/\sqrt{59} = \pm 0.255$
   d. Since n>30, use Formula 13-1.
      CV: $r_s = \pm 2.575/\sqrt{79} = \pm 0.290$

9. Let the DJIA values be x, and the sales values be y.
   The original scores are already ranks.
   Place the $R_x$ values in list L1 and the $R_y$ values in list L2.
   Use STAT TESTS LinRegTTest on lists L1 and L2.

   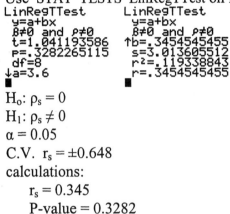

   $H_o$: $\rho_s = 0$
   $H_1$: $\rho_s \neq 0$
   $\alpha = 0.05$
   C.V. $r_s = \pm 0.648$
   calculations:
   $r_s = 0.345$
   P-value = 0.3282

   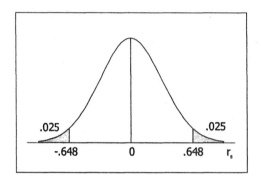

   conclusion:
   Do not reject $H_o$; there is not sufficient evidence to reject the claim that $\rho_s = 0$.
   There does not appear to be a correlation between the DJIA and the numbers of cars sold.

11. Let the salary values be x, and the stress scores be y.
    The original scores are already ranks.
    Place the $R_x$ values in list L1 and the $R_y$ values in list L2.
    Use STAT TESTS LinRegTTest on lists L1 and L2.

    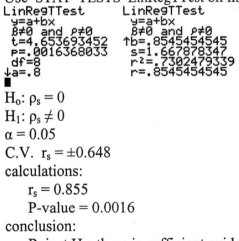

    $H_o$: $\rho_s = 0$
    $H_1$: $\rho_s \neq 0$
    $\alpha = 0.05$
    C.V. $r_s = \pm 0.648$
    calculations:
    $r_s = 0.855$
    P-value = 0.0016

    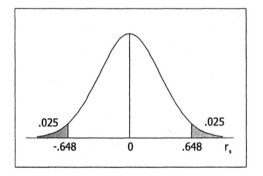

    conclusion:
    Reject $H_o$; there is sufficient evidence to reject
    the claim that $\rho_s = 0$ and conclude that $\rho_s \neq 0$ (in fact, that $\rho_s > 0$).
    Yes, it does appear that salary increases as stress increases.

13. Let the chirp values be x and the temperatures be y.
Replacing the x and y values with their ranks yields the following.
   chirps/min    2  7  6  1  8  5  4  3
   temperature   1  8  6  3  7  5  2  4

Place the $R_x$ values in list L1 and the $R_y$ values in list L2.
Use STAT TESTS LinRegTTest on lists L1 and L2.

```
LinRegTTest          LinRegTTest
y=a+bx               y=a+bx
β≠0 and ρ≠0          β≠0 and ρ≠0
t=4.076197323        ↑b=.8571428571
p=.0065300173        s=1.362770288
df=6                 r²=.7346938776
↓a=.6428571429       r=.8571428571
```

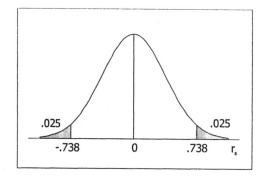

$H_0$: $\rho_s = 0$
$H_1$: $\rho_s \neq 0$
$\alpha = 0.05$
C.V. $r_s = \pm 0.738$
calculations:
   $r_s = 0.857$
   P-value = 0.0065
conclusion:
   Reject $H_0$; there is sufficient evidence to reject the claim that $\rho_s = 0$ and conclude that $\rho_s \neq 0$ (in fact, that $\rho_s > 0$).
Yes, conclude that there is a relationship between the number of chirps in a minute and the temperature – the higher the number of chirps, the higher the temperature.

15. Let the audience values be x and the sales values be y.
Replacing the x and y values with their ranks yields the following.
   impressions   9   1   2.5  7.5  6  5  2.5  7.5  4
   sales         7  4.5  4.5   8   3  1   2    9   6

Place the $R_x$ values in list L1 and the $R_y$ values in list L2.
Use STAT TESTS LinRegTTest on lists L1 and L2.

```
LinRegTTest          LinRegTTest
y=a+bx               y=a+bx
β≠0 and ρ≠0          β≠0 and ρ≠0
t=1.793860936        ↑b=.563559322
p=.1159211398        s=2.413108622
df=7                 r²=.3149302094
↓a=2.18220339        r=.5611864301
```

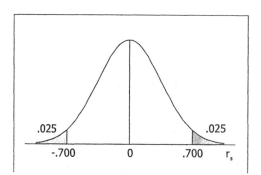

$H_0$: $\rho_s = 0$
$H_1$: $\rho_s \neq 0$
$\alpha = 0.05$
C.V. $r_s = \pm 0.700$
calculations:
   $r_s = 0.561$
   P-value = 0.1159
conclusion:
   Do not reject $H_0$; there is not sufficient evidence to reject the claim that $\rho_s = 0$.

Rank Correlation  SECTION 13-6  263

This page gives the information necessary for working exercises #17 and #18.

| | | exercise #17 | | | | | | exercise #18 | | | |
|---|---|---|---|---|---|---|---|---|---|---|---|
| # | nic | $R_n$ | tar | $R_t$ | car | $R_c$ | act | $R_a$ | 5dp | $R_5$ | 1dp | $R_1$ |
| 1 | 1.2 | 24.0 | 16 | 23.5 | 15 | 19.0 | 80 | 28.0 | 80 | 32.0 | 78 | 24.5 |
| 2 | 1.2 | 24.0 | 16 | 23.5 | 15 | 19.0 | 77 | 23.0 | 80 | 32.0 | 75 | 16.5 |
| 3 | 1.0 | 15.0 | 16 | 23.5 | 17 | 26.0 | 81 | 31.0 | 79 | 28.5 | 81 | 33.0 |
| 4 | 0.8 | 8.5 | 9 | 8.0 | 6 | 3.0 | 85 | 35.0 | 80 | 32.0 | 85 | 35.0 |
| 5 | 0.1 | 1.0 | 1 | 1.0 | 1 | 1.0 | 73 | 15.0 | 79 | 28.5 | 76 | 21.0 |
| 6 | 0.8 | 8.5 | 8 | 5.5 | 8 | 6.0 | 73 | 15.0 | 82 | 34.0 | 75 | 16.5 |
| 7 | 0.8 | 8.5 | 10 | 10.0 | 10 | 8.0 | 80 | 28.0 | 76 | 21.0 | 79 | 27.0 |
| 8 | 1.0 | 15.0 | 16 | 23.5 | 17 | 26.0 | 72 | 13.0 | 73 | 12.0 | 74 | 13.0 |
| 9 | 1.0 | 15.0 | 14 | 16.5 | 13 | 14.0 | 83 | 34.0 | 77 | 24.0 | 75 | 16.5 |
| 10 | 1.0 | 15.0 | 13 | 14.0 | 13 | 14.0 | 81 | 31.0 | 83 | 35.0 | 80 | 30.0 |
| 11 | 1.1 | 20.0 | 13 | 14.0 | 13 | 14.0 | 75 | 18.5 | 77 | 24.0 | 75 | 16.5 |
| 12 | 1.2 | 24.0 | 15 | 19.0 | 15 | 19.0 | 78 | 25.5 | 79 | 28.5 | 79 | 27.0 |
| 13 | 1.2 | 24.0 | 16 | 23.5 | 15 | 19.0 | 80 | 28.0 | 74 | 15.0 | 80 | 30.0 |
| 14 | 0.7 | 5.5 | 9 | 8.0 | 11 | 10.5 | 71 | 11.0 | 75 | 18.0 | 70 | 9.5 |
| 15 | 0.9 | 11.0 | 11 | 11.0 | 15 | 19.0 | 73 | 15.0 | 76 | 21.0 | 72 | 12.0 |
| 16 | 0.2 | 2.0 | 2 | 2.0 | 3 | 2.0 | 78 | 25.5 | 78 | 26.0 | 79 | 27.0 |
| 17 | 1.4 | 28.5 | 18 | 28.5 | 18 | 28.5 | 75 | 18.5 | 76 | 21.0 | 75 | 16.5 |
| 18 | 1.2 | 24.0 | 15 | 19.0 | 15 | 19.0 | 63 | 3.5 | 75 | 18.0 | 67 | 3.0 |
| 19 | 1.1 | 20.0 | 13 | 14.0 | 12 | 12.0 | 63 | 3.5 | 77 | 24.0 | 64 | 2.0 |
| 20 | 1.0 | 15.0 | 15 | 19.0 | 16 | 23.5 | 70 | 9.0 | 71 | 8.5 | 69 | 7.0 |
| 21 | 1.3 | 27.0 | 17 | 27.0 | 16 | 23.5 | 77 | 23.0 | 74 | 15.0 | 77 | 23.0 |
| 22 | 0.8 | 8.5 | 9 | 8.0 | 10 | 8.0 | 82 | 33.0 | 73 | 12.0 | 81 | 33.0 |
| 23 | 1.0 | 15.0 | 12 | 12.0 | 10 | 8.0 | 81 | 31.0 | 75 | 18.0 | 81 | 33.0 |
| 24 | 1.0 | 15.0 | 14 | 16.5 | 17 | 26.0 | 76 | 20.5 | 79 | 28.5 | 80 | 30.0 |
| 25 | 0.5 | 3.0 | 5 | 3.0 | 7 | 4.5 | 77 | 23.0 | 74 | 15.0 | 78 | 24.5 |
| 26 | 0.6 | 4.0 | 6 | 4.0 | 7 | 4.5 | 76 | 20.5 | 71 | 8.5 | 76 | 21.0 |
| 27 | 0.7 | 5.5 | 8 | 5.5 | 11 | 10.5 | 74 | 17.0 | 70 | 6.5 | 76 | 21.0 |
| 28 | 1.4 | 28.5 | 18 | 28.5 | 15 | 19.0 | 66 | 6.0 | 73 | 12.0 | 70 | 9.5 |
| 29 | 1.1 | 20.0 | 16 | 23.5 | 18 | 28.5 | 66 | 6.0 | 72 | 10.0 | 69 | 7.0 |
| 30 | | 435 | | 435 | | 435 | 62 | 2.0 | 69 | 5.0 | 68 | 4.5 |
| 31 | | | | | | | 71 | 11.0 | 70 | 6.5 | 75 | 16.5 |
| 32 | | | | | | | 68 | 8.0 | 68 | 4.0 | 71 | 11.0 |
| 33 | | | | | | | 66 | 6.0 | 67 | 3.0 | 68 | 4.5 |
| 34 | | | | | | | 71 | 11.0 | 61 | 1.0 | 69 | 7.0 |
| 35 | | | | | | | 58 | 1.0 | 64 | 2.0 | 56 | 1.0 |
| | | | | | | | | 630 | | 630 | | 630 |

Place the $R_n$ values in L1, the $R_t$ values in L2, and the $R_c$ values in L3.

C.V. = ±0.368 from Table A-9

a. Use STAT TESTS LinRegTTest on L1 and L2.
```
LinRegTTest              LinRegTTest
y=a+bx                   y=a+bx
β≠0 and ρ≠0              β≠0 and ρ≠0
t=12.01025253            ↑b=.9226310484
p=2.433677E-12           s=3.421738057
df=27                    r²=.8423322364
↓a=1.160534274           r=.9177865964
```

b. Use STAT TESTS LinRegTTest on L1 and L3.
```
LinRegTTest              LinRegTTest
y=a+bx                   y=a+bx
β≠0 and ρ≠0              β≠0 and ρ≠0
t=5.694494265            ↑b=.7405493952
p=4.7495036E-6           s=5.792548857
df=27                    r²=.5456630887
↓a=3.891759073           r=.7386901168
```

Place the $R_a$ values in L1, the $R_5$ values in L2, and the $R_1$ values in L3.

C.V. = ±1.960/$\sqrt{34}$ = ±0.336 from Formula 13-1

a. Use STAT TESTS LinRegTTest on L1 and L2.
```
LinRegTTest              LinRegTTest
y=a+bx                   y=a+bx
β≠0 and ρ≠0              β≠0 and ρ≠0
t=4.045268841            ↑b=.575435883
p=2.9582893E-4           s=8.482628103
df=33                    r²=.3314993456
↓a=7.642154106           r=.5757597985
```

b. Use STAT TESTS LinRegTTest on L1 and L3.
```
LinRegTTest              LinRegTTest
y=a+bx                   y=a+bx
β≠0 and ρ≠0              β≠0 and ρ≠0
t=12.72030386            ↑b=.9094488189
p=2.846014E-14           s=4.263454708
df=33                    r²=.8306008134
↓a=1.62992126            r=.9113730375
```

17. Refer to the information on page 263.

a. $H_0$: $\rho_s = 0$
$H_1$: $\rho_s \neq 0$
$\alpha = 0.05$
C.V.  $r_s = \pm 0.368$
calculations:
$r_s = 0.918$
P-value = 2.43E-12 ≈ 0
conclusion:
Reject $H_0$; there is sufficient evidence to support the claim that $\rho_s \neq 0$ (in fact, that $\rho_s > 0$).
There a significant correlation between nicotine and tar – and researchers need not measure both of them, thus saving laboratory expenses.

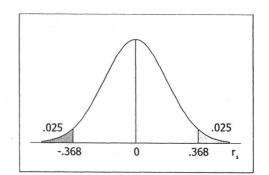

b. $H_0$: $\rho_s = 0$
$H_1$: $\rho_s \neq 0$
$\alpha = 0.05$
C.V.  $r_s = \pm 0.368$
calculations:
$r_s = 0.739$
P-value = 4.75E-6 = 0.000005
conclusion:
Reject $H_0$; there is sufficient evidence to support the claim that $\rho_s \neq 0$ (in fact, that $\rho_s > 0$).
There a significant correlation between nicotine and carbon monoxide – and researchers need not measure both of them, thus saving laboratory expenses.

c. Tar is the better choice, since it has a higher correlation with nicotine.

19. This exercise requires four tasks: the scatterplot, the nonparametric test for Spearman's correlation (this section), the parametric test for Pearson's correlation (section 10-2), a comparison of the nonparametric and parametric tests.
Place the x (times) values in list L1 and the y values (word counts) in list L2.
Determine the x ranks ($R_x$) and the y ranks ($R_y$) as listed below.
    time  1  2  3  4  5  6  7  8  9  10  11
    words  1  2  3  4  5  6  7  8  9  10  11
Place the $R_x$ values in list L3 and the $R_y$ values in list L4.
Use STATPLOT on lists L1 and L2, and the ZOOM 9 to get the scatterplot.
Use STAT TESTS LinRegTTest on lists L3 and L4 for the nonparametric test
Use STAT TESTS LinRegTTest on lists L1 and L2 for the parametric test.

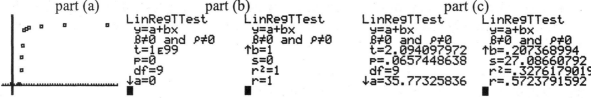

a. The scatterplot is given in the first screen display above. It reveals the non-linear pattern typical for learning curves.

b. See the second and third display screens above.
   $H_o$: $\rho_s = 0$
   $H_1$: $\rho_s \neq 0$
   $\alpha = 0.05$ [assumed]
   C.V. $r_s = \pm 0.618$
   calculations:
   $r_s = 1.000$
   P-value = 0.0000
   conclusion:
   Reject $H_o$; there is sufficient evidence to reject the claim that $\rho_s = 0$ and conclude that $\rho_s \neq 0$ (in fact, that $\rho_s > 0$).

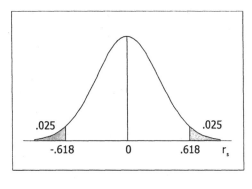

c. See the fourth and fifth display screens above.
   $H_o$: $\rho = 0$
   $H_1$: $\rho \neq 0$
   $\alpha = 0.05$ [assumed] and df = 9
   C.V. $t = \pm t_{\alpha/2} = \pm 2.262$
   calculations:
   $t_r = (r - \mu_r)/s_r = 2.094$
   P-value = $2 \cdot P(t_9 > 2.094) = 0.0657$
   conclusion:
   Do not reject $H_o$; there is not sufficient evidence to reject the claim that $\rho \neq 0$.

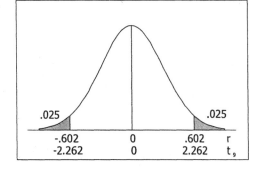

d. The nonparametric test is able to detect the nonlinear relationship.

## 13-7 Runs Test for Randomness

1. No. The runs test for randomness requires the order in which the subjects were selected, and shuffling the papers does not preserve that order.

3. No. The runs test determines only whether the order of the subjects in the sample is random according to a particular criterion – not whether the sample is a convenience sample, a voluntary response sample, etc.

NOTE: In each exercise, the item that appears first in the sequence is considered to be of the first type and its count is designated by $n_1$.

5. $n_1 = 4$ (# of Y's)           G = 8 (# of runs)
   $n_2 = 8$ (# of N's)           CV: 3,10 (from Table A-10)
   Yes. According to the runs test, the sequence is random in that it does not have significantly fewer or significantly more runs than expected by chance – but there appears to be a pattern of one Y followed by three N's.

7. $n_1 = 6$ (# of M's)           G = 7 (# of runs)
   $n_2 = 6$ (# of F's)           CV: 3,11 (from Table A-10)
   Yes. According to the runs test, the sequence is random in that it does not have significantly fewer or significantly more runs than expected by chance.

266   CHAPTER 13   Nonparametric Statistics

9. Since $n_1 = 13$ and $n_2 = 7$, use Table A-10.
   $H_0$: the sequence is random
   $H_1$: the sequence is not random
   $\alpha = 0.05$
   C.V. G = 5, 15
     calculations:
       G = 11
     conclusion:
       Do not reject $H_0$; there is not sufficient evidence to reject the claim that the values occur in a random sequence.

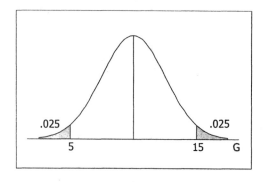

11. Since $n_1 = 14$ and $n_2 = 19$, use Table A-10.
    $H_0$: the sequence is random
    $H_1$: the sequence is not random
    $\alpha = 0.05$
    C.V. G = 11, 23
      calculations:
        G = 20
      conclusion:
        Do not reject $H_0$; there is not sufficient evidence to reject the claim that the values occur in a random sequence.

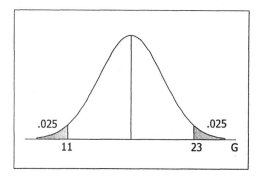

    The test for randomness does not address the relative abilities of the two leagues, but $n_1 = 14 \approx 19 = n_2$ suggests that neither league is dominant over the other.

13. The median of the 17 values arranged in numerical order from $x_1 = 20{,}595$ to $x_{17} = 36{,}448$ is $x_9 = 26{,}995$. Passing through the values chronologically and assigning A's and B's as directed (and ignoring values equal to the median), yields the sequence
        B B B B B B B B A A A A A A A A
    Since $n_1 = 8$ and $n_2 = 8$, use Table A-10.
    $H_0$: the sequence is random
    $H_1$: the sequence is not random
    $\alpha = 0.05$
    C.V. G = 4, 14
      calculations:
        G = 2
      conclusion:
        Reject $H_0$; there is sufficient evidence to reject the claim that the values occur in a random sequence.

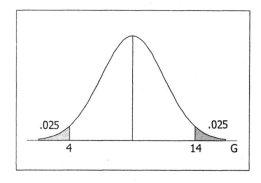

    It appears that since 1987 there has been an upward trend in the number of indoor movie theaters.

15. The genders of the 54 bears (1=male, 2=female) are as follows.
  1 1 1 1 2 2 1 1 2 2 1 1 2 1 2 1 1 2 1 1 2 2 1 1 1 1 1 2 2 1 1 2 2 1 1 1 1 1 2 1 1 1 1 2 2 1 1 1 1 2 2 2 1 1
  Since $n_1 = 35$ and $n_2 = 19$, use the normal approximation.
  $\mu_G = 2n_1n_2/(n_1 + n_2) + 1 = 2(35)(19)/54 + 1 = 25.630$
  $\sigma^2_G = [2n_1n_2(2n_1n_2 - n_1 - n_2)]/[(n_1 + n_2)^2(n_1 + n_2 - 1)]$
  $= [2(35)(19)(1276)]/[(54)^2(53)] = 10.981$
  $H_0$: the sequence is random
  $H_1$: the sequence is not random
  $\alpha = 0.05$
  C.V. $z = \pm z_{\alpha/2} = \pm 1.96$
  calculations:
  $G = 23$
  $z_G = (G - \mu_G)/\sigma^2_G$
  $= (23 - 25.630)/\sqrt{10.981}$
  $= -2.630/3.314 = -0.794$
  P-value $= 2 \cdot P(z<-0.79) = 2(0.2148) = 0.4296$
  conclusion:
  Do not reject $H_0$; there is not sufficient evidence to reject the claim that the values occur in a random sequence.

17. The sequence of the 100 digits in O's and E's is follows.
  O E O O O E E O O O E O O O O E O E E E E E O O
  E O E O O O E E E E O O O O E O O O O O O O O E
  O E E E O O E O E E O O E O E O E O E E E E E E
  E E E O O E E E E O E E E O O E E O O O E E O O
  Since $n_1 = 49$ and $n_2 = 51$, use the normal approximation.
  $\mu_G = 2n_1n_2/(n_1 + n_2) + 1$
  $= 2(49)(51)/100 + 1 = 50.980$
  $\sigma^2_G = [2n_1n_2(2n_1n_2 - n_1 - n_2)]/[(n_1 + n_2)^2(n_1 + n_2 - 1)]$
  $= [2(49)(51)(4898)]/[(100)^2(99)] = 24.727$
  $H_0$: the sequence is random
  $H_1$: the sequence is not random
  $\alpha = 0.05$
  C.V. $z = \pm z_{\alpha/2} = \pm 1.96$
  calculations:
  $G = 43$
  $z_G = (G - \mu_G)/\sigma^2_G$
  $= (43 - 50.980)/\sqrt{24.727}$
  $= -7.980/4.973 = -1.605$
  P-value $= 2 \cdot P(z<-1.60) = 2(0.0548) = 0.1096$
  conclusion:
  Do not reject $H_0$; there is not sufficient evidence to reject the claim that the odd and even digits in $\pi$ occur in a random sequence.

19. The minimum possible number of runs is G = 2, and occurs when all the A's are together and all the B's are together (e.g., AABB).
The maximum possible number of runs is G = 4, and occurs when the A's and B's alternate (e.g., ABAB).
Because the critical values for $n_1=n_2=2$ are  C.V. G = 1,6
the null hypothesis of the sequence being random can never be rejected at the 0.05 level. This means that it is not possible for such a small sample to provide 95% certainty that a non-random phenomenon is occurring. Table A-10 could have given that information more clearly by using an asterisk, as in Table A-7 and Table A-8, to indicate those sample sizes for which it is not possible to achieve statistical significance no matter what the data.

## Statistical Literacy and Critical Thinking

1. A nonparametric test is a test not requiring that the sample data come from a population that has a particular distribution (usually the normal distribution) defined by certain parameters.

2. There is no difference. "Nonparametric" and "distribution-free" are different names for the same category of tests.

3. A rank is a number assigned to a value according to its order in a list of values sorted by magnitude. The smallest value is typically assigned rank 1, the next smallest is assigned rank 2, etc.

4. The efficiency of a nonparametric test refers to the sample size the corresponding parametric test requires to obtain the same precision that a nonparametric reaches with a sample of size 100 – whenever all the requirements of the parametric test are met. The efficiency of a non-parametric test is a number less than 1.00 and can be interpreted as "percent of effectiveness" in the following sense: *The efficiency of the sign test for detecting the difference between matched pairs is 0.63. This means that when the parametric assumptions are met (i.e., when the original populations have normal distributions), the sign test requires a sample of size 100 to detect the same differences that the parametric test can detect with a sample of size 63. And so there is a sense in which the nonparametric test is using only 63% of the information available in the data.*

# Review Exercises

1. Use the sign test of section 13-2. Let the first trial be group 1.

   | child | A | B | C | D | E | F | G | H | I | J | K | L | M | N | O |
   |---|---|---|---|---|---|---|---|---|---|---|---|---|---|---|---|
   | $1^{st} - 2^{nd}$ | 0 | + | + | + | + | + | + | − | − | + | + | + | + | + | + |

   n = 14: 12 +'s and 2 −'s
   $H_0$: median difference = 0
   $H_1$: median difference ≠ 0
   α = 0.05
   C.V.  $x = x_{L,\alpha/2} = 2$
   $x = x_{U,\alpha/2} = 14 - 2 = 12$
   calculations:
   x = 2 (using less frequent count)
   x = 12 (using + count)
   P-value = $2 \cdot P(x \leq 2)$
   = $2 \cdot \text{binomcdf}(14, 0.5, 2) = 0.0129$

   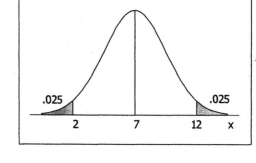

   conclusion:
   Reject $H_0$; there is sufficient evidence to reject the claim that the median difference is 0 and to conclude that the median times are different (in fact, that the times for the first trial are greater).

2. Use the Wilcoxon signed-rank test of section 13-3.
   Put the first trial values in list L1 and the second trial values in list L2.
   Use L1 − L2 → L3. Use PRGM SRTEST on list L3.

   | $1^{st} - 2^{nd}$ | 0 | 13 | 5 | 15 | 15 | 126 | 28 | −2 | −5 | 31 | 3 | 51 | 3 | 14 |
   |---|---|---|---|---|---|---|---|---|---|---|---|---|---|---|
   | R | − | 6 | 4.5 | 8.5 | 8.5 | 14 | 10 | −1 | −4.5 | 11 | 2.5 | 13 | 2.5 | 7 |

   ΣR− = 5.5     n = 14 non-zero ranks
   ΣR+ = 99.5    check: ΣR = n(n+1)/2 = 14(15)/2 = 105
   ΣR  = 105
   $H_0$: median difference = 0
   $H_1$: median difference ≠ 0
   α = 0.05
   C.V.  $T = T_{L,\alpha/2} = 21$
   $T = T_{U,\alpha/2} = 105 - 21 = 84$
   calculations:
   T = 5.5 (using smaller ranks)
   T = 99.5 (using positive ranks)

   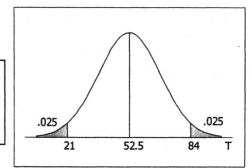

   conclusion:
   Reject $H_0$; there is sufficient evidence to reject the claim that the median difference is 0 and to conclude that the median times are different (in fact, that the times for the first trial are greater).

270  CHAPTER 13  Nonparametric Statistics

3. Use the rank correlation test of section 13-6.
   Let the business school values be x, and the law school values be y.
   The original scores are already ranks.
   Place the $R_x$ values in list L1 and the $R_y$ values in list L2.
   Use STAT TESTS LinRegTTest on lists L1 and L2.

   ```
   LinRegTTest           LinRegTTest
   y=a+bx                y=a+bx
   β≠0 and ρ≠0           β≠0 and ρ≠0
   t=2.844367334         ↑b=.7090909091
   p=.0216659233         s=2.26434819
   df=8                  r²=.5028099174
   ↓a=1.6                r=.7090909091
   ```

   $H_o$: $\rho_s = 0$
   $H_1$: $\rho_s \neq 0$
   $\alpha = 0.05$
   C.V. $r_s = \pm 0.648$
   calculations:
      $r_s = 0.709$
      P-value $= 0.0217$

   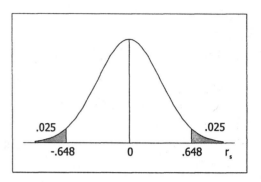

   conclusion:
      Reject $H_o$; there is sufficient evidence to reject the claim that $\rho_s = 0$ and conclude
      that $\rho_s \neq 0$ (in fact, that $\rho_s > 0$).
   Yes. There appears to be a correlation between the business school rankings and the law
   school rankings.

4. NOTE: It is arbitrary whether one assigns the ranks 1,2,...,n from low to high or from high to
   low. Usually the ranks are assigned from low to high so that low values have low ranks and
   high values have high ranks. To preserve the direction of the relationship, however, the same
   system should be used for both x and y. Since the schools were ranked with lower numbers
   indicating better ranks (i.e., rank #1 is better than rank #8), rank the MCAT scores from high
   to low so that the lower numbers also indicate better ranks for the y variable 0.
   Use the rank correlation test of section 13-6.
   Let the ranks be x, and the MCAT scores be y.
   The original x scores are already ranks; use y ranks 2 3 1 6 4 5 8 7 [see NOTE above].
   Place the $R_x$ values in list L1 and the $R_y$ values in list L2.
   Use STAT TESTS LinRegTTest on lists L1 and L2.

   ```
   LinRegTTest           LinRegTTest
   y=a+bx                y=a+bx
   β≠0 and ρ≠0           β≠0 and ρ≠0
   t=3.377557511         ↑b=.8095238095
   p=.01490266676        s=1.553286327
   df=6                  r²=.6553287982
   ↓a=.8571428571        r=.8095238095
   ```

   $H_o$: $\rho_s = 0$
   $H_1$: $\rho_s \neq 0$
   $\alpha = 0.05$
   C.V. $r_s = \pm 0.738$
   calculations:
      $r_s = 0.810$
      P-value $= 0.0149$

   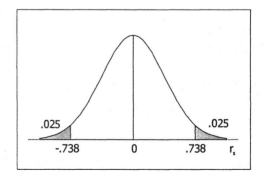

   conclusion:
      Reject $H_o$; there is sufficient evidence to reject the claim that $\rho_s = 0$ and conclude that
      $\rho_s \neq 0$ (in fact, that $\rho_s > 0$).
   Yes, the schools with the better rankings are the schools with the better MCAT scores.

5. Use the sign test of section 13-2.
   Let x = the number of females hired.
   claim: p < 0.5 [this is the charge that there is bias against women (i.e., in favor of men)]
   n = 40: 15 +'s and 25 −'s
   [the data are in the proper direction to proceed]
   $H_0$: p = 0.5
   $H_1$: p < 0.5
   α = 0.05
   C.V. $z = -z_α = -1.645$
   calculations:
   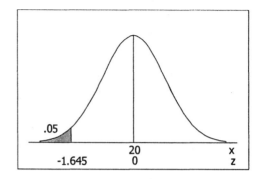
      x = 15
      P-value = P(x≤15)
         = binomcdf(40,0.5,15) = 0.0769
   conclusion:
      Do not reject $H_0$; there is not sufficient evidence to support the charge of bias against women (i.e., in favor of men).

6. Use the Wilcoxon rank-sum test of section 13-4. Let the beer drinkers be group 1.
   Place the 12 beer values in L1 and the 14 liquor values in L2.
   Use PRGM RSTEST on lists L1 and L2.
   ```
   GROUP A USED              19.4422221
   SUM OF RANKS R=    Z SCORE R
              89.5              -3.728997624
   MEAN R=            TEST STAT U=
              162                11.5
   STD DEV R=         MEAN U=
        19.4422221              84
   ```
   $H_0$: median$_1$ = median$_2$
   $H_1$: median$_1$ ≠ median$_2$
   α = 0.05
   C.V. $z = ±z_{α/2} = ±1.960$
   calculations:
      $z = (R - μ_R)/σ_R = -3.729$
      P-value = 2·P(z<−3.72900) = 1.92E-4 = 0.0002
   conclusion:
      Reject $H_0$; there is sufficient evidence to reject the claim that median$_1$ = median$_2$ and to conclude that median$_1$ ≠ median$_2$ (in fact, that beer drinkers have a lower BAC than liquor drinkers.
   By this criterion, it appears that liquor drinkers are more dangerous.

7. Use the runs test of section 13-7.
   The sequence of the 40 digits in O's and E's is follows.
   O O E O O O O O E E E O E E O E O E E O
   O O O E E E E O O E O O O O O E O E E E
   Since $n_1$ = 22 and $n_2$ = 18, use the normal approximation.
      $μ_G = 2n_1n_2/(n_1 + n_2) + 1$
         = 2(22)(18)/40 + 1 = 20.800
      $σ^2_G = [2n_1n_2(2n_1n_2 - n_1 - n_2)]/[(n_1 + n_2)^2(n_1 + n_2 - 1)]$
         = [2(22)(18)(752)]/[(40)^2(39)] = 9.545

$H_o$: the sequence is random
$H_1$: the sequence is not random
$\alpha = 0.05$
C.V. $z = \pm z_{\alpha/2} = \pm 1.96$
 calculations:
  $G = 18$
  $z_G = (G - \mu_G)/\sigma^2_G$
   $= (18 - 20.800)/\sqrt{9.545}$
   $= -2.800/3.089 = -0.906$
  P-value $= 2 \cdot P(z < -0.90629) = 0.3648$

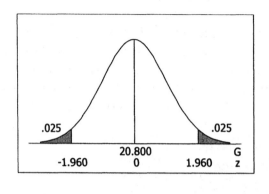

conclusion:
 Do not reject Ho; there is not sufficient evidence to reject the claim that the odd and even digits in the lottery occur in a random sequence.

8. Use the Kruskal-Wallis test of section 13-5.
 Place the data in 5x4 matrix [A], with the 5 subcompact values in column1, the 5 compact values in column 2, etc. NOTE: This is the transpose of the arrangement in the exercise.
 Use PROG KWTEST.
 For a review, the exercise is also worked by hand.

| sub | R | com | R | mid | R | full | R |
|---|---|---|---|---|---|---|---|
| 422 | 2 | 584 | 4 | 181 | 1 | 804 | 8 |
| 519 | 3 | 946 | 11 | 629 | 6 | 971 | 12 |
| 595 | 5 | 984 | 13 | 645 | 7 | 996 | 14 |
| 885 | 10 | 1051 | 15 | 880 | 9 | 1085 | 17 |
| 1063 | 16 | 1193 | 18 | 1686 | 20 | 1376 | 19 |
|  | 36 |  | 61 |  | 43 |  | 70 |

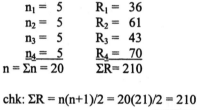

$n_1 = 5 \quad R_1 = 36$
$n_2 = 5 \quad R_2 = 61$
$n_3 = 5 \quad R_3 = 43$
$n_4 = 5 \quad R_4 = 70$
$n = \Sigma n = 20 \quad \Sigma R = 210$

chk: $\Sigma R = n(n+1)/2 = 20(21)/2 = 210$

$H_o$: the populations have the same median
$H_1$: at least one median is different
$\alpha = 0.05$ and df $= 3$
C.V. $H = \chi^2_\alpha = 7.815$
calculations:
 $H = \{12/[n(n+1)]\} \cdot [\Sigma(R_i^2/n_i)] - 3(n+1)$
  $= \{12/[20(21)]\} \cdot [(36)^2/5 + (61)^2/5$
   $+ (43)^2/5 + (70)^2/5] - 3(21)$
  $= \{0.0286\} \cdot [2353.2] - 63$
  $= 4.234$

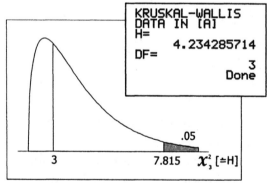

P-value $= \chi^2 \text{cdf}(4.234, 99, 3) = 0.2373$
conclusion:
 Do not reject $H_o$; there is not sufficient evidence to support the claim that leg injury measurements for the four car weight categories are not all the same.
 No. By this measurement, the data do not show that heavier cars are safer in a crash.

# Cumulative Review Exercises

Place the 9 before scores in list L1 and the 9 after scores in list L2.
The following steps and displays are used in various of the exercises.
Use STAT CALC 1-Var Stats on L1. (exercise #1)
Use STATPLOT on L1 and L2, and then use ZOOM 9. (exercise #2)
Use STAT TESTS LinRegTTest on L1 and L2. (exercise #3)

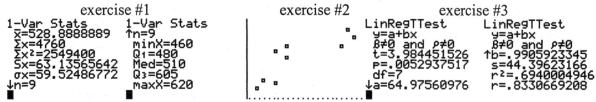

1. Refer to the display screens above.
   mean: $\bar{x} = 528.9$     range: $R = 620 - 460 = 160$
   median: $\tilde{x} = 510$     standard deviation: $s = 63.1$
   variance: $s^2 = (63.13565642)^2 = 8986.1$

2. The scatterplot is given above in the third display screen. Yes, there appears to be a correlation between the before and after scores. The correlation appears to be a positive one – meaning that those who got the lower scores before the program tended to have the lower scores after the program, and those who got the higher scores before the program tended to have the higher scores after the program.

3. Refer to the display screens above.
   $r = 0.8331$
   $H_o: \rho = 0$
   $H_1: \rho \neq 0$
   $\alpha = 0.05$ [assumed] and df = 7
   C.V. $t = \pm t_{\alpha/2} = -2.365$
   calculations:
   $t_r = (r - \mu_r)/s_r = 3.984$

   P-value = $2 \cdot P(t7>3.984) = 0.0053$
   conclusion:
   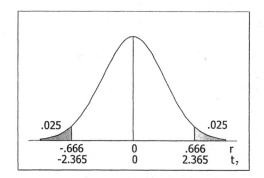

   Reject $H_o$; there is sufficient evidence to conclude that $\rho \neq 0$ (in fact, that $\rho > 0$).
   No, a significant linear correlation does not mean the preparation course is effective – it means only that you can predict the after score from the before score (and vice-versa). If every student gets exactly 10 points <u>lower</u> on the after test, for example, there would be a perfect positive correlation of r = 1.00.

4. Rank the 9 before scores and the 9 after scores to obtain the following table of ranks.
   Before  1   2   3.5   3.5   5.5   5.5   7   9   8
   After   1   3   2     5     4     6.5   6.5 8   9
   Place the 9 before ranks in list L3 and the 9 after ranks in list L4.
   Use Use STAT TESTS LinRegTTest on L3 and L4.

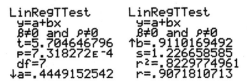

274   CHAPTER 13   Nonparametric Statistics

$H_o$: $\rho_s = 0$
$H_1$: $\rho_s \neq 0$
$\alpha = 0.05$ [assumed]
C.V. $r_s = \pm 0.700$
calculations:
  $r_s = 0.907$
  P-value = 7.32E-4 = 0.0007
conclusion:
  Reject $H_o$; there is sufficient evidence to conclude that $\rho_s \neq 0$ (in fact, that $\rho_s > 0$).
Yes, there is a positive correlation between the ranking and the MCAT score.

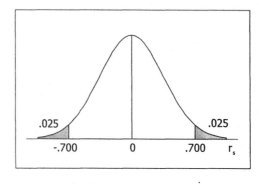

5. The following steps and displays are used in various of the remaining exercises.
   Use L1 − L2 → L5, and then use STAT TESTS TTest on list L5.
   original claim: $\mu_d = 0$
   $H_o$: $\mu_d = 0$
   $H_1$: $\mu_d \neq 0$
   $\alpha = 0.05$ [assumed]
   and df = 8
   C.V. $t = \pm t_{\alpha/2} = \pm 2.306$
   calculations
   $t_{\bar{d}} = (\bar{d} - \mu_{\bar{d}})/s_{\bar{d}} = -4.334$
   P-value = $2 \cdot P(t_8 < -4.334) = 0.0025$
   conclusion:
     Reject $H_o$; there is sufficient evidence to reject the claim that $\mu_d = 0$ and conclude that $\mu_d \neq 0$ (in fact, that $\mu_d < 0$ – i.e., that the before scores are lower).

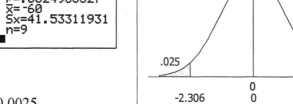

6. Subtract (Before − After) to create a row of differences as given below.
   The signs of the differences are used in this exercise.
   The signed ranks of the differences are used in exercise #7.

   | difference (B − A): | -20 | -40 | -10 | -120 | -80 | -120 | -30 | -40 | -80 |
   | sign of difference: | − | − | − | − | − | − | − | − | − |
   | signed rank of difference: | -2 | -4.5 | -1 | -8.5 | -6.5 | -8.5 | -3 | -4.5 | -6.5 |

   The above table indicates the following.
   n = 9 pairs: 0 +'s and 9 −'s
   $H_o$: median difference = 0
   $H_1$: median difference ≠ 0
   $\alpha = 0.05$ [assumed]
   C.V. $x = x_{L,\alpha/2} = 1$
        $x = x_{U,\alpha/2} = 9 - 1 = 8$
   calculations:
     x = 0
     P-value = $2 \cdot \text{binomcdf}(9,0.5,1) = 0.0039$

   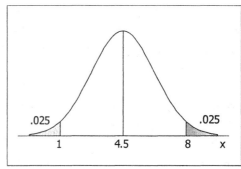

   conclusion:
     Reject $H_o$; there is sufficient evidence to reject the claim that there is no difference between the before and after scores and to conclude median difference ≠ 0 (in fact, that median difference < 0 – i.e., that the before scores are lower).

7. By calculator, use PRGM SRTEST on list L5 of differences created in exercise #5.
   By hand, refer to row of signed ranks of the differences created in exercise #7
   and calculate the following preliminary statistics.

   $\Sigma R- = 45$   $\quad$ n = 9 non-zero ranks
   $\Sigma R+ = 0$   $\quad$ check: $\Sigma R = n(n+1)/2 = 9(10)/2 = 45$
   $\Sigma R\ = 45$

   $H_o$: median difference = 0
   $H_1$: median difference $\neq$ 0
   $\alpha = 0.05$
   C.V. $T = T_{L,\alpha/2} = 6$
   $\quad\ T = T_{U,\alpha/2} = 45-6 = 39$
   calculations:
   $\quad$ T = 0

   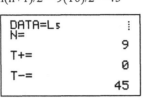
   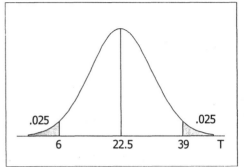

   conclusion:
   Reject $H_o$; there is sufficient evidence to reject the claim that there is no difference between the before and after scores and to conclude median difference $\neq$ 0 (in fact, that median difference < 0 – i.e., that the before scores are lower).

8. The main question of interest is whether the coaching program is effective in raising math SAT test scores. Four of the preceding exercises (viz., #1-#4) do not address that question, and the three exercises (viz., #5-#7) that do are not equally effective. The following paragraphs summarize the relevance of each of the preceding exercises to this question.

   a. exercise #1 – descriptive statistics for the before scores. If these results were compared to similar results for the after scores, one could make a crude comparisons between the two populations. Such comparisons would not take advantage of the individual pairings, and any differences between the before and after scores would probably be hidden by the greater differences between individual students.

   b. exercise #2 – scatterplot. Visualize the best straight line through the points. The tightness of the points around that line indicate the consistency of the effect of the program (the program is consistent from student to student if the points are tightly clustered around the line), and the slope of the line indicates the effect (the program is effective if the slope is significantly greater than 1.00). But such detailed examination requires formal regression analysis, not merely a scatterplot.

   c. exercise #3 – linear correlation using the raw scores. This measures only how well the after scores can be predicted from the before score (or vice-versa), and does not address the issue of which score is higher.

   d. exercise #4 – linear correlation using the ranks. This measures only how well the after scores can be predicted from the before score (or vice-versa), and does not address the issue of which score is higher. It is more effective than exercise #3 in detecting a relationship that is not linear in the raw scores but that is linear in ranks.

   e. exercise #5 – matched pairs using the difference in the raw scores. This is the most effective measure. It addresses the question directly and uses all the available information – both the actual raw scores and the individual pairings.

   f. exercise #6 – matched pairs using only the sign of the difference in the raw scores. This addresses the question, but in using only the sign of the change and not the magnitude of the change it does not use all the available information. In general, it is the least efficient of the three appropriate tests in exercises #5, #6 and #7.

   g. exercise #7 – matched pairs using the signed rank of the difference in the raw scores. This addresses the question, but in using only the (signed) rank of the change and not the actual magnitude of the change it does not use all the available information.

# Chapter 14

# Statistical Process Control

## 14-2 Control Charts for Variation and Mean

1. Process data are values in chronological sequence that measure a characteristic of a controllable activity.

3. A control chart is a plot of values over time, with a center line representing the standard value and upper and lower lines indicating limits beyond which values are considered unacceptable. An R chart is a control chart that plots the ranges of consecutive samples of size n in order to monitor the variation of a process. An $\bar{x}$ chart is a control chart that plots the means of consecutive samples of size n in order to monitor the central tendency of a process.

5. a. and b. Within statistical control.
   c. No. The variation is too large. The runs chart indicates a variation of about ±1.2 ounces, which is about ±10% of the target value, and this is not acceptable for 12 ounce cans of soda.

7. a. and b. Not within statistical control, because there is one unusually low value.
   c. No. The dramatic departure of the extremely low value from the acceptable values suggests that there was a problem with the machine or the operator – and that steps to identify the cause of the departure are in order.

NOTE: In this section, k = number of sample subgroups
          n = number of observations per sample subgroup

Although the $\bar{\bar{x}}$ line is not specified in the text as a necessary part of a run chart, such a line is helpful. It is included in figures generated by many statistical software packages, and it is included in this manual.

9. There are k=25 samples of size n=7 each.  $\quad$ UCL = $D_4 \bar{R}$ = 1.924(54.96) = 105.74
   $\bar{R} = \Sigma R/k = 1374/25 = 54.96$  $\quad\quad\quad\quad\quad\quad$ LCL = $D_3 \bar{R}$ = 0.076(54.96) = 4.18

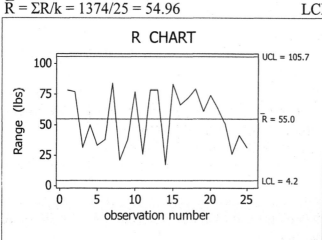

NOTE: The manual follows the recommendation in the text and uses Minitab to generate all the figures in Chapter 14. If Minitab is not available, construct figures by hand that include the labeling unavailable on the TI-83/84 Plus.

The process variation appears to be within statistical control.

11. There are k·n = 20·5 = 100 total observations on the run chart.
    For convenience, the problem is worked in milligrams (i.e., the original weights x 1000).
    $\bar{\bar{x}} = \Sigma \bar{x}/k = \Sigma x/(k \cdot n) = 570980/100 = 5709.8$

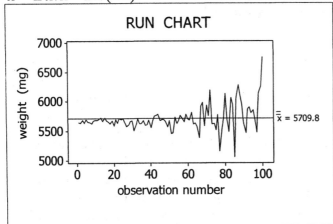

Yes, there appears to be a pattern that suggests the process is not within statistical control – viz., the variability is increasing. In practical terms, this suggests that the equipment is not holding its tolerance and that the process needs to be shut down for an adjustment.

13. For convenience, the problem is worked in milligrams (i.e., the original weights x 1000).
    There are k=20 samples of size n=5 each.  $\bar{\bar{x}} = \Sigma \bar{x}/k = 114196.0/20 = 5709.8$
    $\bar{R} = \Sigma R/k = 7292/20 = 364.6$
    UCL = $\bar{\bar{x}} + A_2 \bar{R}$ = 5709.8 + (0.577)(364.6) = 5709.8 + 210.4 = 5920.2
    LCL = $\bar{\bar{x}} - A_2 \bar{R}$ = 5709.8 – (0.577)(364.6) = 5709.8 – 210.4 = 5499.4

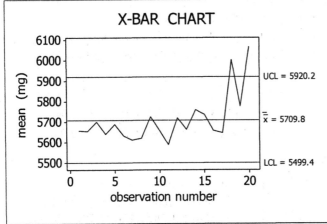

The process variation is not within statistical control. To be out of control, the process must meet at least one of three criteria – and this process actually meets all three!
    (1) There is a pattern to the $\bar{x}$ values – in this case, an upward drift.
    (2) There is a point beyond the upper or lower control limit – in this case, there are 2 points above the UCL.

    (3) There are at least 8 consecutive points on the same side of the center line – in this case, the first 8 points are below the center line.
    Yes, corrective action is needed.

**278 CHAPTER 14 Statistical Process Control**

NOTE: Application of the techniques of this section in exercises #14, #15 and #16 is interesting – but probably not appropriate. It is questionable whether the Boston rainfall data meet the definition of process data. In general, process data are values in chronological sequence that measure a characteristic of a controllable activity. If the activity (rainfall) is not controllable, it is really irrelevant to discuss whether or not it is within statistical control.

The following chart summarizes the information necessary for exercises #15 and #16.

| week | $\bar{x}$ | R | week | $\bar{x}$ | R | week | $\bar{x}$ | R | week | $\bar{x}$ | R |
|---|---|---|---|---|---|---|---|---|---|---|---|
| 1 | 0.019 | 0.05 | 14 | 0.000 | 0.00 | 27 | 0.036 | 0.14 | 40 | 0.067 | 0.24 |
| 2 | 0.027 | 0.10 | 15 | 0.097 | 0.40 | 28 | 0.027 | 0.11 | 41 | 0.003 | 0.02 |
| 3 | 0.101 | 0.71 | 16 | 0.237 | 0.87 | 29 | 0.010 | 0.05 | 42 | 0.003 | 0.02 |
| 4 | 0.186 | 0.64 | 17 | 0.067 | 0.47 | 30 | 0.017 | 0.12 | 43 | 0.181 | 0.68 |
| 5 | 0.016 | 0.05 | 18 | 0.054 | 0.24 | 31 | 0.083 | 0.44 | 44 | 0.243 | 1.48 |
| 6 | 0.091 | 0.64 | 19 | 0.036 | 0.14 | 32 | 0.044 | 0.18 | 45 | 0.290 | 1.28 |
| 7 | 0.051 | 0.30 | 20 | 0.237 | 0.92 | 33 | 0.164 | 0.64 | 46 | 0.139 | 0.96 |
| 8 | 0.001 | 0.01 | 21 | 0.049 | 0.27 | 34 | 0.134 | 0.85 | 47 | 0.116 | 0.79 |
| 9 | 0.039 | 0.16 | 22 | 0.003 | 0.01 | 35 | 0.009 | 0.03 | 48 | 0.109 | 0.41 |
| 10 | 0.121 | 0.39 | 23 | 0.000 | 0.00 | 36 | 0.017 | 0.12 | 49 | 0.017 | 0.08 |
| 11 | 0.181 | 0.78 | 24 | 0.101 | 0.71 | 37 | 0.050 | 0.26 | 50 | 0.000 | 0.00 |
| 12 | 0.029 | 0.17 | 25 | 0.097 | 0.33 | 38 | 0.057 | 0.40 | 51 | 0.106 | 0.74 |
| 13 | 0.416 | 1.41 | 26 | 0.000 | 0.00 | 39 | 0.017 | 0.12 | 52 | 0.146 | 0.43 |

NOTE: The year begins and ends on Wednesday. Removing the last Wednesday gives 52 weeks of data that begin on Wednesday and end on Tuesday – and each row in Data Set 10 is that type of week. To construct a run chart for those k·n = 52·7 = 364 days, one would have to move the Monday and Tuesday columns to the end (i.e., following Sunday) before reading through the data by rows – but this is irrelevant to calculating $\bar{x}$ and R for each Wednesday to Tuesday week.

15. There are k=52 samples of size n=7 each.   $UCL = D_4 \bar{R} = 1.924(0.3915) = 0.753$
    $\bar{R} = \Sigma R/k = 20.36/52 = 0.3915$       $LCL = D_3 \bar{R} = 0.076(0.3915) = 0.030$

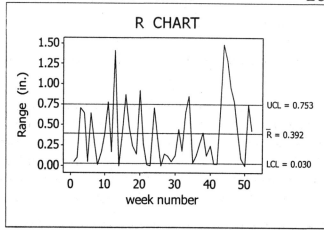

The process variation is not within statistical control – there are points above the upper control limit and below the lower control limit. While only one point above or below the appropriate control limit is sufficient to declare the process out of control, these data have multiple points beyond each limit. A point above the upper control limit represents a large range for that sample – which indicates a potential problem within the time frame of that particular sample, as well as the fact that the variance from sample to sample exhibits more than acceptable fluctuation. A point below the lower control limit represents a small range for that sample – which does not indicate a problem with that particular sample, but does indicate that the variance from sample to sample exhibits more than acceptable fluctuation. See also the NOTE at the top of the page.

17. There are k=20 samples of size n=4 each.    UCL = $B_4 \bar{s}$ = 2.266(9.6925) = 21.963
$\bar{s} = \Sigma s/k$ = 193.85/20 = 9.6925    LCL = $B_3 \bar{s}$ = 0.000(9.6925) = 0.000

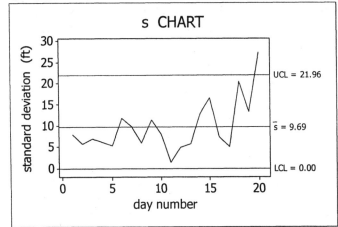

This is very similar to the R chart given in the text, and both charts indicate that the process variation is not within statistical control. For small n's in the sub-samples (n=4 in this example), there will be a very high correlation between the values of R and s – and so the two charts will be almost identical, but with different labels on the vertical axis.

## 14-3 Control Charts for Attributes

1. A p chart is a control chart to monitor the proportion of items having a particular attribute over time. It is used to determine whether the corresponding process is within statistical control.

3. When the formula determines the lower control limit in a p chart to be less than zero, use the LCL = 0.00. In any problem with upper and lower physical boundaries, the control limits should be truncated to match those boundaries – as having a limit outside the realm of possibility is not informative, and adjusting the control limits to match the physical boundaries may help to identify errors resulting in otherwise impossible values. In the given scenario, the UCL and LCL should be 0.250 and 0.000 respectively.

5. This process is within statistical control. Since the first third of the values are generally less than the overall mean, the middle third are generally more than the overall mean, and the final third are generally less than the overall mean, however, one may wish to check future analyses to see whether such a pattern of clustering tends to repeat itself. The final seven values being less than the overall mean is just one away from meeting the criterion for being declared out of statistical control whenever 8 consecutive values fall above or below the center line.

7. This process is out of statistical control. There is an upward tend, and there is a point above the upper control limit.

280 CHAPTER 14 Statistical Process Control

9. $\bar{p} = (\Sigma x)/(\Sigma n) = (155+152+\ldots+139)/(10)(10{,}000) = 1462/100{,}000 = 0.01462$

$\sqrt{\bar{p}\bar{q}/n} = \sqrt{(0.01462)(0.98538)/10000} = 0.001200$

UCL $= \bar{p} + 3\sqrt{\bar{p}\bar{q}/n} = 0.01462 + 3(0.001200) = 0.01462 + 0.00360 = 0.01822$

LCL $= \bar{p} - 3\sqrt{\bar{p}\bar{q}/n} = 0.01462 - 3(0.001200) = 0.01462 - 0.00360 = 0.01102$

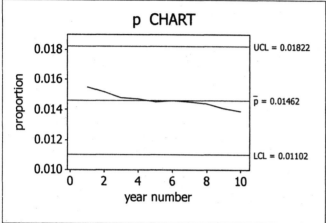

The process is out of statistical control because there is a downward trend. One possible explanation for a decreasing U.S. birth rate might be an increasing U.S. life expectancy. The 10,000 persons were selected each year from the entire population – as life expectancy increases, the proportion of persons in the child-bearing years (and hence the relative number of births to the entire population) would be expected to decrease.

11. $\bar{p} = (\Sigma x)/(\Sigma n) = (8+6+\ldots+8)/(12)(1{,}000) = 86/12000 = 0.00717$

$\sqrt{\bar{p}\bar{q}/n} = \sqrt{(0.00717)(0.99283)/1000} = 0.002667$

UCL $= \bar{p} + 3\sqrt{\bar{p}\bar{q}/n} = 0.00717 + 3(0.002667) = 0.00717 + 0.00800 = 0.01517$

LCL $= \bar{p} - 3\sqrt{\bar{p}\bar{q}/n} = 0.00717 - 3(0.002667) = 0.00717 - 0.00800 = -0.00084$,

truncated at 0

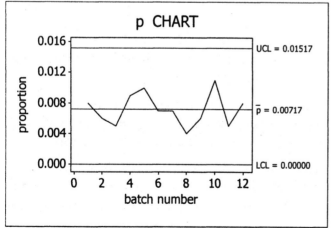

The process is within statistical control. Assuming that the level of $0.00717 = 0.717\%$ defects is acceptable, no corrective action is required.

13. $\bar{p} = (\Sigma x)/(\Sigma n) = (3+3+\ldots+5)/(52)(7) = 127/364 = 0.3489$

$\sqrt{\bar{p}\bar{q}/n} = \sqrt{(0.3489)(0.6511)/7} = 0.18015$

UCL $= \bar{p} + 3\sqrt{\bar{p}\bar{q}/n} = 0.3489 + 3(0.18015) = 0.3489 + 0.5404 = 0.8893$

LCL $= \bar{p} - 3\sqrt{\bar{p}\bar{q}/n} = 0.3489 - 3(0.18015) = 0.3489 - 0.5404 = -0.1915$, truncated at 0

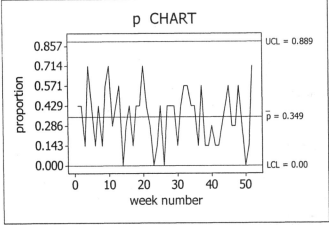

The 52 sample proportions are 3/7, 3/7,…, 5/7.
The vertical labels correspond to 0/7 = 0.000, 1/7 = 0.143, 2/7 = 0.286, etc.
The process is within statistical control.

**Statistical Literacy and Critical Thinking**

1. A process is designed to result in a product that meet certain specifications. Because conditions change over time, a process meeting those specifications at one given point in time may not always be meeting those specifications. For that reason it is important to monitor the process over time. Oftentimes detection of a changing pattern can lead to implementation of the necessary adjustments before the product strays so far from the specifications that the process output is unsatisfactory.

2. Statistical process control is the monitoring of data over time in order to detect trends or problems that indicate the production process is out of statistical control and/or that lead to the production of a product that does not meet prescribed specifications.

3. If a bottling plant runs a process that is not monitored, product not meeting the proper specifications could be leaving the plant and entering the market. The resulting financial and/or public relations loss in dealing with the defective products could be devastating.

4. In general, a process can go out of control because of a change in mean or a change in variation. The fact that the mean is within statistical control does not indicate that the variation is within statistical control, and vice-versa. For this reason, one should monitor a process with both an $\bar{x}$ chart and an R chart.

282  CHAPTER 14  Statistical Process Control

**Review Exercises**

Refer to the following tables for exercises #1-#3.

|        | kwh1 | kwh2 | kwh3 | $\bar{x}$ | R |
|--------|------|------|------|-----------|------|
| year1a | 3375 | 2661 | 2073 | 2703.00 | 1302 |
| year1b | 2579 | 2858 | 2296 | 2577.67 | 562 |
| year2a | 2812 | 2433 | 2266 | 2503.67 | 546 |
| year2b | 3128 | 3286 | 2749 | 3054.33 | 537 |
| year3a | 3427 | 578  | 3792 | 2599.00 | 3214 |

For exercise #1, use the 15 kwh values by row.
For exercise #2, use the range values in the R column.
For exercise #3, use the mean values in the $\bar{x}$ column.

1. The run chart, using the 15 kwh values by row, is given below.

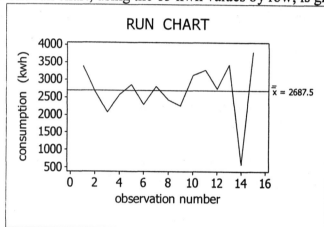

The existence of the extremely low value suggests that the process is not within statistical control – i.e., there appears to be a value that cannot be explained by normal variation, and that was caused by some identifiable circumstances.

2. There are k=5 samples of size n=3 each.  
$\bar{R} = \Sigma R/k = 6161/5 = 1232.2$

UCL = $D_4 \bar{R}$ = 2.574(1232.2) = 3171.7  
LCL = $D_3 \bar{R}$ = 0.000(1232.2) = 0

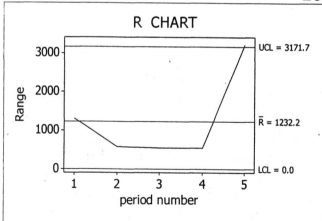

The process variation is out of statistical control – because there is a point above the upper control limit.

3. There are k=5 samples of size n=3 each.  $\bar{\bar{x}} = \Sigma\bar{x}/k = 13427.67/5 = 2687.5$
$\bar{R} = \Sigma R/k = 6161/5 = 1232.2$

UCL = $\bar{\bar{x}} + A_2\bar{R}$ = 2687.5 + (1.023)(1232.2) = 2687.5 + 1260.5 = 3948.1
LCL = $\bar{\bar{x}} - A_2\bar{R}$ = 2687.5 - (1.023)(1232.2) = 2687.5 - 1260.5 = 1427.0

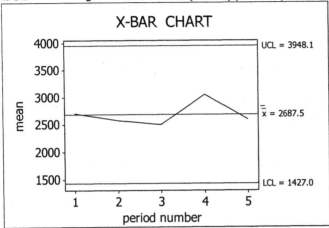

The process mean is within statistical control.

4. $\bar{p} = (\Sigma x)/(\Sigma n) = (270+264+\ldots+343)/(13)(100,000) = 4183/1,300,000 = 0.00322$
$\sqrt{\bar{p}\bar{q}/n} = \sqrt{(0.00322)(0.99678)/100000} = 0.000179$
UCL = $\bar{p} + 3\sqrt{\bar{p}\bar{q}/n}$ = 0.00322 + 3(0.000179) = 0.00322 + 0.00053 = 0.00375
LCL = $\bar{p} - 3\sqrt{\bar{p}\bar{q}/n}$ = 0.00322 - 3(0.000179) = 0.00322 - 0.00053 = 0.00268

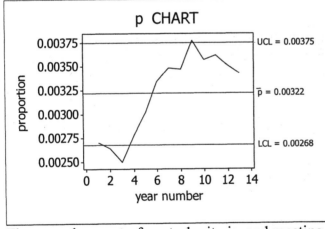

There are three out-of-control criteria, and meeting any one of them means that the process is not within statistical control. This process meets all three criteria.
   (1) There is a pattern – in this case, an upward trend.
   (2) There is a point outside the upper or lower control limit – in this case, there are 2 points below the LCL and 1 point above the UCL.
   (3) The run of 8 rule applies – in this case, the last 8 points are all above the center line.

284   CHAPTER 14   Statistical Process Control

5. $\bar{p} = (\Sigma x)/(\Sigma n) = (4+2+\ldots+14)/(14)(100) = 82/1400 = 0.0586$

$\sqrt{\bar{p}\bar{q}/n} = \sqrt{(0.0586)(0.9414)/100} = 0.02348$

UCL $= \bar{p} + 3\sqrt{\bar{p}\bar{q}/n} = 0.0586 + 3(0.02348) = 0.0586 + 0.0704 = 0.1290$

LCL $= \bar{p} - 3\sqrt{\bar{p}\bar{q}/n} = 0.0586 - 3(0.02348) = 0.0596 - 0.0704 = -0.0119$, truncated at 0

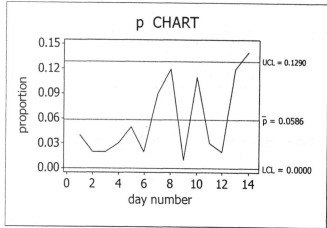

The process is not within statistical control for two reasons – there is a pattern of increasing variation, and there is a point beyond the upper control limit.

## Cumulative Review Exercises

1. $\bar{p} = (\Sigma x)/(\Sigma n) = (6+4+\ldots+2)/(10)(200) = 48/2000 = 0.0240$

$\sqrt{\bar{p}\bar{q}/n} = \sqrt{(0.0240)(0.9760)/200} = 0.01082$

UCL $= \bar{p} + 3\sqrt{\bar{p}\bar{q}/n} = 0.0240 + 3(0.01082) = 0.0240 + 0.0325 = 0.0565$

LCL $= \bar{p} - 3\sqrt{\bar{p}\bar{q}/n} = 0.0240 - 3(0.01082) = 0.0240 - 0.0325 = -0.0085$, truncated at 0

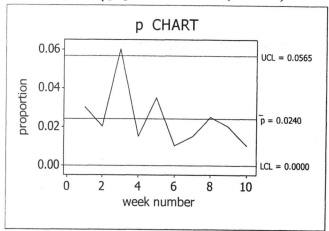

The process is not within statistical control because there is a point beyond the upper control limit.

2. As a review, the exercise is worked by hand.
   $\alpha = 0.05$ and $\hat{p} = x/n = 48/2000 = 0.0240$

   $$\hat{p} \pm z_{\alpha/2}\sqrt{\hat{p}\hat{q}/n}$$
   $0.0240 \pm 1.96\sqrt{(0.0240)(0.9760)/2000}$

   $0.0240 \pm .0067$

   $0.0173 < p < 0.0307$

   Using STAT TESTS 1-PropZInt,
   x: 48
   n: 2000
   C-Level: 0.95
   $0.0173 < p < 0.0307$

3. As a review, the exercise is worked both by hand and using STAT TESTS 1-PropZTest.

   original claim: $p > 0.01$
   $\hat{p} = x/n = 48/2000 = 0.024$

   $H_o$: $p = 0.01$
   $H_1$: $p > 0.01$
   $\alpha = 0.05$
   C.V. $z = z_\alpha = 1.645$
   calculations:
   $z\hat{p} = (\hat{p} - \mu_{\hat{p}})/\sigma_{\hat{p}}$
   $= (0.024 - 0.01)/\sqrt{(0.01)(0.99)/2000}$
   $= 0.014/0.00222 = 6.295$
   P-value = $P(z > 6.295) = 1.56\text{E-}10 \approx 0$

   conclusion:
   Reject $H_o$; there is sufficient evidence to support the claim that $p > 0.01$.

   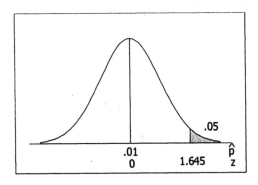

4. a. $P(A) = P(A_1 \text{ and } A_2 \text{ and}\ldots\text{and } A_8)$
      $= P(A_1) \cdot P(A_2) \cdot \ldots \cdot P(A_8) = (\frac{1}{2})^8$
      $= 1/256 = 0.0039$

   b. $P(B) = P(B_1 \text{ and } B_2 \text{ and}\ldots\text{and } B_8)$
      $= P(B_1) \cdot P(B_2) \cdot \ldots \cdot P(B_8) = (\frac{1}{2})^8$
      $= 1/256 = 0.0039$

   c. $P(A \text{ or } B) = P(A) + P(B) - P(A \text{ and } B)$
      $= 0.0039 + 0.0039 - 0$
      $= 0.0078$

FINAL NOTE: Congratulations! You have completed statistics – the course that everybody likes to hate. I trust that this manual has helped to make the course a little more understandable – and that you leave the course with an appreciation of broad principles, and mot merely memories of manipulating formulas. I wish you well in your continued studies, and that you achieve your full potential wherever your journey of life may lead.